Competing Through
Supply Chain Management

Chapman & Hall
Materials Management/Logistics Series
Eugene L. Magad, Series Editor
William Rainey Harper College

Total Materials Management: Achieving Maximum Profits through Materials/Logistics Operations, Second Edition
by Eugene L. Magad and John M. Amos

International Logistics, by Donald Wood, Anthony Barone,
Paul Murphy and Daniel Wardlow

Global Purchasing: Reaching for the World, by Victor Pooler

MRP II, by John W. Toomey

Distribution: Planning and Control, by David F. Ross

Purchasing and Supply Management: Creating the Vision,
by Victor Pooler and David Pooler

Competing through Supply Chain Management: Creating Market-Winning Strategies through Supply Chain Partnerships,
by David F. Ross

Practical Handbook of Warehousing, 4e,
by Kenneth B. Ackerman

Competing Through Supply Chain Management

Creating Market-Winning Strategies Through Supply Chain Partnerships

David Frederick Ross
Manager of Education for the Interactive Group, Inc.,
Chicago, Illinois

CHAPMAN & HALL

I(T)P® International Thomson Publishing
New York • Albany • Bonn • Boston • Cincinnati • Detroit • London • Madrid • Melbourne
Mexico City • Pacific Grove • Paris • San Francisco • Singapore • Tokyo • Toronto • Washington

JOIN US ON THE INTERNET   WWW: http://www.thomson.com
EMAIL: findit@kiosk.thomson.com

thomson.com is the on-line portal for the products, services and resources available from International Thomson Publishing (ITP).

This Internet kiosk gives users immediate access to more than 34 ITP publishers and over 20,000 products. Through *thomson.com* Internet users can search catalogs, examine subject-specific resource centers and subscribe to electronic discussion lists. You can purchase ITP products from your local bookseller, or directly through *thomson.com*.

Visit Chapman & Hall's Internet Resource Center for information on our new publications, links to useful sites on the World Wide Web and an opportunity to join our e-mail mailing list. Point your browser to: **http://www.chaphall.com**

A service of   I(T)P

Cover Design: Andrea Meyer, Emdash Inc.

Copyright © 1998 Chapman & Hall

Printed in the United States of America

For more information, contact:

Chapman & Hall
115 Fifth Avenue
New York, NY 10003

Chapman & Hall
2-6 Boundary Row
London SE1 8HN
England

Thomas Nelson Australia
102 Dodds Street
South Melbourne, 3205
Victoria, Australia

Chapman & Hall GmbH
Postfach 100 263
D-69442 Weinheim
Germany

International Thomson Editores
Campos Eliseos 385, Piso 7
Col. Polanco
11560 Mexico D.F.
Mexico

International Thomson Publishing-Japan
Hirakawacho-cho Kyowa Building, 3F
1-2-1 Hirakawacho-cho
Chiyoda-ku, 102 Tokyo
Japan

International Thomson Publishing Asia
221 Henderson Road #05-10
Henderson Building
Singapore 0315

1 2 3 4 5 6 7 8 9 10 XXX 01 00 99 98 97

Library of Congress Cataloging-in-Publication Data

Ross, David Frederick, 1948-
     Competing through supply chain management : creating market - winning strategies through supply chain partnerships / David Frederick Ross.
        p.   cm. -- (The Chapman & Hall materials management/logistics series)
     Includes bibliographical references and index.
     ISBN 0-412-13721-6
     1. Business logistics--Cost effectiveness. 2. Delivery of goods-Management. 3. Business networks. I. Title.
     II. Series.
     HD38.5.R675 1997
     658.7'2—dc21   97-2953
                                                          CIP

British Library Cataloguing in Publication Data available

To order this or any other Chapman & Hall book, please contact **International Thomson Publishing, 7625 Empire Drive, Florence, KY 41042.** Phone: (606) 525-6600 or 1-800-842-3636. Fax: (606) 525-7778, e-mail: order@chaphall.com.

For a complete listing of Chapman & Hall's titles, send your requests to **Chapman & Hall, Dept. BC, 115 Fifth Avenue, New York, NY 10003.**

In memory of little Ryan Joseph Ross
the bravest of the brave
who lives in my heart

I took pains to determine the flight of crook-taloned birds, marking which were of the right by nature, and which of the left, and what were their ways of living, each after his kind, and the enmities and affections that were between them, and how they consorted together.

AESCHYLUS, *Prometheus Vinctus*, 488–492

Table of Contents

Preface

This book is a work of *business strategy.*

Since around 1990, companies across a wide variety of industries have become increasingly interested in exploring the opportunities for competitive advantage that can be gained by leveraging the core competencies and innovative capabilities to be found among networks of business partners. Although companies have always acknowledged the importance of the relationships that existed between themselves and their customers and suppliers, it has only been recently that creating and nurturing channel alliances has been recognized as a critical source of strategic advantage. Once a backwater of business management, creating "chains" of customers and suppliers has arisen as perhaps today's most important competitive strategy.

What has caused this awareness of the "interconnectiveness" of enterprises? What does Supply Chain Management (SCM) mean and how is it to be implemented? What impact will the increasing dependence on channel partnerships have on the fabric of today's business environment? What are the possible opportunities as well as the liabilities of channel alliances? Are the benefits of channel partnerships focused primarily on operations issues, or do they hold out the opportunity for the realization of fresh sources of market-winning product and service value?

The supply chain focus of today's enterprise has arisen in response to several critical business requirements. First, today's best companies have come to realize that the effective management of the supply channel constitutes the final frontier in the search for new sources of cost reduction and process improvement. Over the past decade, the application of computerized information tools, the utilization of management techniques such as Just-In-Time (JIT), Total Quality Management (TQM), and Business Process Reengineering (BRP), and the implementation of employee empowerment and cross-functional management philosophies have activated highly agile, lean product design and manufacturing functions capable of superlative quality and service. Sustaining the collective momentum of these management paradigms has required companies to turn outward to their channels of supply and distribution in search of untapped opportunities for cost and cycle time reduction and process agility.

Second, interest in SCM has been stimulated by the realization that closely integrated channels of suppliers and customers can provide today's enterprise with unique sources of competitive competencies. Previous management models focused on employing quality and improvement methods that sought to increase marketplace value by leveraging the capabilities to be found in *internal* business processes. In contrast, SCM shifts attention to the previously unseen opportunities that appear when companies seek to converge the innovative competencies and unique resources of their *external* chains of customers and suppliers in the pursuit of radically new sources of competitive advantage.

Third, few companies today still maintain a dependence on vertical integration to provide them with competitive advantage. Over the past few years companies have continued to divest themselves of non-profitable businesses and functions for which they had weak core competencies, preferring to use channel partners who specialize in these business areas. In such an environment, managing supply chain partners has become the key to market leadership.

Fourth, the growth of international competition has open new markets previously inaccessible even just a few years ago. Equipped with today's newest information and communication technologies and able to leverage the tremendous capabilities occurring in global logistics management, customers are no longer limited to national sources of products and services. The ability to assemble closely networked supply chains provides even the smallest company with the capability to maximize customer satisfaction and accessibility at the lowest possible cost.

Fifth, enterprise reengineering and operations streamlining have forced companies to look seriously at their supply chain partners. The growing dependence on the supply chain has been accentuated by the following changes to traditional business practices:

- The growth of information sharing between vendors and customers
- The rise of process-focused teams replacing traditional departmental functions
- The shift in the marketplace from the mass-production of standardized products to flexible operations providing customized products
- Increased reliance on purchased materials and outside processing with a simultaneous reduction in the number of suppliers
- Greater emphasis on organizational and process agility
- Rise of employee empowerment management techniques that require the implementation of rules-based, real-time decision support systems [1].

Far from being a peripheral issue, effectively managing the supply chain has become critical to competitive survival. As Ralph Drayer, Vice president for Product Supply/Customer Service at Procter & Gambel, recently put it, "Winning in the marketplace of the 90s is going to require a far different kind of relationship—one that recognizes that the ultimate winners will be those who understand the interdependence of the retailer/manufacturer business systems and who work together to exploit opportunities to deliver superior customer value [2]." Effectively managing supply chains means that companies must create market-wining partnerships with other companies that promote the competitive advantage of the whole channel system. They must also know how to utilize technology to coordinate internal activities with those of their trading partners. Networking information and synchronizing the capacities and resources to be found in each channel node permit the exploration of new regions of competitive space and unassailable marketplace leadership.

Such a fundamental refocusing of business strategy has been brought about by five major business dynamics driving the marketplace today. Each dynamic provides a different perspective centered around a common theme: *How can companies realize fundamentally new avenues of marketplace advantage by closely integrating the core competencies and capabilities for innovative thinking to be found among their supply chain business partners?*

The first and most potent dynamic driving the movement to SCM is the veritable revolution occurring in the growth of the power of the "voice of the customer." In the past, companies sought to create and deliver standardized, mass-market products to customers who had relatively very little purchase choice. Today's customer, on the other hand, requires tailored combinations of high-quality products and services that will provide them with unique value and solutions assisting them to realize their own competitive strategies. Responding to this marketplace challenge constitutes the second of today's business dynamics and it can be described as the activity of constructing agile product design, manufacturing, and delivery processes that continually provide superior product and service quality, yet can be quickly and cost-effectively configured to meet individual customer needs. Succeeding in this dynamic means that firms must not only continually search for new ideas and processes that win the customer's order but that they must also possess the competencies and resources to realize market-wining innovations enabling them to create whole new markets beyond the boarders of their existing customers.

The third marketplace dynamic centers around the tremendous breakthroughs occurring in information and communications technologies. Today, as the value of accurate information grows and the speed of communications accelerates, *information* has increasingly begun to be seen as the fundamental source of wealth. This means that companies aspiring to market leadership must view technology not only as a critical management tool that shortens cycle times and increases the productivity of business functions through automation but also as a key enabler providing the enterprise with the opportunity to activate highly competitive organizational cultures and channel structures in the search for new sources of marketplace leadership. Information technology has made possible the interactive networking of congruent functions across the supply channel and the linkage of once separate companies into single, competitive supply chain systems.

The competitive advantages provided by information technologies has been further enhanced by the dramatic changes occurring in the fourth dynamic, the marketing and distribution channel environment. This dynamic has several dimensions. The first can be characterized by the globalization of the world economy, the rapid introduction of new products and services on an international scale, and increased demands for quick-response delivery. The second dimension can be found in the explosion of strategic alliances and partnerships on a global scale. The close interlinkage of companies along a supply channel system has enabled the formation of interenterprise "virtual" organizations capable of leveraging the skills, physical resources, and innovative knowledge of a matrix of productive capacities originating from different locations in the supply network. The ability of companies to exploit the peer-to-peer networking of marketers, designers, manufacturers, and distributors provided by today's information technologies will facilitate the creation of new forms of competitive-enriching collaboration and enhance the growth of supply chain alliances.

Finally, the fifth business dynamic, supply channel logistics integration, provides the last source for the growth of SCM. In the past, logistics was perceived primarily as an operations activity focused around product delivery and cost management. In contrast, today's best companies utilize logistics as a strategic, cross-functional, interenterpise management activity whose mission is to both plan and coordinate all inventory and delivery activities as well as to realize new opportunities for competitive advantage. As companies increasingly seek to more closely integrate logistics operations across the supply channel, dependence on the SCM strategic philosophy will deepen as they explore concurrently new ways to satisfy customers, develop new products, explore new competitive regions, implement new technology tools, and broaden and enrich their channel relationships.

It is the thesis of this book that the concept of SCM constitutes today's most potent and influential business strategy because it provides an effective way for companies to manage the marketplace dynamics detailed above. Prior competitive strategies focused on leveraging and reengineering the product and market values, productive processes, and core competencies to be found *within* the boundaries of the enterprise. While such strategic initiatives were critical in refocusing competitive objectives away from the narrow business paradigms of the past, they have become insufficient as competitive strategies in today's era of global collaboration, intensified competition, ever-shortening product life cycles, and increased dependence on supply chain partners. SCM is built around a set of simple competitive values:

- Companies work best when they not only cooperate but when they enter into full partnership with their supply channel allies
- Companies can withstand the onslaught of global competition and flourish only when they combine their destines with channel collaborators
- Companies can not hope to gather on their own without channel partners the necessary resources and core competencies permitting them to design, manufacture, market, and distribute those products and services that preempt the competition
- Companies will succeed in direct portion to their ability to network through information and communications technologies the tremendous repositories of skills and innovative capacities found collectively within their people resources
- Companies can not hope to realize the power of process improvement initiatives unless they seek to extend them beyond the boundaries of their own organizations into the supply chain system as a whole
- Companies can not hope to realize strategic visions that generate whole new regions of competitive space without converging the collective resources and innovative capabilities found among their supply chain allies

Each chapter in the book attempts to explore and elaborate on the different facets of these SCM strategic values. The first chapter focuses on defining SCM and detailing its essential elements. The basis of the discussion that unfolds is that the SCM philosophy provides today's global enterprise with the strategic capabilities to succeed despite the dramatic and fundamental changes occurring to today's marketplace. The chapter concludes with an exploration of the relationship of logistics management and SCM in today's business environment.

Chapter 2 is wholly devoted to a detailed discussion of the five marketplace dynamics driving the development of SCM. Each dynamic—customer service, product and process design, information and communications technologies, channel management, and integrated logistics management—is explored in depth. The objective is to illustrate how collectively these dynamics are requiring companies to be dramatically more creative, flexible, and focused on productive competencies than they have ever been in the past.

Chapter 3 seeks to explore the origins and development of the SCM concept. After a brief discussion on the development of logistics management and its basic functions and objectives, the remainder of the chapter examines the stages marking the rise of SCM. Four stages are identified: decentralized logistics management, centralized logistics management, logistics integration, and the emergence of SCM.

Chapter 4 is concerned with how companies can utilize the strategic and operations components of SCM to fashion market-winning business strategies. The discussion focuses on exploring the nature of today's marketplace and how business strategies centered around the SCM philosophy provide for the creation of effective channel partnerships and the development of new forms of product/ service combinations and new competitive space. The chapter concludes with a seven-step methodology for SCM strategic development to guide planners in strategy formulation.

Chapter 5 focuses on detailing the physical supply channel strategies and functions necessary for the effective implementation of SCM. After defining supply channel management, its historical development, and principle business mission, the chapter discusses supply channel functions in detail. Two areas are explored: basic channel functions, such as inventory movement and functional performance, and channel marketing functions, such as title, transaction, and information flows. The chapter concludes with an overview of the challenges involved in developing effective global strategies.

Chapter 6 is concerned with a detailed examination of the elements necessary for the effective management of supply chain inventories. Discussion begins with an overview of the flow and functions of supply pipeline inventories and how channel buffer stocks can provide a key source of strategic advantage and service value. Next, the chapter turns to an analysis of the principles of supply chain inventory management. Among the topics discussed are computerized tools for inventory management, planning and ordering methods, continuous inventory replenishment techniques, and supplier management.

Chapter 7 seeks to explore the application of quality and improvement management concepts, termed Supply Chain Quality Management (SCQM), to the management of today's global supply channel system. After a short review of the development of quality management for the past 15 years, attention shifts to an analysis of SCQM processes and how channel quality can be utilized to create superior customer service value. The chapter concludes by detailing the requirements necessary for the effective implementation of a channelwide program for continuous SCQM process improvement.

Exploring how changes to the work force and information technologies are impacting SCM is the subject of chapter 8. The analysis begins with an overview of the changes to work force values and objectives in the "Age of Supply Chain Management." Key topics discussed are meeting the challenge of SCM leadership, creating the learning organization, developing new forms of organization, and team-based management styles. After this analysis, the chapter then shifts to a discussion of how information and communications technologies are fundamentally reshaping the tactical and strategic functions of both the work force and the supply channel network.

Chapter 9 concludes the book with a discussion of the management and organizational elements necessary to effectively implement SCM. The goal is to detail the principles that must be followed in any application of SCM. After revisiting the principles of the SCM philosophy and the steps required for effective implementation, the chapter concludes with a series of questions managers can use when determining how SCM can help their businesses achieve competitive success and what direction their implementations should take.

Acknowledgments

Acknowledgments are always one of the most difficult tasks in writing a book. There are so many influences direct and indirect on an author's ideas that it is virtually impossible to render the kind of thanks that is so necessary. The author is greatly indebted to the many students, professionals, and companies he has worked with over the years who have contributed their ideas and experiences. They have provided the laboratory where the author could test the hypotheses and determine the fundamental principles upon which this book is based.

The author would especially like to thank Mr. Eugene Magad, editor of Chapman & Hall's *Materials Management/Logistics Series*, for his scholarly assistance and encouragement. I would also like to thank my friends in the editorial office at Chapman & Hall. Ms. Margaret Cummins, Acquisitions Editor, was especially helpful in her enthusiasm for the project. Ms. Mary Ann Cottone, Senior Managing Editor, at Chapman & Hall rendered her expertise in processing the manuscript through to completion. The author would also like to express his gratitude to Ms. Nancy Sherman and the entire staff at the Oakton Community College Library. Finally, I would like to express my thanks to my personal friends and relatives, and most especially my loving wife Colleen and my son Jonathan, whose strength and devotion somehow enabled me to continue during the dark period after the passing of our little son Ryan.

Notes

1. Rhonda R. Lummus and Karen L. Alber, *Supply Chain Management: Balancing the Supply Chain with Customer Demand*. Falls Church, VA: APICS, 1997, pp. 3–5.
2. Ralph Drayer, "The Emergence of Supply Chain Management in the North America." Excerpt from a speech to suppliers of Procter & Gambel Co., October 1994.

Competing Through
Supply Chain Management

1

Meeting the Challenge of Supply Chain Management

One of the most important topics in the study of the management of contemporary manufacturing and distribution is Supply Chain Management (SCM). Over the past decade, the business literature as well as the popular press have been filled with books and articles describing the dramatic changes occurring in productive processes and organizational structures emanating from the radical breakthroughs taking place in management methods, the implementation of business process reengineering techniques, the globalization of the marketplace, and the explosion in information and communication technologies. Today, academics, consultants, and practitioners alike have begun to explore how these often divergent threads can be woven together to form compelling new strategies, providing companies with exciting marketplace opportunities and the capability to uncover whole new competitive regions in the search for marketplace advantage.

Increasingly, at the center of this emerging dialogue stands SCM. In fact, SCM has today become such a "hot topic" that it is difficult not to pick up a periodical relating to manufacturing, distribution, customer management, or transportation without seeing an article about SCM or SCM related topics. What is more, seminars, roundtable discussions, and scholarly presentations on SCM can be heard regularly in universities and professional societies. This preoccupation with SCM has not been limited to academic settings. An increasing number of Fortune 500 companies have managers with "supply chain" in their title, and, recently, more than one company has created logistical operations groups entrusted with "supply chain" functions. For example, Becton Dickinson and Company, a global manufacturer and distributor of medical supplies and devices and diagnostic systems, formed a new operating division called BD Supply Chain Services early in 1995. This organization is mandated with complete responsibility for all logistics channel services, including distributor strategy and management, physical distribution, international distribution services, transportation, invoicing, credit, and collections [1].

Despite the press given to SCM and the nascent organizations being formed, there exists a good deal of confusion as to what it is, how it can be practically applied, and what are its expected benefits. Whereas most of the literature focuses on some critical operational aspect of SCM, such as transportation, supplier partnering, or customer service, none has sought a comprehensive definition that illuminates fundamental principles and places it squarely within the management context of today's business environment. As a result, considerable confusion has grown surrounding the SCM concept. Part of the problem resides in the fact that SCM, like other management philosophies such as Just-In-Time (JIT), Enterprise Resource Planning (ERP), and Total Quality Management (TQM), can be defined in several ways and possesses a matrix of possible applications. However, the confusion stems primarily from the commonplace error of equating SCM and the array of operational functions constituting modern logistics and supply channel management. As will become apparent, SCM is more than an set of techniques designed to deliver product to the customer faster while reducing supply channel costs. Rather, it is a comprehensive, dynamic, growth-oriented, and competitive-winning management approach to thriving in a business environment driven by global change and uncertainty.

This opening chapter is focused on defining SCM and exploring the competitive challenges and marketplace opportunities that have shaped and continue to drive its development. The chapter begins by offering a detailed definition of SCM. The argument that unfolds centers around the postulate that SCM provides the enterprise with the ability not only to compete in today's marketplace but also to successfully respond to the fundamental changes that are occurring as a result of the shift from a mass-market economy to one dominated by new concepts of products and services, fragmenting global marketplaces, new organizational challenges, and the enabling power of information technologies. Once a working definition of SCM has been established, the chapter then continues with an analysis of the operational and strategic elements of SCM and how they converge to form an indispensable management paradigm assisting today's enterprise in the quest for competitive advantage. At this point, the chapter proceeds to examine the business environment that has enabled the emergence of SCM. After a brief discussion focusing on the problems associated with using conventional approaches to managing organizations and supply channels, the chapter concludes with an exploration of the relationship of logistics management and SCM in today's business environment.

Understanding Supply Chain Management

At no time in history has the effective management of supply chain systems been as important as it is in today's business climate. In the past, the production and distribution of products and information through the marketplace was considered of secondary importance in comparison to the much more critical strategies of managing marketing, sales, and finance. Today, as companies find themselves under constant pressure to source and create high-quality, configurable products and services, deliver them to market faster, increase process flexibility in order to respond to continuous and radical shifts in marketplace requirements, reduce product and operating costs, and search for ways to educate and train their work forces to master the challenges they must face on a daily basis, they have been increasingly turning to the strengths to be found in their supply chain partners to enrich their competitive capabilities. As companies focus internal efforts on building core competencies, reengineering wasteful processes, and applying quality management techniques, they have also been looking outward to their channel alliances to gain access to sources of unique competencies, physical resources, and customer-winning process value.

Clearly, the competitive requirement that not only individual enterprises but whole channels of supply ceaselessly search for new ways to converge their operational and strategic strengths to become more *creative* and more *responsive* to the marketplace has become the hallmark of today's leading companies. As many businesses have experienced, achieving superlative service and cost objectives on the company level alone is often insufficient in keeping one step ahead of the competition. Increasingly, it has become apparent that today's enterprise must refocus its efforts away from conventional business paradigms centered around transaction management and parochial performance metrics, and toward strategies that recognize that to achieve competitive advantage companies must work together across enterprise boundaries and optimize interchannel processes and innovative capabilities that preempt the competition and open whole new areas of competitive space. Companies that succeed in exploiting their supply chains are able to leverage and focus productive functions better while extending the reach of their products and services to meet national and international demand. On the other hand, poorly defined relationships with customers and suppliers, "disconnection" between channel systems resulting in unnecessary costs and redundancies, and hidden supply pipeline capacity constraints can mean disaster for even the best run company and its products.

Opening Questions

For the past half-decade, practitioners, academics, and consultants have been using the term *supply chain management* to describe this emerging management paradigm. However, although there is solid consensus that SCM forms one of the keystones to achieving marketplace leadership in today's new world of exploding technologies and "virtual" companies, there is considerable "fuzziness" when it comes to defining exactly what the concept means, how it applies to businesses, what the benefits are, and how it is to be implemented. What is more, as one logistics scholar points out, discussion of SCM is often shrouded in complex jargon, further clouding understanding and blocking avenues for practical application. The following questions, some fundamental, others seeking workable definitions and practical guidelines, inevitably come to mind whenever serious thought is given to SCM.

- What is SCM?
- Is SCM a fundamentally new idea, or is it simply a fancy new word describing the logistics concept for the late 1990s?
- Is SCM a strategic management paradigm or a set of techniques targeted primarily at operational objectives?
- What is the relationship between SCM and logistics; between SCM and supply channel management?
- Is SCM only for distributors and wholesalers?
- How do suppliers, customers, and supply channel partners fit in?
- What is the relationship between SCM and today's information and communication technologies?
- What benefits can be expected by companies embracing the SCM concept?
- What performance metrics are necessary to measure SCM success?
- Does SCM really have anything to offer enterprises seeking not only to survive but to flourish in today's global marketplace, or is it just today's newest management buzzword?

As will become apparent, SCM is one of today's most important management concepts, enabling not only individual companies but entire business systems to repel the challenges of aggressive competitors, the displacement caused by new technologies and forms of organization, and the onslaught of goods, sometimes from the other side of the word, that commoditize products and processes once thought unique and uncopyable.

Defining Supply Chain Management

As is the case with any management philosophy, especially one that is still in the process of evolving, available definitions of SCM can be characterized as possessing a wide spectrum of different meanings and equally numerous applications. Some definitions of SCM are structured around operational issues. For example, Ellram [2] defines SCM in general terms as an integrated management approach for planning and controlling the flow of materials from suppliers through the distribution channel to the end user. Another logistics scholar considers SCM a management method focused primarily on facilitating the outbound flows of inventory and information. The goal of the concept is to structure interchannel partner integration that will produce the "synchronization of all channel activities in a manner which will create the greatest net comparative value for the customer." [3] Others consider SCM a channel management philosophy. Cooper [4] feels that SCM provides companies with a boundary-spanning channel focus where "all the steps of a product's movement, regardless of corporate, political, or geographical boundaries, from raw material supply through final delivery to ultimate user to satisfy a particular customer group" are planned and supervised. LaLonde defines SCM "as the delivery of enhanced customer and economic value through synchronized management of the flow of physical goods and associated information from sourcing through consumption." [5] Finally, Walton and Miller [6] state that the "strategic integration of trading partners is the Supply Chain Management concept."

What is clear from these definitions is that instead of describing SCM in a lucid phrase that can be used as easily in the boardroom as on the plant floor, they, in fact, describe a diverse business philosophy consisting of a complex matrix of concepts and practical applications that can be employed to respond to a myriad of problems that impact the enterprise both horizontally within its own operations and vertically outside in the channel. On one level, SCM is concerned with strategic issues such as the integration of internal and external business processes, the development of close linkages between channel partners, and the management of products and information as they move across organizational and enterprise boundaries. However, SCM can also be understood as a tactical tool that can be applied to the management of ongoing operational activities such as customer service, control of inbound and outbound flows of materials and information, and the elimination of channel inefficiencies, costs, and redundancies extending from raw materials acquisition, through manufacturing, distribution, and consumption, and final return back through the channel by way of recycling or disposal.

Undoubtedly, much of the reason for the confusion surrounding SCM stems from the fact that the concept did not arrive on the business scene as a complete body of knowledge, replete with a set of proven implementation steps. In fact, as was the case with the JIT and TQM crusades which took the business world by storm in the 1980s, SCM has been described from a variety of vantage points and has been used to manage a matrix of business processes. For example, at a roundtable discussion of SCM conducted at the APICS Convention in October, 1996, participants defined SCM as a way to manage logistics assets, an expansion of JIT purchasing theory, a method to work more closely with suppliers, a technique that enables companies to include customers in logistics decisions, and a new logistics model made possible by today's information and communications technologies. In addition to this lack of basic concurrence regarding principles and scope, understand SCM has been clouded by the fact that it is still evolving and is being architected from many directions as today's best companies search for fresh strategic business philosophies enabling them to leverage the value-generating processes to be found within their networks of channel partners. In this sense, SCM is in the unique position of providing supply chain systems with the capability to generate whole new approaches to product and service innovation as well as being shaped by the practical requirements of identifying and winning today's most profitable marketplace opportunities.

Furthermore, a concise definition of SCM has been obscured by the fact that actual implementation of operational and strategic aspects of SCM have been occurring for years. As is often the case with new management responses to radically changing market conditions, elements of SCM have been put into practice long before the theory has been elucidated in detail. Like a tropical plant, SCM has grown quickly and uncultivated, twisting and flowering to meet one set of business needs, and then, suddenly, sprouting whole new branches to sustain the weight of radical new marketplace initiatives. SCM has arisen today not because it represents the latest management paradigm but because it is how real companies are solving the problems of coping and succeeding in today's intensely competitive and uncertain times.

Another impediment to the formulation of a comprehensive definition of SCM is the confusion that exists between SCM and modern logistics and supply channel management. Today, the effective positioning of the logistics organization has become a fundamental source of competitive advantage by creating value for the customer, driving down operations costs, actualizing the marketing and sales effort, and facilitating operations flexibility. A sound logistics strategy is fundamental in keeping the organization focused on marketplace objectives by synchronizing customer opportunities with actual enterprise operational constraints, capabilities, opportunities, and options available.

Of particular relevance is the trend of current logistics management to assume responsibility not only for internal logistics functions but also for the challenges involved in integrating and coordinating the flow of materials and information along the entire supply channel. The goal is the orchestration of an external logistics strategy that links all channel constituents together as a single operational entity. By integrating inventory, production, and distribution resources on an enterprise level as well as channel level, companies have the opportunity to realize optimal customer value while minimizing total supply chain costs. This outward-looking aspect of logistics management has been defined by many experts as *supply chain management*. In fact, it is not unusual to see books and articles using the terms *logistics* and *supply chain management* as if they were synonymous, or definitions where SCM is considered a subset or purely an extension of the integrated logistics concept. For example, Christopher states that, "It must be recognized that the concept of supply chain management, while relatively new, is in fact no more than an extension of the logic of logistics." [7] In a similar vein, Gopal and Cypress, when defining key logistics terms, state, "The term *logistics management* is often used interchangeably with *supply chain management*." [8]

In a similar vein, the term *supply channel management* has also been confused with SCM. The distinction between the two, however, is clear and the proper usage unequivocal. As will be described shortly, SCM is fundamentally a *philosophy* of channel management which seeks the synchronization and convergence of intraenterprise and interchannel operational and strategic capabilities into a unified, compelling marketplace force. In contrast, the term *supply channel management* refers not to a concept but, rather, to the actual strategic objectives and the structure of channel business functions, institutions, productive values, and physical operations that define the way a particular channel system moves goods and services to market through the supply pipeline. SCM provides today's market leaders with innovative competitive strategies that enable them to leverage the capabilities of coalitions of companies to reinvent whole industries and create new competitive space; supply channels, on the other hand, provide the actual business functions, cooperative relationships, and the daily management of logistics, demand management, and financial transactions upon which a particular application of SCM is founded.

The result of the confusion concerning the proper definitions of SCM, supply channel management, and logistics has been that instead of clarity, management teams, consulting professionals, and academics have become confused by trying to decipher what exactly is meant by SCM. Is SCM simply nothing more than the advanced application of the operations tools to be found in modern logistics management? Is SCM purely an operational tactic that seeks to extend the benefits of logistics and supply channel process integration beyond the internal boundaries of the organization? Is SCM synonymous with channel partner management? Is SCM, on the other hand, a fundamentally new management paradigm, or just today's newest buzzword? Like Alice's Cheshire cat, SCM appears to mean one thing in reference to one idea, and then again something entirely different in another context, depending on who and what is being spoken about.

A Supply Chain Management Definition

Based on the discussion of the definition of SCM up to this point, it can be said that SCM consists of two dynamics. The first describes SCM as an operations management technique that enables companies to move beyond simply optimizing logistics activities only, to one where all enterprise functions—marketing, manufacturing, finance—are optimized by being closely integrated to form the foundation of a common business system. Enterprise integration at this level enables managers to connect and synchronize the day-to-day performance of key value-enhancing activities that provide decisive competitive advantage. These operations activities can be divided into four functional groups. The first homogeneous set of activities, *inbound logistics*, includes sales forecasting, inventory planning, sourcing and purchasing, and inbound transportation. *Processing activities* form the second group. These consist of production, value-added processing, work-in-process inventory management, and finished goods warehousing. The third group consists of *outbound activities,* including finished goods inventory management, customer order management, and outbound and intracompany transportation. The final group can be found in the performance of *support activities,* including logistics systems planning, logistics engineering, and logistics control [9]. The first dynamic of SCM ensures the continuous alignment of departmental tactical objectives, the optimization of all operations functions, and the continuous creation of customer service value from the enterprise perspective.

The second dynamic of SCM can be seen in the extension of integrated logistics management to the performance of interchannel logistics activities. The objective of SCM at this level is to closely interface, if not merge, the logistics functions of an organization with the identical functions being performed by logistics counterparts found in outside supply channel partners. One possible example is the capability of SCM to merge the logistics functions of a multidivisional organization. An even more dramatic application is the ability of companies to link internal logistics functions with those of outside suppliers and customers to fashion a fully integrated channel system. Other examples would be the ability of inventory planners to look directly at their suppliers' inventories via computer-to-computer networking or the capability of manufacturers to synchronize production capacities to match the collective planned inventory requirements of their customers. This dynamic supports the concept that in today's business environment, no company can compete independently nor possesses by itself all of the competencies and knowledge necessary to maintain marketplace leadership. Instead, by integrating logistics functions, supply chain partners, in effect, acknowledge that they are inextricably bound with networks of other supply chain partners in the performance of superior manufacturing and distributive processes designed to create unique customer value.

What is being proposed is that to the two traditional dynamics of SCM must be added a third and much more powerful dynamic. Although the above discussion illuminates several critical dimensions of the SCM concept, it is, in actuality, organized around a common theme. Basically, SCM is viewed primarily as an *operations management activity* focused on accelerating the flow of inventory and information through the supply pipeline, optimizing internal business functions and synchronizing them with those of channel outside partners, and providing the mechanism to facilitate continuous channel-wide cost-reduction efforts and increased productivity. Although these aspects are critical elements of SCM, they, nevertheless, represent only a fraction of the potential residing within the SCM concept. To these operational elements must be added a crucial *strategic dynamic.* SCM is about accelerating delivery times and reducing costs; it is also about utilizing new management methods and the power of information technologies to achieve order-of-magnitude breakthroughs in products and services that target the unique requirements of the marketplace. The operational aspects of SCM provide today's enterprise with the ability to stay even with the competition in the struggle for marketplace advantage. On the other hand, the strategic capability of SCM to fashion a shared vision with channel system partners, form coevolutionary and mutually beneficial channel alliances, and manage complex relationships with suppliers and customers enables today's innovative enterprise to lead market direction, spawn new associated businesses, and explore radically new opportunities.

It is the emerging strategic capabilities of SCM that is the central focus of this book. In the ensuing chapters, the value-creating capacities of SCM both within the organization and in the supply channel will be explored. This value-enhancing competitiveness of SCM will be seen as characterizing today's agile organizations that tirelessly search the marketplace as well as the entire supply channel to uncover new sources of innovation and customer enrichment. Furthermore, SCM provides for the engineering of cooperative design processes among allied partners who have joined resources and special competencies to ensure rapid product development and roll-out and the quick configuration of processes that enable the manufacture and delivery of individualized products and services on a global scale. At the core of SCM can be found the enabling power of today's information and communication technologies. Technology enables channel partners to be networked together, which, in turn, enables the orchestration of decisive decision-making processes that span multiple layers of converging supply channels focused on providing the best product and service alternatives to the customer. The competitive goal of SCM is to coevolve capabilities found in each business system around the collective competencies of each constituent to generate new sources of product and service value, new processes and technologies, and new forms of vertical integration and scale economies that will not only sustain survival but also engineer continuous market system dominance.

Based on these factors, a meaningful definition of SCM can be constructed. The definition is as follows:

> Supply chain management is a continuously evolving management philosophy that seeks to unify the collective productive competencies and resources of the business functions found both within the enterprise and outside in the firm's allied business partners located along intersecting supply channels into a highly competitive, customer-enriching supply system focused on developing innovative solutions and synchronizing the flow of marketplace products, services, and information to create unique, individualized sources of customer value.

Supply chain management is such a popular topic in current literature and professional seminars because it directly addresses the fundamental issues confronting today's enterprise. SCM is an open-ended philosophy for managing companies and supply chains in which they participate. SCM enables whole supply channels to leverage the following spectrum of reciprocal developments that are radically transforming today's marketplace:

- The explosion in information and communications technologies that are revolutionizing the very nature of products and how they are developed, marketed, sold, and transported
- The shift in the supply channel from a concern with pushing standardized, mass-produced products to a *pull system* able to quickly respond to customer requirements for uniquely configured products and services
- The development of business partner alliances and virtual organizations that are formed and then mutate to realize the unique opportunities to be found in today's ever-changing global marketplace
- The design of channel strategies that unify the material procurement, product design, production processes, business operations, market distribution, and customer wants and needs performed up and down the supply channel
- The creation of new organizational structures that are uniquely positioned to leverage information and communications technologies, superior skills, productive capacities, and the expertise of people both within the enterprise and among allied channel partners in the pursuit of superior competitive advantage.

Elements of Supply Chain Management

The SCM definition stated above expands the meaning and capabilities of SCM considerably and enables it to be considered from several fresh perspectives. The traditional definitions described earlier are more or less centered around detailing SCM as purely an extension of logistics operations management. Their focus is on the effective performance of the day-to-day activities associated with the optimization of distribution and manufacturing processes and accelerating the flow of inventory and information through the channel system.

By adding the strategic dynamic, the concept of SCM takes on a whole new dimension. To begin with, SCM provides the conceptual foundation enabling the separate companies constituting a supply channel network to compete as a single *unified* competitive entity. This means considerably more than just sharing inventory records and sending data via a computerized Electronic Data Interchange (EDI) system link-up. SCM supports the convergence of the individual marketing, product design, production, and logistics plans and activities found within each organization into a coherent value-enhancing supply channel system positioned to respond decisively to the rapidly changing needs of a global marketplace. The range of SCM activities extends from the establishment of channel alliances and supporting information and communications technologies through business forecasting, product design collaboration, pricing and promotions, demand and supply planning, sourcing and procurement, manufacturing, and logistics management. The goal is the creation of a constantly regenerating channel business system that provides channel partners with the capability to continuously search for new avenues, not just to improve existing products and processes but also to facilitate the generation of whole new markets and innovative solutions to meet individual customer requirements.

Opening Discussion

Activating the strategic elements of SCM requires companies to significantly broaden their understanding of the basis of the SCM concept. Formerly, utilizing SCM was the responsibility of a firm's logistics department. However, by elevating SCM to the status of a strategic management philosophy, it has now become perhaps an enterprise's foremost competitive strategy. Realizing the potential to be found in this new perception of SCM requires a full understanding of the following fundamental principles:

1. *SCM provides for a strategic view of the supply channel.*

    Although the operations or logistics functions of a firm have traditionally been the focus of the application of SCM, what is being argued is that the most significant characteristic of SCM can be found in its ability to provide companies with an external strategic orientation; that is, in today's business climate of shrinking product life cycles and shifting market niches, the ability of companies to leverage and fuse the core competencies and physical resources, marketing and production processes, information technologies, and logistics capabilities of their supply chain partners has become the fundamental source of marketplace advantage today. This means that world-class market leaders achieve and sustain dominance by exploiting the productive and innovative capabilities of a closely interwoven network of supporting channel systems that enable them to realize order-of-magnitude breakthroughs in product design, delivery, customer service, cost management, and value-added services that are far beyond the scope of their own capacities. Strategic SCM permits whole supply chains to realize unique marketplace value that squeezes out the competition.

2. *Achieving SCM objectives is not simply the result of something that occurs only inside of a company.*

    Although the application of process reengineering and quality management techniques designed to shrink costs, eliminate redundancies, and accelerate throughput and productivities are fundamental elements of the SCM concept, the implementation of SCM is not synonymous with initiatives designed to improve logistics functions solely. In addition, SCM is not simply about broadening and enriching the existing relationships that have been developed with a company's immediate suppliers and customers. Finally, SCM is not the same as *vertical integration*, which involves ownership of the business entities found up and down the supply pipeline. All of these management activities focus inward and are designed to increase a firm's ability to be competitive on its own. The real value of SCM, however, is to be found not in its *internal* but in its *external* orientation. Companies pursuing SCM have the capability to dramatically improve their productive and innovative competencies by converging not only logistics and manufacturing operations but also the strategic elements of finance, marketing, operations planning, and product development possessed by the universe of supply chain systems to which they belong. Through the formation of strategic alliances, the coupling of interchannel functions through information technology, and the open access to complementary core channel competencies, SCM enables whole supply chains to engineer unified supply networks capable of providing exceptional competitive value. By sharing and combining resources and facilities, risk and infrastructure cost to compete, and human skills and technological tools, supply chain members have the flexibility to rapidly configure new business approaches to meet the challenges of today's intensely competitive global marketplace to a degree unattainable by single companies acting on their own.

3. *The SCM strategy is completely customer driven.*

In the mass-production era, manufacturers produced standardized products that were presented to the customer base, who, in turn, were expected to make their choices motivated by promotion, price, and the variety available. Logistics' role in this business era was to push product through the supply channel to the customer in as cheap and as rapid a manner as was possible. In contrast, in today's marketplace, products and accompanying services are being driven by the customer, who expects to receive product configurations and services matched to their unique needs. This reversal of the flow of demand from a push to a *pull* of goods and services has resulted in a radical reformulation of both market expectations and the productive and innovative capabilities of traditional marketing, production, and distribution functions. By networking both the operations and strategic capacities of suppliers, manufacturers, wholesalers, and retailers along the entire supply channel, SCM facilitates cooperative product design, flexible production processes, quality deployment, information transfer, and inventory flows that enable companies to respond with the unique product configuration and mix of services demanded by the customer. Today, SCM plays a dual role: first, as a *communicator* of customer demand that extends from the point of sale all the way back to the supplier, and second, as a *physical flow process* that engineers the timely and cost-effective movement of goods through the entire supply pipeline.

The Supply Chain Management Business Philosophy

The power of SCM can best be comprehended when it is considered a strategic business philosophy. From the outset, however, it must be said that implementing SCM is not identical with Business Process Reengineering (BPR). Because BPR requires the whole company to explore and focus on improving its core value-added processes, beginning with suppliers and ending with delivery to the customer, it is not surprising that a BPR improvement initiative can get confused with SCM. In addition, SCM is more than just engineering processes that more efficiently synchronize the flow of inventory through the supply pipeline or the activation of information systems to network channel members. Also, it is not simply about cutting costs, improving efficiency, or coercing channel partners into unfair relationships. Finally, it is not just an expanded version of integrated logistics management, TQM, or ERP. Rather, SCM is a holistic, enterprisewide view of how the productive resources and talents of an allied group of businesses can be blended to form a single channel system possessing the flexibility to successfully respond to any marketplace opportunity with superior competitive advantage. Strategic SCM can best be described as a boundary-spanning, channel-unifying, dynamic, and growth-oriented competitive concept.

Supply chain management is a *boundary-spanning* management philosophy that requires companies to search for competitive advantage by looking beyond the frontiers of their own organizations. Earlier management theories, such as JIT, Lean Manufacturing, MRP II, ERP, and TQM, focused on productivity and cost-reduction objectives that were to be realized within the four walls of the enterprise. SCM, on the other hand, recognizes that to seize marketplace leadership companies need to compete, not as individual firms but as active members of a *unified* supply chain, managed as if it were a single business entity. No single company today can be organized nor possess the blend of skills, talents, technologies, and information to respond successfully by itself to the speed of change and the uncertainty of today's global marketplace. The essence of SCM can perhaps best be described as the continuous formation and permutation of the operating entities of a single company or of a group of independent companies into temporary alliances that, by leveraging the core competencies of each member, can capitalize on highly profitable marketplace opportunities. What makes SCM such a potent marketplace force today is its ability to provide a seamless channel structure which is physically dispersed and consists of different competencies yet functions as a coherent customer-satisfying resource whose boundaries appear invisible to the customer [10].

Supply chain management is a *unifying* force that enables the entire supply chain to act as a single competitive entity. Although it represent a tremendous advancement over the *disconnected* channels of the past, logistics functional integration cannot respond fast enough to the speed of change and the requirements for interconnecting the human, physical, and informational resources neces-

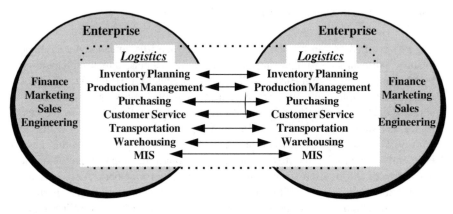

Figure 1.1. Logistics channel integration.

sary to achieve dominance in today's global marketplace. As illustrated in Figure 1.1, integrated channel organizations still maintain their own internal functions, performance measurements, and distinctive business strategies. The result is that the timely flow of marketplace information is delayed as it passes serially through each channel partner on its way through the supply pipeline. In contrast, supply chain *unification* means that all channel business entities must abandon their independent distinctiveness and function as a single supply chain process, as shown in Figure 1.2. Unification means that not only logistics activities, such as inventory planning, purchasing, and transportation, but also core business processes, such as marketing, sales, and product development performed by each channel partner, function as correlative processes led by interchannel process

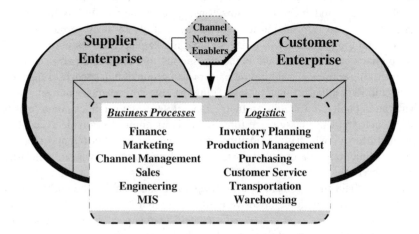

Figure 1.2. Unified channel functions.

teams. Through the convergence of information systems networking and telecommunications technologies, channel partners can now bypass the encumbrances of dealing sequentially with the transfer of critical marketplace information and the traditional movement of inventory from collection point to collection point on their way up and down the supply channel.

Supply chain management is a *dynamic and open-ended approach* to marketplace competitiveness. Similar to JIT and TQM, SCM is not a business formula or a computer system, but a continuous channel-level management process of shaping and reshaping intracompany and intercompany performance, information technology tools, products and services, and organizational and personal excellence to exploit the ever-changing contexts of customer opportunities. Also, because it is a philosophy of management rather than a detailed set of operations activities, SCM is never totally attained by any company or group of companies, nor can the elements of success enjoyed by one supply channel be transferred to another channel with the expectation of identical levels of performance. By its very nature, SCM is dynamic because today's global business environment is intrinsically dynamic. SCM enables companies to harness the incertitude produced by the speed and relentlessness of change characteristic of today's marketplace by assisting them to find a basic congruence of strategic values, converge channel core competencies, remove redundancies, shrink costs, and accelerate the velocity of information and product among supply channel partners. Finally, SCM provides channel executives with the capability to continually reposition marketing strategies, the role of channel members, the mix of products and services, productive and distributive processes, and human and technological resources to respond decisively to the marketplace with overwhelming competitive superiority.

Supply chain management is, above all, a business philosophy that enables individual companies as well as channel members to achieve high levels of *productivity, profit,* and *growth.* By following the SCM philosophy, channel constituents have the ability to develop virtually endless permutations of channel structures, each possessed of superior core competencies and the potential for innovative product and service breakthroughs targeted at realizing distinct marketplace opportunities and new competitive space. In the past, successful companies relied on the development of fixed, intricate channels of supply where standardized products would be distributed based on the least-cost principle. Today, competitive advantage belongs to those supply channels that can activate concurrent business processes and core competencies that merge infrastructures, share risk and costs, leverage the shortness of today's product life cycles, and reduce time to market, and that gain and anticipate new vistas for competitive leadership.

Implementing SCM

Converting the principles of SCM into a detailed program for competitive advantage is perhaps one of the most difficult challenges for even the best of organizations. In addition, there is no easy formula to follow to guarantee the successful implementation of SCM. Each company may have a different experience depending on their goals and points of emphasis. BASF Corp., the North American representative of the BASF Group with $5 billion in sales and 16,000 employees, has discovered 5 success factors in developing their SCM initiative. The first is the selection of "world-class" supply channel partners, ranging from logistics to computer software support, that assist the company to leverage state-of-the-art services and technologies. The second success factor is the development and promotion of the company's people resources. Company management feels that education and training are critical processes that foster awareness, encourage the creation of new skills and knowledge, and prepare the organization to meet the challenges of SCM. The third factor is found in BASF's commitment to the development of an operational structure consisting of cross-functional teams and the breakdown of the organization by functional area. The creation of channels of communication concerning information and ideas forms the fourth critical factor. As the organization focuses increasingly on customer value creation by communicating concurrently across functions, business units, and channel partners, BASF has become extremely sensitive to technology enablers that provide for effective decision-support. The final success factor, management leadership, is the keystone in setting goals and direction as well as providing education and training, guiding the whole process.

Although BASF has just begun to tap the unlimited competitive potentiality of SCM, the company has been able to report outstanding successes. In those areas where the SCM concept has been applied, customer service levels have improved 54–99%, on-time delivery to at least 97%, reduced inventory investment by 30%, and reduced administrative effort by 50%. As the company deepens its commitment to the work force and its efforts to apply integrated, computerized information technologies as part of the corporate SCM plan (labeled "Vision '96"), other benefits, cost reductions, and productive gains are expected [11].

The Emergence of Supply Chain Management

The group of ideas and market forces that have been slowly converging over the past half-decade into the concept of strategic SCM is the result of dramatic changes occurring in the nature of business competition, the rise of new global markets, and the veritable revolution in the way the enterprise is structured and how business processes are planned and executed. Current literature and business seminars are filled with descriptions of the critical changes driving such potentially threatening forces as the continued fracturing of markets, development of more flexible manufacturing processes, collapsing product life cycles, global networks, mass customization, corporate reengineering, and others. Detailing the content of these changes is a fairly easy task. These key business "revolutions" and how they are molding the shape of strategic SCM are described in Chapter 2. What is more difficult to define, however, is how these interconnected dynamics collectively are transforming the marketplace and subtly engineering dramatic changes in the methods by which companies maneuver for competitive advantage. And, although investigating today's business dynamics does shed some light on what is happening, it is obvious that merely tallying these market forces does not add up to an understanding of the whole.

The Trouble with Conventional Organizations

Much of the problem in understanding today's global business changes resides in the fact that most of us still conceive of competition and the marketplace according to the traditional paradigms instilled in the business schools and handed down in the folklore of business management. The scenario goes something like this. Competition is played out on the battlefield of markets, where individual companies employ the weapons of product and service quality, price, and delivery in the struggle for supremacy. Successful companies compete by designing, marketing, manufacturing, and distributing families of low-cost, mass-produced goods and services that enable them to win out against potential rivals. Profits provide capital for product and process improvement. Marketshare is then solidified and extended by listening to the "voice of the customer" and refining superlative logistics functions that further shrink costs and cycle times.

Trouble begins when powerful or innovative competitors, either within or outside of the established product and marketing system, challenge the existing paradigm. These challenges can take the form of the appearance of new products and processes, access to innovative knowledge bases and critical core competencies, development of new systems of customer value, and shifts in the critical mass among the players constituting the market system. Outwardly, the signs of these challenges usually appear in the inability of former leaders to control the market, the arrival of new copy-cat entrants, growing commoditization of once unassailable product and service offerings, loss of channel bargaining power, and eventual erosion of profit margins.

The traditional response of management to these formidable challenges has been to turn inward and closely examine those products and processes that had once proved so successful. One strategy is to abandon those products and service offerings whose margins have become razor-thin and reside on the periphery of corporate competencies. Similarly, whole markets, once considered proprietary reserves, can be deserted and left to new competitors who have carved out seemingly impregnable positions. Top management can also attempt to become creative, seeking to reengineer wasteful practices and redundancies out of the organization while streamlining process flows. The most widely used response is simply to downsize. The first to go is the company's work force, followed by ever-widening cuts in productive and administrative assets. Although the round of brutal cost cutting enables management to maintain margins for a while, the holding action is pyrrhic and self-defeating. Denuded of critical work force competencies and shrinking cash flows, the downward spiral caused by the failure to invest in new processes and products accelerates, the savage assault of competitors on margins intensifies, and stakeholders withdraw in search of more lucrative investments. In the end, the business collapses as surviving competitors devour what is left of the remains of a once proud industry leader that failed to adapt to new marketplace realities.

Although the above drama presents a grim picture which too many companies in the steel, automotive, electronics, machine tool, and other industries know all too well, it is a story caused not by greed or laziness on the part of workers and management, but is the result of a failure to understand the dynamics of today's business environment. The arrival of new competitors, new information and mechanical technologies, deregulation, and the challenge of responding to a global marketplace consisting of customers who demand more from their suppliers at less cost have exposed the fatal weaknesses of companies that feel that they can still depend on the tired paradigms of the past. The following problems facing conventional organizations have slowly begun to emerge.

- *Reliance on "fixed" industry structures.* Mass-production era organizations view the marketplace as composed of immutable *industries*, such as the automotive industry, the health care industry, the retail industry, and others. Each industry is considered a closed business system, inhabited by rival companies who struggle with one another for local supremacy.

- *Presence of functional boundaries that inhibit process innovation.* The ability of companies to create the innovative intraenterprise and interenterprise breakthroughs necessary to engineer responsive systems to meet the thrusts of sophisticated competitors and the needs of today's global customer is frustrated by organizations founded on rigid business functional hierarchies. Conventional organizations tend to focus on local performance measurements and piecemeal improvements associated with narrow business functional areas.

- *Dependence on vertical and horizontal integration.* Conventional organizations attempt to maintain control of supply channels, production processes, and distribution by integrating different business sectors together. The goal is to maintain flexibility and industry independence and to safeguard product and process competencies.

- *Focus on economies of scale and scope.* Conventional organizations are structured to exploit the economies of scope and scale made possible by vertical integration and centralized hierarchical management methods. The goal is to engineer products and processes that ensure lowest unit cost for standardized products and services to be sold to a mass market.

- *Inability to provide the customer with a single point of contact.* Traditional organizations divide servicing the customer into several spheres. Marketing and sales focus on pretransactional activities associated with isolating markets and executing, negotiating, and closing activities. The responsibility of production and distribution operations is to fulfill time, place, and delivery utilities. Finally, postsales is the reserve of separate customer service, training, and warranty groups.

Many companies today have been bewildered by the marketplace forces that have made their traditional operating paradigms obsolete. Global competition, fragmenting markets, short product life cycles, shrinking development cycle times, exploding product lines, mass product customization, and the need to continuously reinvent the organization are but a few of the dynamics that have shaken once powerful companies to their very foundations.

Today's Business Ecosystems

The problem with the conventional business paradigm is that it assumes each company is an island and that cooperation with other organizations is self-defeating. Such a view ignores the fact that every enterprise is, in reality, part of a much larger matrix of intersecting business systems composed of intricate, mutually supportive webs of customers, products, and information played out on a global scale. The idea that companies can develop and become successful independent of the business systems to which they belong is a myth. Today, companies around the globe have begun to understand that survival requires cooperation as well as competition with other channel members, sometimes including competing channels, that together shape the destiny of the whole system.

Moore [12] aptly describes this dynamic in biological terms as a *business ecosystem* and perceives the central organic mechanism as the intense coevolution of companies within a channel system coalescing around commonly shared strategic visions and mutually supportive competencies. Basically, Moore feels that instead of developing in isolation, today's most successful companies achieve marketplace leadership by identifying and being identified as mutually supporting members of a distinct business ecosystem. Although each member of the system struggles to achieve its particular business goals, ultimate success can only be realized by drawing on the core competencies, physical resources, and innovative capabilities of other channel system members. In other words, today's market system is being driven not by individual companies but by clusters of allied business partners who contest, and sometimes cooperate, with other business ecosystems in the search for competitive advantage.

The concept of supply chain coevolution provides several key benefits. To begin with, coevolution lets companies focus on what they do best, while still providing them with access to a broad range of channel resources and processes that are beyond the scope of their internal competencies. Microsoft, for example, dominates a business ecosystem composed of industries ranging from personal computers and consumer electronics to information and communications technologies. To support its growth, Microsoft's business channel system depends on an extensive network of suppliers, such as Hewlett-Packard and Intel, and fosters the evolution of thousands of customers in other business ecosystems. Also, by building mutually sustaining business communities, each member of the supply chain can develop much stronger defenses to ward off potential competitors than would be the case with enterprises that develop in isolation. Moreover, channel coevolution is a reciprocal process that accelerates innovation by underscoring the advantages found when allied companies join capabilities and work cooperatively and competitively to generate new products and productive processes, create new forms of customer value, and, in the process, drive the next round of innovation.

Finally, supply chain coevolution enables the development of a broad sense of shared vision among the members of the business ecosystem community. Shared vision accelerates the process of activating the potential of a new competitive idea, rousing the innovative power not only of a company and the supply chain system to which it belongs but also of other intersecting systems attracted by the inherent possibilities. An example of such visioning is the business, communications, and entertainment potential to be found in the Internet. The Internet has spawned whole new industries and joint developments ranging from advertising and services, to software, to technical articles and other literature. This sense of visioning and shared imagination provides for an extended view of the marketplace. Instead of a purely enterprise focus, today's business environment must be conceived of as an enormous macro-marketing system composed of mutually supportive channel systems—individual companies, channels of customers and suppliers, competitors, government and labor unions—that both draw on the resources and contribute their own competencies to the community at large. Figure 1.3 attempts to portray this concept.

Today, companies have begun to realize that if they are to succeed in the feverishly competitive world of contemporary global business, they need to organize themselves and develop strategies to exploit the opportunities of the supply chain systems found in their business environments. Two essential dynamics have emerged. The first involves continuously looking for innovative ways not just to penetrate existing markets but to create new sources of value that, in turn, germinate whole new markets. Such an objective requires a complete reevaluation of the traditional concept of what constitutes *competitiveness*. Value-enhancing competitiveness is characterized by agile marketing and operations organizations that focus on constantly creating and delivering customer-enriching, individualized combinations of products and services at the time and in the quantities demanded by the marketplace. It requires the engineering of cooperative product design processes where enterprises utilize the special competencies of a

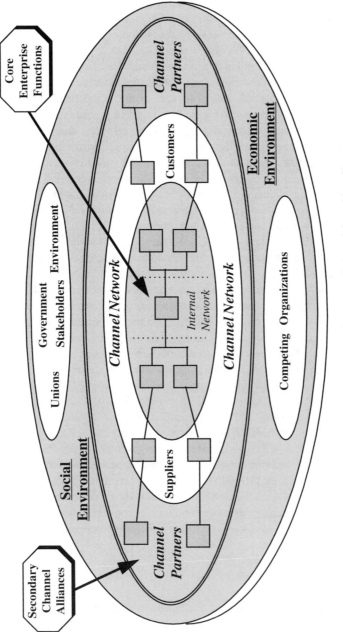

Core
Enterprise
Functions

Secondary
Channel
Alliances

Figure 1.3. Supply channel systems. (Adapted from Note 2)

network of business partners for the rapid development, concurrent management, and eventual disposal of products at the end of their life cycles. Finally, at the core of customer value-creating organizations stands the ability of the work force to activate breakthrough thinking and latent resources both within the enterprise and among channels of cooperating companies.

The second essential critical dynamic necessary for companies seeking marketplace leadership is the development of strategic relationships with other organizations in the search for the vision as well as the competencies and resources necessary to sustain competitive survival. To an unprecedented degree, today's companies are entering into partnerships, joint ventures, and collaborations of every imaginable kind, including the structuring of "virtual" organizations. The goal is to accelerate product development time to market while reducing cost structures so as to exceed customer expectation and preempt the competition. Achieving this goal requires the creation of panchannel processes that provide for the coevolution of customer value across companies inhabiting the same or intersecting business ecosystems by merging similar capabilities, designing teams for the joint development of new processes and technologies, and structuring new forms of vertical integration and economies of scale that draw on the core competencies within or among associated enterprises.

Today's most creative companies, such as Intel, US Robotics, L.L. Bean, Wal-Mart, and others, have learned how to continuously create customer value by leading economic coevolution. The role of their top management groups is to search for and initiate centers of innovation, and then to orchestrate a network of supporting enterprises from which new forms of competition and new businesses can evolve. Such activities are not just for major companies such as IBM, Hitachi, Xerox, and other global giants. For example, NESI (New England Suppliers Institute), a public organization composed of member companies and government and educational partners, was created to improve relationships between small and large New England manufacturers. Its goal is to create business partnering alliances aimed at developing stronger regional customer and supplier bases. Among the activities of the association can be found assistance in achieving quality goals, such as ISO 9000 certification, sharing of "best practices," the formation of virtual manufacturing networks, and seminars, roundtables and focus groups on topics such as customer-supplier relationships, pull systems for small manufacturers, EDI, total-cost-reduction techniques, and supplier development models [13].

Role of Supply Chain Management

As stated above, the fundamental challenges before today's enterprise have less to do with the structuring of more effective management methods or the ceaseless search to eliminate costs, and more to do with understanding the dynamics of the intersecting business ecosystems to which the company belongs and exploring the potential for new forms of competitive advantage gained through supply chain coevolutionary processes. No team of senior managers would seriously maintain in the current environment that they can guide their company to success by running their organization in abject isolation and ignoring the interlocking web of ideas, technologies, and processes exploding all around them. In fact, for over a decade, managers have been learning how to leverage the strengths of business partners up and down the supply channel. Beginning with the creation of preferred supplier and customer agreements, today's logistics executives have been cautiously yet decisively moving toward the establishment of closely integrated networks of channel partners. Indeed, some companies, such as IBM, Toshiba, Siemens, and others, have tapped into the tremendous enabling power of technology and new management thinking to tear down internal and external functional boundaries, leverage capital and human capacities, and merge core competencies to establish "virtual" organizations capable of responding immediately to any marketplace opportunity.

What is being maintained in this text is that this nascent management paradigm forms the essence of the SCM concept. As pointed out earlier, the operations-oriented definitions of SCM are too limited in scope to assist today's enterprise in leveraging the opportunities for marketplace advantage to be found in the channel systems to which they belong. On the other hand, the strategic dimension of SCM provides companies with the vision as well as the day-to-day operational principles to exploit the competitive possibilities of a global business environment bursting with coalitions, partnerships, and consortia of all sorts. Strategic SCM provides executives with the ability to deal with today's coevolving networks of global information and supply channels, structure their organizations to meet the challenges of creating decisive customer value, actualize breakthrough product and service opportunities, and create competitive-enhancing, interlocking supply chain communities focused on the development of shared visions and possessed of the unlimited potential for innovation.

In the future, managers can expect their concentration on traditional management paradigms and metrics to diminish and involvement with SCM strategies to dramatically increase. Effectively utilizing SCM means that managers must seek to merge the supply channels to which they belong into innovative, value-enhancing customer-focused channel systems. This involves understanding critical channel partner competencies, identifying centers of innovative potential, and mapping how well they are being realized. Finally, managers must continuously engineer new business relationships that leverage the capabilities of the very best companies and structure superlative channel system product and delivery processes that provide continuous and unassailable customer value.

Supply Chain Management and Logistics

This chapter has focused on detailing SCM as a strategic supply chain management philosophy that seeks to utilize the special competencies both within organizations and the channel systems in which they participate in the search for sustained marketplace leadership. At this point, it might be useful to redefine SCM's relationship to *logistics management*. It has already been discussed that integrated logistics management constitutes the tactical side of the SCM concept. In addition, there can be no denying that in the emergence of modern logistics can be found the seedbed of SCM. As the role of logistics has expanded from a preoccupation with warehousing and transportation to today's concern with integrating the logistics operations of the entire supply channel, SCM has been instrumental in merging marketing and manufacturing with distribution functions to provide the enterprise with new sources of competitive strength. In addition, the application of SCM can been seen in the integration of logistics activities among supply chain partners in the pursuit of shorter cycle times and reduced channel costs.

Now that a concise definition of SCM has been introduced, it is possible to clarify the role of logistics management. Several key questions come to mind:

- What is the definition of logistics management?
- What are the organizational elements of the logistics function?
- How does logistics add value?
- How does SCM and logistics support one another?
- What impact will the growth of SCM have on the future of logistics?

Defining Logistics

Over the past 25 years, the management science of logistics has progressed from a purely operational activity to a competitive weapon enabling the enterprise to pursue new avenues of marketplace advantage. As an operations function and as a business concept, logistics management has evolved over the past four decades. Originally, the role of logistics management centered around providing value-added warehousing and transportation functions that enabled companies to deliver product in support of marketing and financial objectives in as timely and cost-efficient a manner as possible. Today, modern logistics has become in itself a key enterprise competitive resource, creating marketplace value by responding to ever-higher levels of customer expectation, engineering operations that produce superlative quality while reducing costs, and highlighting misalignment of key enterprise elements, such as product positioning, channel integration, work force competencies, and business strategies that can inhibit enterprise growth.

These objectives have been detailed in several popular logistics definitions. The most often quoted has been formulated by the Council of Logistics Management.

> Logistics is the process of planning, implementing, and controlling the efficient flow and storage of raw materials, in-process inventory, finished goods, services,

and related information from point of origin to point of consumption (including inbound, outbound, internal, and external movements) for the purpose of conforming to customer requirements. [14]

This definition portrays logistics as a integrative process that links the flow of materials and information as they move from channel function to channel function.

A more channel/value-oriented definition can be found in Novack, Rinehart, and Wells. [15]

> Logistics involves the creation of time, place, quantity, form, and possession utilities within and among firms and individuals through strategic management, infrastructure management, and resource management, with the goal of creating products/services that satisfy customers through the attainment of value.

In a similar vein, Christopher [16] defines the mission of logistics management as the planning and coordinating of

> all those activities necessary to achieve desired levels of delivered service and quality at lowest possible cost. Logistics must therefore be seen as the link between the marketplace and the operating activity of the business. The scope of logistics spans the organization, from the management of raw materials through to the delivery of the final product.

These and other definitions imply that logistics creates competitive value by optimizing on operations cost and productivity, better capacity and resource utilization, inventory reduction, and closer integration with suppliers. Of equal importance is the ability of logistics to assist in the creation of marketplace leadership through customer service and timely product and service delivery. Furthermore, the effective and efficient performance of these objectives is achieved by the integration of all functional processes performed by supply channel members. These processes include decisions relating to channel design and structures, resource allocation, the human organization, operations, and finance. Logistics creates competitive advantage by flawlessly executing customer service objectives, achieving conformance to quality standards, and increasing marketplace value.

Perhaps the best way to understand logistics is to divide it into two separate, yet closely integrated functions as illustrated in Figure 1.4. The first function is termed *materials management* and is identified with the *incoming flow* of information and material into the enterprise. Materials management can be defined as the collection of business functions supporting the cycle of materials flow from the planning, purchase, and control of inventory, to manufacturing and delivery of finished goods to the distribution channel system. Specific activities in materials management can be separated into *product acquisition*, such as supplier management, purchasing, receiving, and quality assurance, and *inventory management*,

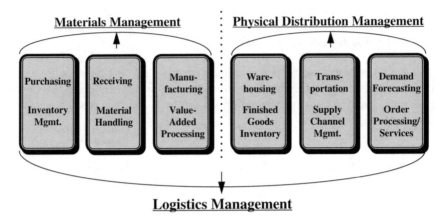

Figure 1.4. Logistics management.

such as receiving, materials handling and storage, work-in-process inventories, value-added processing, and delivery of goods to the distribution channel.

The second function comprising logistics is *physical distribution*. This function is normally associated with the warehousing and movement of finished goods and service parts through the distribution channel to meet customer order fulfillment and delivery requirements. Often physical distribution is so closely tied to customer service, forecasting, and channel management that, in some companies, is actually part of the marketing/sales department. Physical distribution activities can be described in detail as including warehousing, transportation, finished goods handling and control, value-added processing, customer order administration, warehouse site/location analysis, product packaging, shipping, and returns goods management. Although it can be rightfully argued that separating logistics into the above two categories is somewhat artificial, the objective of the exercise is not to describe an organizational structure but to communicate concrete value-added activities that detail the flow of products and information as they move up and down the channel supply system.

Logistics Quality and Value

Whereas efficient logistics functions enable companies to respond to customer requirements, conformance to quality standards link logistics output with the level of service quality expected by the customer. Essentially, logistics quality consists of three key elements: logistics productivity, logistics service performance, and performance measurement systems. Logistics productivity is demonstrated by the establishment of meaningful productivity standards, the ability to manage logistics costs, the integration of quality management processes, and the broadening of logistics service levels. Logistics performance can be assessed through metrics associated with customer service goals, such as product availability, order cycle time, logistics system flexibility, depth of service information, and breadth of postsales service support.

Finally, logistics adds competitive advantage by adding value to the customer that extends far beyond the simple performance of product transaction activities. These value-added functions consist of the following five key dynamics [17]:

- *Transportation.* This activity adds value by ensuring the timely delivery of quality materials, components, and finished goods that enhance customer capabilities to respond effectively to their marketplace opportunities.

- *Operations.* The ability to effectively execute internal and channel-wide operations, such as production, warehousing, distribution, and delivery that enable companies to continuously reduce costs, increase profits, and engineer flexible organizations that position them to respond to any marketplace opportunity.

- *Inventory.* Controlling inventory costs while increasing serviceability locally within each company and externally throughout the supply channel pipeline has a direct impact on logistics performance. Inventory management methods that constantly search for methods to reduce inventory enable companies to better leverage financial resource that accentuate customer service needs.

- *Information.* Today's information and communication technologies enable the linkage of whole supply channels, make more effective the planning of logistics resources, and facilitate customer order processing and service management. Information management increases competitiveness by shrinking order cycle times, reducing stocked and in-transit inventories, and facilitating planning and operations activities.

- *Special Functions.* Logistics provides competitive advantage by meeting governmental regulatory requirements, such as the execution of reverse logistics activities, and pollution and energy conservation requirements. In addition, integrated logistics assists in the enhancing marketing activities such as sales and special promotions.

Redefining Logistics

The evolution of strategic SCM has elevated logistics management to a new level of critical importance in today's enterprise. It is the role of SCM to assist in the generation of new marketplaces and seek out new forms of product and service innovation that enable individual companies, if not whole channel systems, to achieve unassailable competitive advantage. By providing the capability to tap into the almost limitless reserve of skills and resources found within the supply chain system, SCM can activate the visioning and productive powers of an extended community of businesses to realize continuous order-of-magnitude breakthroughs in products, processes, and services. Actualizing the possibilities of strategic SCM requires superior, highly integrated logistics operations that are aligned and structured to meet the quick response needs of a high-velocity, continuously changing global marketplace. And, although marketing and sales, manufacturing, and finance will be required to find new, dramatic solutions to the challenge of strategic SCM, it is logistics that will be entrusted with establishing the groundwork where channel alliances will be germinated and the cycle of innovation, design, production, distribution, and eventual obsolescence played out. Effectively performing such a role will require logistics managers to critically rethink the nature of both their own roles and responsibilities as well as the structure and daily operational objectives of logistics functions.

Supply chain management presents individual managers with fresh opportunities and new career directions. Instead of a singular focus on achieving day-to-day operational targets, logistics executives will be required to spend increasingly larger amounts of time structuring channel partnerships and searching for channel network opportunities. This will mean that the ability to manage marketing, design/development, and logistics activities *outside* of the firm will eventually become more important to logistics professionals than the ability to manage *within* their own companies. Skills associated with negotiation, team-building, the capacity for innovation, the development of consensus with channel partners, and the application of information and communications tools will vastly overshadow the traditional operations activities associated purely with managing warehousing and transportation. Moreover, as the marketplace becomes more international, logistics managers can expect to see responsibilities associated with managing on a global scale to equally grow [18]. The expansion in the scope of leadership necessary to realize the full potential of SCM is summarized in Table 1.1.

Supply chain management also requires radical thinking that challenges traditional paradigms not only about logistics organizational structure and design but also about how channel partners are to be integrated, the level of participation of suppliers and customers in product and process design and logistics decisions, employee ownership and involvement, and the use of information and communications technologies. Activating the power of logistics in the *era of supply chain management* will necessitate changes in the way logistics resources are organized and how they are to be effectively managed. The ability of logistics to stimulate innovative thinking will enable people throughout the internal organization and outside in the channel network to focus specialized skills and talents on targeted business processes, as well as facilitate the growth of cultural values fostering teamwork, value-added activity, continuous improvement, and total customer service.

Table 1.1. SCM management focus.

Management Scope	Traditional Focus	SCM Focus
Management process	Products, sales, revenues	Interorganizational processes, extended processes, investment in channel innovation
Key performance targets	Departmental objectives, processes and product specs	Innovative and value-adding capabilities of the entire channel
Business goals and objectives	Consistency of performance, departmental alignment, key benchmark metrics	Alignment of channel objectives and goals, shared competitive channel vision
Business Relationships	Focus on internal structures and organizational values	Structured channel partnerships, coevolving processes and objectives
Business Process Improvements	Reductions in costs and defects; rate of improvements in products and processes	Rate of progress of entire channel; rate of channel value creation and innovation

Research Opportunities

As a management science, the concept of SCM is in its infancy. Although the purpose of this book is to attempt a concise definition of the operational and strategic aspects of SCM, it has barely scratched the surface of the vast amount of topics open for further research. Thus far, the subject has received an altogether uneven treatment. Although it is true that articles, research papers, and popular seminars purporting to delineate some critical aspect of SCM abound, all have failed to provide a consistent, structured body of knowledge. In fact, SCM has become such a popular term for describing any type of business activity relating to dealing with customers and suppliers that it has begun to be devoid of meaningful content. Some proponents claim that SCM is nothing more than changing one's viewpoint about how one manages customers and assets by including outside suppliers and partnerships. Others perceive it as a method for optimizing supply channel performance across multiple enterprises. The American Production and Inventory Control Society (APICS) has even created a new certification exam for its CPIM program entitled *Basics of Supply Chain Management,* in which the term SCM is really used as a catch-all term to cover the fundamental operations activities of a manufacturer as well as to introduce the supply channel as a set of interlinked processes used to manage a firm's inventory and information flows [19].

What is being argued is that educators and industry experts have spent a great deal of time discussing those aspects of SCM that are eminently *researchable* and eminently *discussable,* thereby passing over topics that really *need* to be researched and discussed. Such an attempt to more fully detail the content of SCM can only come about when practitioners and consultants work closely with researchers and educators in exploring new SCM topics while validating primary assumptions. The following areas of SCM afford significant opportunity for future productive research:

1. SCM Theory
    - **Developing SCM strategies.** Although the operations aspects of SCM have come under close scrutiny from a variety of directions, the subject of SCM strategy has received little or no attention. Significant research needs to be undertaken on the integration of SCM with company marketing, financial, manufacturing, and logistics strategies. Research should include defining the procedures for developing intercompany SCM strategies and what form an effective SCM strategic plan should take. In addition, much research needs to be performed on methodologies for the successful implementation of strategic SCM.
    - **Integration of SCM with other management philosophies.** As SCM matures as one of today's most critical management philosophies, it needs to be integrated with other contemporary philosophies such as JIT, ERP, TQM, business process reengineering, and agile manufacturing. The goal is to show how SCM provides today's enterprise with effective tools to manage intercompany strategies that are complementary with other high-profile value-creating management paradigms.
    - **SCM and global trade.** Probably one of today's most fertile areas for the application of SCM can be found in the growth of the global marketplace. The crisp execution of international trade requires the development of superlative channels of supply that promote the smooth flow of products and information across international boundaries. SCM enables international companies to closely integrate strategic competencies and resources that provide interchannel process teams with the ability to pursue the best opportunities for global marketplace advantage.
    - **Requirements and costs of SCM.** Shifting from an enterprise to an interenterprise management focus will have a profound impact on companies moving toward SCM. What will implementing SCM require in the way of a skilled and knowledgeable work force? What organizational changes will be required to operate with other "virtual" supply chain partners? What will it cost to implement the

information technologies necessary to network the supply chain? Other research in this area includes investigating how companies will have to alter traditional paradigms governing marketing, manufacturing, and logistics decisions when implementing SCM.

2. SCM Operations Elements

    -   **Defining SCM operations strategies.** Although the operations aspects of SCM have been explored in great detail over the past few years, the research has been disjointed and oftentimes confusing. What is needed is a concise delineation of the areas of SCM operations functions ranging from purchasing to customer service. The goal is the creation of an SCM information repository for companies to reference when undertaking SCM improvement projects.

    -   **Applying information and communication technology** (ICT). Perhaps one of the most uncharted areas of SCM is the application and implementation of computerized tools to facilitate SCM operations functions. Without a doubt, the capabilities of ICT systems are at the very heart of SCM and their continued development directly impacts the development of the SCM concept. Research needs to explore issues ranging from the utilization of ICT for data collection to complex networking tools such as the Internet and electronic notes that bind supply chain members closely together.

    -   **Designing effective supply channels.** The supply channel is the physical manifestation of the SCM concept. Supply channels define the relationship partner companies have with each other as well as the actual flow of products and information as they pass up and down the supply pipeline. Research in this area must focus on such questions as: What is the channel's competitive mission? How should the channel be designed to facilitate SCM concepts? What is to be the nature of the integration of product development, manufacturing, and logistics functions that are to govern how companies interact with each other? A fertile area for research is exploring how the application of SCM principles impacts the performance of basic channel operations and marketing functions.

    -   **SCM process teams.** Realizing SCM's potential requires channel members to construct effective process teams. How should these process teams be developed and structured? What rules should govern team leadership, decision making, and the selection of improvement initiatives? How do companies select team members? Further research into defining the mechanisms necessary to detail team performance and capability for innovation is required.

3. Using SCM on a daily basis.

- **Managing supply channel inventories.** The prime purpose of channel inventory is to provide network partners and external customers with products in the quantities and at the time and place required. Critical questions for research are the following: What is the optimal level of inventory vertical integration and what channel inventory positioning alternatives should be pursued by each channel member? What is the desired balance between inventory and customer service necessary to meet the marketing objectives of the entire supply chain? What is the desired balance between inventory investment and associated overhead costs? How can the causes that produce pipeline cycle and safety stock inventories be eliminated? A further area for research is developing a channel inventory planning process that optimizes channel service levels and facilitates inventory flow-through.

- **Synchronizing channel ordering systems.** One of the fundamental requirements in supply chain inventory management is the coordination and integration of the inventory ordering systems used on all levels of the supply channel. The goal of this line of inquiry is to research the various ordering techniques like Order Point/Economic Order Quantity (EOQ) and Distribution Requirements Planning (DRP) to tightly link channel replenishment requirements. How can inventory planners across the channel make optimum use of channel ordering systems? What are the limits of these systems and what new techniques can expand the scope of channel inventory management? Another important area for research in this area is the implementation of continuous replenishment techniques and their impact on channel organizations and productivities.

- **Customer order management.** Today's integrated supply channel requires world-class customer order management systems. These systems should provide for the synchronization of inventory supply and customer demand. What automation tools should be implemented to improve order management? How are companies to establish the customer service targets to be used by the entire supply channel? How are supply channels to manage the physical distribution requirements of the channel? Much research needs to be done in the area of defining customer service quality gaps on the channel level.

4. Quality Management

- **Developing channel quality standards.** The creation and maintenance of superlative channel service quality and value is critical to strategic SCM. Who is responsible within the supply chain for the establishment of channel quality standards? What steps are necessary to establish an effective channel quality management program? What are the dimensions of a supply chain's quality program? What quality analysis techniques are best fit for the supply chain? Who will be responsible for developing effective quality management teams and how will they be applied to improve supply channel productivities?

5. SCM Education

- **Curricula design and available literature.** As this book is being written, educational materials designed to teach the elements of SCM are largely undeveloped. Few college courses teach courses where SCM is even mentioned, and professionals societies have not advanced beyond the concept of integrated logistics management. A minimal number of case studies have been written on the topic and sources have been limited to trade-magazine articles. This area is wide open to creative research and development. Much work is needed in curricula design, development of teaching materials, classroom examples and exercises, and the writing of case studies.

Summary and Transition

Perhaps the greatest challenge before today's enterprise is managing the impact of the enormous changes taking place in the development and marketing of products on a global basis, the need to constantly reengineer effective organizational structures and values, and the capability to activate the unlimited possibilities found in information and communications technology enablers. In the past, companies approached the marketplace by offering an array of standardized products and services, seeking competitive advantage through strategies focused on low-price, vertical channel dominance, internal efficiencies and lowest-cost processes, and mass-marketing approaches. As the final years of the 20th century conclude, however, companies have increasingly found themselves having to grapple with new concepts of what constitutes customer value, new requirements for flexible product design and manufacturing processes, the growth of "virtual" companies and integrated partnerships, and continuously fragmenting niche marketplaces demanding customized products and services to satisfy a "marketplace of one."

Of all these changes, perhaps the most important is the realization that every enterprise is, in reality, part of a much larger matrix of intersecting business systems composed of intricate, mutually supportive webs of customers, products, and information played out on a global scale. Market leaders understand that in order not only to survive but to succeed in today's business environment, they need to continuously activate the synergy that occurs when the capabilities aı. competencies found within their own companies are merged with those of other channel partners who, together, constitute a unique business ecosystem. Companies can no longer count on developing breakthrough products, services, and marketplace avenues in isolation. Only by converging the competencies possessed by channel partners into a unified competitive entity can leading-edge organizations hope to deliver the value their customers demand. The emerging networks of channel alliances and collaboration dotting today's business landscape, driven by the tremendous power of technology and new management thinking, have permitted companies to bypass the limitations of traditional internal and external functional boundaries, leverage capital and human capabilities, and merge channel partner centers of expertise to establish virtual organizations capable of responding decisively to any marketplace opportunity.

What is being maintained in this book is that this nascent management paradigm forms the essence of *supply chain management* (SCM). SCM provides today's managers with the strategic vision as well as the day-to-day operational principles to activate new areas of competitive advantage. Today, companies are finding that just focusing on designing more effective channel operations functions is insufficient to achieve and maintain superior levels of market leadership. What is needed is a comprehensive channel management philosophy that can be translated into strategies focused around merging the customer-satisfying functions and capacities occurring in each business supply channel node to form a unified, seamless structure with the strength to turn away the advances of even the most powerful of competitors. SCM is the term for this new strategic paradigm and it is defined as:

> a continuously evolving management philosophy that seeks to unify the collective productive competencies and resources of the business functions found both within the enterprise and outside in the firm's allied business partners located along intersecting supply channels into a highly competitive, customer-enriching supply system focused on developing innovative solutions and synchronizing the flow of marketplace products, services, and information to create unique, individualized sources of customer value.

Supply chain management has become today's most important concept for competitive advantage because it enables companies organized along a supply channel to exploit the new realities transforming the marketplace. As will be detailed in the next chapter, there are five critical dynamics driving today's marketplace. The discussion will surround each of these dynamics and how SCM can assist companies to succeed in each.

Notes

1. Robert A. Novack, C. John Langley, and Lloyd M. Rinehart, *Creating Logistics Value*. Oak Brook, IL: Council of Logistics Management, 1995, p. 216.

2. Lisa M. Ellram, "Supply Chain Management: The Industrial Organization Perspective." *International Journal of Physical Distribution and Logistics Management* 21 (1) (1991), 13–22; Lisa M Ellram and Martha C. Cooper, "Supply Chain Management, Partnerships, and the Shipper-Third Party Relationship," *International Journal of Physical Distribution and Logistics Management* 1 (2) (1990), 1–10.

3. John T. Mentzer, "Managing Channel Relations in the 21st Century," *Journal of Business Logistics* 14 (1) (1993), 31; C. John Langley, Jr. and Mary C. Holcomb, "Creating Logistics Customer Value," *Journal of Business Logistics* 13 (2) (1992), 1–28.

4. Martha C. Cooper, "Logistics in the Decade of the 1990's," in *The Logistics Handbook*. (James F. Robeson and William C. Copacino, eds.) New York: The Free Press, 1994, p. 46.

5. Bernard LaLonde, "Small Shipments—Manage the Supply Chain," *Transportation & Distribution* 37 (11) (November 1996), 15.

6. Lisa Williams Walton and Linda G. Miller, "Moving Toward LIS Theory Development: A Framework of Technology Adoption Within Channels," *Journal of Business Logistics* 16 (2) (1995), 117.

7. Martin Christopher, *Logistics and Supply Chain Management*. Burr Ridge, IL: Richard D. Irwin, 1994, p. 12.

8. Christopher Gopal and Harold Cypress, *Integrated Distribution Management*. Homewood, IL: Business one Irwin, 1993, p. 1.

9. Robert A. Novack, C. John Langley, and Lloyd M. Rinehart, *Creating Logistics Value*. Oak Brook, IL: Council of Logistics Management, 1995, pp. 39–40.

10. Steven L. Goldman, Roger N. Nagel, and Kenneth Preiss, *Agile Competitors and Virtual Organizations*. New York: Van Nostrand Reinhold, 1995, p. 204.

11. Bernard H. Flickinger and Thomas E. Baker, "Supply Chain Management in the 1990s," *APICS: The Performance Advantage* 5 (2) (February 1995), 24–28.

12. James F. Moore, *The Death of Competition*. New York: HarperCollins, 1996, pp. 15–16.

13. Sherry R. Gordon, "Changing the Structure of Business," *APICS: The Performance Advantage*, 5 (5) (May 1995), 36–39.

14. Council of Logistics Management, Oakbrook, IL, 1985.

15. Robert A. Novack, Lloyd M. Rinehart, and Michael V. Wells, "Rethinking Concept Foundations in Logistics Management," *Journal of Business Logistics*, 13 (2) (1992), 233–267.

16. Christopher, *Logistics and Supply Chain Management*, p. 9.

17. Robert A. Novack, et al., *Creating Logistics Value*, pp. 40–45.

18. See Bernard J. LaLonde, "Evolution of the Integrated Logistics Concept," in *The Logistics Handbook*, pp. 11–12, and David F. Ross, "Managing in a New Era," *APICS: The Performance Advantage* 6 (7) (July 1996), 62–63.

19. Stephen N. Chapman, "The New CPIM Module—The Basics of Supply Chain Management" 7 (1) (January 1997), pp. 18–20.

2

The Challenge of Today's Business Environment

Supply chain management (SCM) is a response to the overwhelming speed of the changes shaping today's business environment. Change and the acceleration of change on a global basis have become the foremost topics of the late 1990s, and its vernacular dominates today's management literature. SCM is founded on a philosophy of managing change and the impact of change on the fabric of the topography of business, enterprise, and supply channel cultures and structures, and how information technology is continuously reshaping the contours of tomorrow's marketplace playing field. In the past, procurement and market distribution were conceived as purely operational functions. One logistics authority in the late 1960s defined the supply channel succinctly as "moving goods at least cost," and detailed channel objectives as minimizing stock-outs, solidifying relations with customers, increasing delivery discounts, improving market coverage, freeing marketing people to create demand, and reducing inventory levels and freight and warehouse costs [1].

In contrast to those days when a preoccupation with reducing operations costs was the sole objective in achieving effective distribution management, today's enterprise must manage complex inbound and as well as outbound supply channel systems. The ability of companies to leverage the competencies of both their internal organizations and their channel partners, the explosion in high-value and mass-customized products and services, and the blinding speed characterizing customer delivery have elevated management of the supply channel from the operational backwaters of the past to perhaps today's single most important business function. Only those companies that have the ability to continually reposition their own businesses on the inside and their partners out in the supply channel to meet the challenges of a continually changing global marketplace will survive to the 21st century. In such an environment, marketplace leaders perceive SCM as an aggressive, competitive-enhancing catalyst enabling the activation of radically new management paradigms centering around organizational structure, channel configurability, production and logistics, and market leadership.

In this chapter, five key business dynamics impacting SCM will be detailed. The first, and most important, is satisfying the customer. The continuous development of products and services marked by superior design, high quality, low cost, and configurability demanded by today's marketplace forms the second dynamic. The third dynamic centers around the tremendous enabling power of information and communication technologies. The fourth dynamic focuses on the competitive advantage that can be leveraged by effectively utilizing the capabilities of the internal organization, the core competencies of channel partners, and the opportunities to be found in the global marketplace. Finally, the ability of the entire supply channel to compete successfully along the dimensions of time, customer service, and cost through logistics is the fifth business dynamic. All five business dynamics have a common motif: The relentlessness and swiftness of change caused by the heightening of customer expectations, shrinking product and service life cycles, and exploding information and communication technologies is irrevocable and unforgiving and is forcing even the best of companies to be dramatically more creative, flexible, and focused on productive competencies than they have ever been in the past. As will be seen, SCM provides whole supply channels with operational and strategic paradigms not just to weather, but to succeed despite the intensity of today's competitive environment.

Changing Business Perspectives

In the previous chapter, it was stated that the emergence of *supply chain management* is the result of the convergence of several critical market, product and service, organizational, and information and communication technology developments that are revolutionizing the very structure of today's business environment. Without a clear understanding of these dramatic marketplace changes, the fresh perspectives offered by SCM can be misinterpreted and its ability to create exciting new opportunities for competitive advantage lost. Perhaps the most revealing way to understand the impact of these developments is to counterpoise them with the business concerns of the past two decades. Such a comparison illustrates a rather different management view of the scope and responsibilities of the distributive and logistics functions than what can be found in today's enterprise.

Such a statement is self-evident. The critical marketplace concerns that dominated those times required logistics to perform other roles. Rather than a theoretical science, business management is preeminently centered around the necessity of responding to the current needs of the marketplace by shaping purposeful organizational objectives and realizing fresh competitive opportunities. SCM has emerged because the requirements of today's business environment have rendered past operational paradigms unworkable and driven enterprises to explore fundamentally new approaches to organizing the productive competencies of their organizations to meet the competitive realities of the marketplace.

Historical Perspective

In the mid-1970s, the business objectives of manufacturers and distributors were dominated by concerns over cost management and an internal search for functional definition. Unlike the comparatively prosperous decades of the past, companies in the United States were for the first time experiencing severe challenges from foreign competitors, materials and energy shortages, and bloated and inefficient organizations. For example, in the 1978 edition of *Logistical Management*, Bowersox labeled the decade as "a period of changing priorities" and detailed the challenges facing companies, and the logistics function in particular. The critical topics were as follows: (1) escalating costs and continued shortages in energy, particularly oil, (2) growing material shortages, (3) growing environmental pressures, (4) a shrinking United States economy racked by spiraling inflation and unemployment, (5) the integration of materials management and physical distribution in an effort to better control costs, (6) the capability of utilizing quantitative management tools to facilitate logistics decisions, (7) the elevation of logistics functions from a purely operational to a strategic position in corporate decision making, and (8) the propagation of new concepts of logistics organization and theory to assist companies in the struggle for competitive survival [2].

By the early 1980s, the business environment was in the midst of the worst recession since the Great Depression, accompanied by double-digit interest rates and high unemployment. In addition, US companies were now finding themselves in what was considered a life or death struggle with competitor nations offering high-quality products and driven by radically new management techniques. The response of logistics professionals to this period of marketplace uncertainty was the proposal of fundamentally new approaches targeted at cost reduction and ways to streamline the channel pipeline. For example, LaLonde felt the reconfiguration of the following five key logistics areas as essential to marketplace survival in the 1980s: (1) redefining sourcing/procurement strategies; (2) customizing customer service strategies to fit marketplace needs; (3) realigning the physical network of distribution channels; (4) developing techniques that fostered a closer integration of material flows; and (5) utilizing information technology as a substitute for inventory in the channel pipeline [3].

During the late 1980s and earlier 1990s, new market forces quickened the pace of internal logistics and external channel management change. One of the most significant was the continuation of deregulation, which in turn, spawned a whole new industry of logistics service companies. The acceleration of the application of information and communication technologies to supply channel functions was also changing the face of logistics operations. In conjunction with the implementation of Just-In-Time (JIT) and *lean manufacturing* management techniques, technology tools such as EDI and bar code symbology enabled the closer integration of logistics functions and provided companies with the ability to further compress time and reduce costs throughout the supply channel. Finally, the rising power of the customer significantly impacted the development of the logistics organization. Firms that elevated logistics to a strategic position found their ability to compete had increased significantly. According to a 1987 Arthur Andersen report, these dramatic changes had ushered in a "period of aggressive restructuring" in the distribution industry. Logistics and supply chain management were now being positioned as critical resources in responding to a business environment marked by maturing markets, declining margins, excess capacity, new forms of competition, and the fragmentation of the marketplace [4].

Today's Business Perspective

As discussed in the previous chapter, the climate of today's business environment has required enterprise managers to rethink traditional business structures and conventional methods of approaching the marketplace. This reevaluation process is particularly true of the logistics function. Not only has the operational requirements of the enterprise changed to meet customer demand, but the need for quick response and continuous innovation in product development, services, manufacturing, and delivery have necessitated companies moving beyond the logistics concept to embrace the SCM paradigm. Although a concern with logistics cost management is critical, today's logistics manager is more than ever migrating away from an internal operations focus to an external strategic viewpoint driven by marketplace globalization, the enabling power of technology, increasing use of third-party service providers, developing business alliances and "virtual" organizations, and continued organizational reengineering. Still other changes are reshaping the values, educational requirements, and objectives of today's work force.

Understanding the impact of the challenges confronting today's enterprise is at the core of utilizing SCM for competitive advantage. These challenges have been organized around five critical business dynamics. Collectively, the five dynamics present a dramatic departure from the past, each requiring radically new solutions. Understanding today's customer is the first dynamic. The key to this dynamic is the search for tailored combinations of products, services, and information that will provide the customer with unique value and a solution to their buying needs. Today's second business dynamic, the development of high-quality products and services, is characterized by the ability of producers to construct agile design, manufacturing, and delivery processes that provide superior product and service quality, yet can be quickly and cost-effectively configured to meet individual customer needs. The explosion in *information and communication technologies* (ICT) constitutes the third dynamic. The convergence of ICT and empowerment-based management techniques stimulating quality and productivity has opened new opportunities to reduce channel redundancies and shrink cycle times while uncovering new business perspectives for the alignment of channel operational and strategic activities. Leveraging the opportunities to be found in the opening of the global marketplace and the growth of strategic alliances and partnerships marks the fourth dynamic. Finally, the last business dynamic can be found in the capability of SCM to provide sources of continuous marketplace value by supporting outstanding customer service, reducing cycle times, increasing logistics service quality, and reducing total supply channel costs.

Customer Dynamics

The management of customer service and the ability to respond effectively to the needs and expectations of today's customer have become the dominant objectives pursued at all levels in the supply channel. In fact, it is difficult to read today's business literature without encountering some new model of customer service or detailed case studies illustrating some revolutionary "customer-focused" management style or technology application. The truth of the matter is that today's customers, with their expectations set by "world-class" companies across industries and continents, *are* demanding more from their supply channels in regard to high-quality, customized products and services, quick-response deliveries, and information and communication technologies. In the past, enterprises competed by selling product lines consisting of standardized, mass-produced products. Today, customers are demanding to be treated as unique individuals and expecting their suppliers to provide configured, variable combinations of products, services, and information that are capable of evolving as their needs change.

Understanding Customer Value

The key to satisfying the customer is understanding and continuously increasing what the customer perceives as value. What this means is that companies must be flexible enough to provide their customers the right mixture of products and services that fulfills their needs or enables them to pursue new opportunities. This is quite different from mass-production-era marketing objectives that focused on persuading customers to purchase already-configured products and services. Today's customers are searching for a tailored combination of products, services, and information that will provide them with unique value and a solution to their buying needs. This point has been succinctly summarized by Treacy and Wiersema [5]:

> Customers today want more of the things they value. If they value low cost, they want it lower. If they value convenience or speed when they buy, they want it easier and faster. If they look for state-of-the-art design, they want to see the

art pushed forward. If they need expert advice, they want companies to give them more depth, more time, and more of a feeling that they're the only customer.

Although what is considered value will differ in detail and scope based on the actual products and services wanted, today's customer demands at least one, and oftentimes several, of the following key marketplace attributes from their suppliers.

1. *Quality.* In the past, consumers and industrial customers focused on price as the central motivating factor guiding purchasing decisions. Adding quality, whether it be in the form of added features, use of materials, or superlative service, added cost. Today, nothing less than superior quality in products and services has become an *order qualifier*; it merely provides admission to the game along with other competitors good enough to make the grade. Whether it be performance, reliability, durability, or even aesthetics, customers want increasingly higher-quality, lower-cost products, and they expect it in the automobiles, cameras, apparel, and services they purchase.

2. *Price.* The cost of products and services has always been a key standard of customer value. It used to be that customers had to accept higher prices, as suppliers sought to protect margins as costs to produce goods and services rose. Now, customers want prices to continually decrease, despite the fact that producer costs are on the increase. In today's market, whether it be commodity-type goods or luxury automobiles, companies are finding themselves under a clear mandate to reduce costs rather than increase prices to maintain or improve margins. Higher prices give customers every reason to defy the protection long afforded by reputation and brand by shopping with the competition.

3. *Delivery.* For most businesses, the speed and reliability of delivery is a qualifier in the search for marketplace success; indeed, it is very often an order-losing qualifier. Conditioned by a "fast-food culture" and instantaneous information and communications response, today's customers have very high expectations concerning the element of time-to-delivery and it has become an essential value dimension. In addition, inconveniences and delays caused by incomplete orders, stock-outs, and late shipments are no longer tolerated. Simply, customers now consider time and how it can be constantly diminished as an essential part of the transaction itself. Rigidity in availability, service offerings, and convenience cause customers to search elsewhere for solutions to their product and service needs.

4. *Products.* Customers are expecting more out of the products the marketplace offers them. They are no longer satisfied with generic, standardized goods that are based on a "one-size-fits-all" marketing strategy. They

value products that are characterized by superior design, high quality, low cost, robust functionality, and a broad range of possible choices. In addition, today's products should be easily customized to fit particular requirements and flexible enough to grow through add-ons and upgrades that correspond to the customer's changing needs.

5. *Service.* Premium service is the final value most highly prized by today's customer. Whether it be when renting a car, shopping for apparel, or purchasing a hamburger, customers demand instantaneous, flawlessly executed, and responsive service. In order to compete, today's supplier must have an energizing service vision that sees service leadership as a continuous commitment permeating all the plans and actions of the enterprise. Customers want to do business with producers and merchants who ceaseless search for ways to exceed their expectations the first time and every time, and see the goal of zero defects as their central operational paradigm.

Obviously, the crucial requirement in achieving superlative service is identifying what customers value and then engineering productive and supply processes agile enough to continuously meet the particular needs of individual customers as they arise. Accountingwise, value is easily calculated as the sum of benefits acquired, minus the cost. Outside of an accounting viewpoint, however, customer value can be defined from several perspectives, depending upon the nature and purpose of the products and services desired by the customer. What is considered as value can take many forms, and ranges from purchasing simple commodities whose value consists purely in mere possession, to the acquisition of products and services that provide business solutions enabling customers to pursue strategies that will enrich their organizations far beyond the cost of the product or service.

Customer value, then, should be conceived as a set of attributes whose worth is not necessarily determined by its cost to produce or the price at which it is sold. This is quite different from the view of mass-production-era concepts of marketplace value. In the past, value-added usually meant that a basic product or service was enhanced by the use of expensive materials, specialized machinery, highly skilled labor, or processes that ensured high quality. In this cost-based view of value, the marketplace considered product and service value as consisting principally in the sum of the materials and labor consumed in their creation, independent of the needs of the customer or the added value they could produce when used to accomplish a specific business purpose.

In contrast, in today's global marketplace, value-added is, in the words of Goldman, et al. [6], what increases "the bottom-line enrichment to the customer." Often, it is not important that the product or service solution has a higher cost than possible alternatives. What is important is that customers perceive the value received in terms of how well it will assist them to achieve other, more critical strategic goals, and not merely as the sum of material, labor, and overhead expended by the producer. Today's customer is most willing to pay Federal Express $14 to deliver a letter that the U.S. Post Office would charge less than a dollar, or spend hundreds of thousands of dollars on communications equipment if the results enable them to capitalize on a strategic opportunity.

What is being suggested is that what constitutes value can vary, oftentimes significantly, depending on the requirements of the customer. Some products and services, those that are commodity in nature, have more easily identifiable values such as ownership, availability, low cost, convenience, and a recognized level of quality. The warehouse "club" store chain Price/Costco, for example, is focused on providing a highly selective inventory of high-quality grocery and household products to customers who want to feel that they are purchasing top brands at discount prices. The key to the chain's success is purchasing in large discount quantities and then passing on the savings to the customer. Maintaining the right assortment of products is accomplished through a continuous process whereby product families of leading brands are evaluated, followed by the selection of a few items from each family calculated to be the most appealing to the customer and the elimination of less successful ones. Although each store only stocks about 3500 items, considerably less than the 50,000 items or more offered by competitors, item sales are diligently recorded and closely reviewed to ensure that the products customer really want are on hand. In addition, new items and assortments and promotions are strategically scheduled to continually pique customer interest. Like other discount retailers, Price/Costco competes not just by offering the lowest price but by reducing their customer's overall costs in the form of reduced time spent at the checkout counter, convenience, and operational excellence [7].

For non-commodity-type products, customers usually require the fulfillment of more complex values. Although the customer's focus is ostensibly on product possession or service performance completion, there is normally a wide spectrum of supporting values that are often considered as important as the physical worth of the product itself. For Inacom, a major provider of information management products and services to national and international markets, what customers value is superlative service. In the past, personal computers, peripherals, and software were sold in standardized configurations, forcing the customer to choose from narrow product lines of proprietary brands. Recognizing that what customers really wanted was the ability to configure a customized solution that met their individual computing needs, Inacom has positioned itself as an information systems integrator with the capability of assembling specialized computer systems, networks, software, and communications equipment to customer specification and delivering them in extremely short turnaround times. As computer equipment becomes more and more generic and interchangeable, Inacom's goal is to build the supply chain from the customer back. This means working closely with commercial and consumer input from technology experts, MIS professionals, user groups, and others to create responsive delivery systems that provide an individualized, not simply a manufacturer's only, solution to customer needs [8].

The most complex type of value is found where the customer is searching for a product or service that will provide a solution strategic business problem. When the focus of the transaction is product-based, the value to the customer often is no greater than the cost of the product itself. The goods purchased are usually consumed quickly, with the result that their impact on the enterprise is small and indirect. In addition, the price of the product is normally determined by calculating the sum of the material and processing costs, plus the margin of profit. In contrast, product and service solutions have a long-term impact and are targeted directly at enabling breakthrough processes that will impact the strategic positioning of the company. Solutions provide the customer with unique product-information-service combinations designed to produce distinct bottom-line enrichment value. Unlike the cost of product-based value, solution-based pricing is determined by an agreement on the part of buyer and seller as to the enrichment value of the solution [9].

For Systems Software Associates (SSA), an international supplier of client-server enterprise information systems, providing solution-based customer value stands at the very core of the company's product and service strategy. This fact is belied by the corporate business mission statement: "SSA will provide a competitive advantage to our clients on a worldwide basis through the implementation of our business enterprise information systems." Through a combination of "world-class" software, highly configurable architecture and technology, and a matrix of support services, SSA can provide targeted, multidimensional solutions to a wide spectrum of manufacturing and distribution companies. In addition, SSA is able to leverage a large and highly professional global alliance of business partners that enrich the software giant by providing unique technical, sales, and support competencies far beyond what could be found in comparable vertically integrated companies. Although SSA does physically sell software, its goal is to develop long-term enduring relationships with its customers. In the final analysis, what SSA's customers value is the ability to work with a responsive supplier in the continuous creation of business information software and technology solutions that evolve as the needs of the customer base change to meet new competitive challenges.

Each of the above three companies seeks to compete by signaling out a critical dimension of customer value. Price/Costco has gained and maintains market leadership by focusing on providing the best price for the goods and services they offer their customers. Inacom's strategy, on the other hand, is to offer the best products. Although price is, of course, important, customers choose Inacom because they know they can receive a uniquely configured product custom-fitted to their individual specification. Finally, SSA seeks competitive positioning by offering the best solution to their client's information processing needs. Customers buy SSA's software and technology products and services because they offer a "bottom-line" impact on their ability to be competitive by reducing operations and asset costs, thus enabling them to both increase marketshare and as well as to penetrate new markets.

Product and Service Dynamics

In order to meet the requirements of today's customers, manufacturers and distributors have had to revolutionize the way they had designed, manufactured, and distributed products to the marketplace. As stated earlier, companies in the past competed by selling product lines consisting of standardized, mass-produced products and services. The focus of the entire process centered around the production and distribution of goods at the least unit cost. Efficiency was calculated around minimizing direct manufacturing and distribution costs associated with materials, labor, and overheads. Marketing concentrated on persuading customers to buy products whose value was fixed in the form of standardized pricing. The role of sales was one dimensional: The purchasing transaction was considered as the culmination of the sales process after which neither seller or buyer seldom expected ongoing opportunities for increased value-added products or services.

In today's marketplace, the old paradigms surrounding the production and marketing of products and services have given way to new requirements and opportunities for the development of deep, sustainable relationships among producers, distributors, and customers. As is illustrated in Figure 2.1, the expectations of the customer with regard to traditional marketing strategies have changed

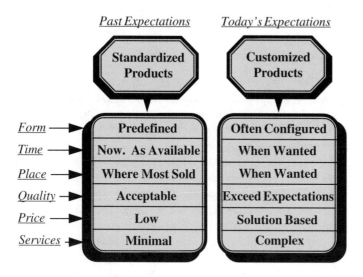

Figure 2.1. Changes in products and services.

considerably. In the illustration, the marketing utilities of form, time, and place have been expanded with the addition of quality, price, and value-added services. Furthermore, the way these marketing utilities had been considered in the past have been contrasted with the requirements of today's customer. The central theme of the figure is to illustrate the radical shift in the marketplace away from standardized to more customized products and services that seek to meet the unique and particular needs of each customer. The implications of the illustration are critical. In the past, companies competed by offering low-cost products that possessed standards of quality and availability barely above those offered by competitors. To gain dominance in today's marketplace, companies need to continuously offer products and services of superior quality that can be configured and priced to meet individualized requirements and that also offer opportunities to continually add value through the creation of mutually benefiting relationships established between supplier and purchaser.

Product and Service Process Developments

Companies focused on competing through leading-edge products and services must continually search for new ideas and processes not only to stay one step ahead of the competition but also to execute breakthrough innovations that enable them to surpass the performance standards expected by their customers. The reality of today's marketplace is that it is increasingly being characterized by difference and not similarity. As traditional markets segment and then segment again, producers and distributors can no longer merely seek to keep pace. They must be able to anticipate, if not drive, the creation of whole new customer expectations if they are to maintain customer loyalties and market leadership. Thus, in a business environment where the range of product and service requirements are expanding at a bewildering rate, the following process developments need to be structured to permit effective response.

- *Design.* The integration of product and service design, manufacturing processes, and distributive functions is perhaps the most fundamental dynamic of a business. Design has been dramatically impacted by the continuously shrinking life cycles of today's products on the one hand, and the acceleration in the rate of new product introduction on the other. The shortness of some product life cycles is staggering, even to the restless buying habits of today's consumer. For example, the life cycle of any one of Panasonic's line of consumer electronic products, such as CD players, TVs, and VCRs, is just 90 days. The market for personal computers is so volatile and dynamic that a product can be obsoleted while still in production. Shelf life may be 6 months, at best. At Intel, new product design and innovation is something of a passion. Normally, the company is supporting three generations of processor chips at any one given time: One that it currently in the market, one that is being beta tested as the replacement, and one that is being designed. The cost for this commitment, in the words of CEO Andy Grove, to make "the fastest chips in the newest applications" was $3.5 billion in 1994 [10].

The smallness of the window of opportunity for today's products directly impacts on how they are to be serviced and delivered. This means that individual enterprises must not only seek to continuously streamline inbound and outbound logistics, but that suppliers also must work very closely with their manufacturers to uncover new opportunities to speed product movement through the channel pipeline. For the Electronics Boutique, a video game and computer software and hardware retailer with 527 stores in the United States, the United Kingdom, and Puerto Rico, response time to new products is absolutely critical. Customers find out, usually electronically, about new products while they are still on the drawing board and normally want availability on the day of release. Considering that the first week or two after a video game's release are its peak sales period, disappointing customers usually results in a direct sales loss, as they simply cross the mall to competitors such as Software Etc., Babbage's, or Toys R' Us. Besides compressing cycle times through superlative logistics, the company has been able to maintain its position by closely working with its suppliers. Recognizing Electronic Boutique's commitment to quick response has resulted in deals with manufacturers, such as Sega, for the advanced release of their hottest new games [11].

- *Cost.* In the past, effective product management meant that products had to be designed with the capability for continuous process improvement and material cost reduction. Today, as product life cycles continue to spiral downward, the ability of manufacturers to squeeze the time it takes from idea conception to sales has added a new dimension to product cost management. Simply, manufacturers must be able to reduce design costs, time to market, and risk of failure continuously if they are to survive in the scramble to keep one step ahead of competitors on the one side, and the expectations of customers on the other. In addition, today's struggle for new product leadership has a *global* basis as manufacturers and distributors seek to leverage worldwide opportunities for materials, components, and finished goods that possess superior quality and can be brought to market quickly.

To be successful in today's environment, both manufacturers and distributors must continue to search for new ways to accelerate product design and delivery as a source of cost competition. One method is the activation of the internal capabilities of the enterprise. Working from the Deming principles of TQM, Rhone-Poulenc Rorer, an international pharmaceutical firm, has schooled its work force in the concepts of continuous improvement and statistical process control, and has provided for the rapid synthesizing of marketing, engineering, technology, and human resource capabilities into focused teams that will facilitate design time to market. Another powerful source of shrinking design-time-to-market costs is the use of intercompany relationships. By closely integrating with channel partners who possess critical competencies, companies can rapidly increase technology transfer and the availability of resources. The development of the PowerPC processor provides an excellent example. Designed jointly by IBM, Apple, and Motorola, the product appeared on the market in less than 2 years, due to the willingness of the collaborators to share proprietary intellectual property [12].

- *Product Range.* The sheer volume of new products making their way to market has not only accelerated, but it has virtually exploded with no limit in sight. As markets have increasingly segmented, producers and distributors have had to react quickly with a constant stream of products and services targeted at each emerging segment. The examples are endless. Nike and Reebok have multiple styles of athletic shoe for every sporting activity and every taste, and new styles arrive each season. The number of Walkman-type radios at a Circuit City or a Best Buy is numbing. Phillips offers more than 800 color TV models; Seiko sells 3000 different kinds of watches. There are so many brands of toothpaste, ranging from the variety of available flavors to whiteners and tartar-free, that even dentists are confused when it comes to recommending one to a patient.

  As the life cycles of automobiles, computers, electronics, business services, and a host of other products continue to decrease, the message for manufacturers and distributors is clear. Companies hoping not just to survive but to flourish in today's marketplace must have superior processes in place to enable them to continuously add value to their suite of products and services that possess features and quality beyond the boundaries of the performance expectations set by their customers. The role of manufacturing and distribution in such an environment is to develop processes flexible enough to reflect the proliferation of the product base, yet achieve cost-effectiveness despite low-volume, customized production and more frequent delivery.

- *Production Processes.* The ability to provide faster deliveries, higher quality, lower prices, and better service can only occur if manufacturing processes are available that enable enterprises to produce products when the customer wants them with near-zero lead time. In the past, supply channels sought to solve the problem of delivery responsiveness by allowing inventories to stockpile. Today, JIT techniques utilizing group technology, design for automation, setup-time reduction, and breakthroughs in production equipment have enabled both large and small companies to utilize mixed model scheduling, synchronized manufacturing, and JIT pull systems to create high-volume production lines capable of manufacturing a wide variety of products that meet the individualized needs of the customer. To stay competitive, companies need to rethink traditional lot-size production. Instead of calculating the most economic lot-size to produce, the real challenge becomes determining the ratio between production cycle time and customer delivery demand. And, as customer requirements for shorter delivery cycles increase, the goal becomes working toward the JIT ideal of producing to a "lot size of one."

Take, for example, the IBM manufacturing sites in Rochester, NY and Santa Palomba, Italy, which construct the world's supply of AS/400 computers. On any given day, dozens of new systems—all configured to unique specifications—make their way to IBM's customers. Assembled in a matter of minutes and tested over a period of hours or days, each new system is touched by hundreds of IBM professionals whose jobs reflect the state of the art in manufacturing today. Each computer is built to order in a continuous-flow manufacturing environment. The process begins when the Customer Order Analysis Tracking System (COATS) receives an order from the field. The system translates the order into a customized bill of material and manufacturing routing. The correct parts and customized assembly instructions are then scheduled to come together at the right work stations at the right time, with no setup time between orders. Utilizing sophisticated client/server applications run on IBM's own equipment, the shop-floor control system provides assembly instructions, part information, data collection, test direction, and failure analysis. Finally, another integrated system tracks all the final steps, including packaging, labeling, loading, shipping, and invoicing [13].

- *Services.* Customers today not only require high-quality, uniquely configured products, but they also expect their purchases to be accompanied with a matrix of value-added services. Many traditional services can be described as being almost "commodity" in nature because they directly accompany the product. Such services as quality, warranties, packaging, rebates, and documentation allow customers to receive additional value with the receipt of the tangible product. In addition, there are other services that are *extrinsic* to the product. Such services as discounting, credit, delivery reliability, transportation, product assortment, training, and others add value by reducing the customer's internal costs, facilitating the flow of business information, and improving productivity.

For many products the associated information and/or services are often more important to the customer than the physical product itself. For example, when customers purchase high-tech products what they really value is not the mere possession of the technology hardware but the ability to gain access to information or other products made possible by

Table 2.1. Dimensions of Quality

Type	Dimensions	Functions
Performance	Product's operating conformance	Design
Features	Secondary operating characteristics	Design
Reliability	Probability of breakdown within a given period	Design
Conformance	Degree to which product matches stated specifications	Manufacturing
Durability	Measure of product's functional life cycle	Design
Serviceability	Ease of servicing (planned or breakdown)	Design and After-sales
Aesthetics	Product's outward appearance	Design
Perceived Quality	Value of product as perceived by the customer	Marketing and Design

Source: D.A. Garvin, "Competing on the Eight Dimensions of Quality," *Harvard Business Review* (November-December 1987), 101–119.

the hardware. The model and components of a personal computer can be made by any one or combination of manufacturers. What the customer wants is the ability to use the wide variety of software available that will run on the computer. For example, Nintendo and Sega game systems are sold at cost, or even below. Both companies have realized that their profits come from the sale of the games, the number of which are expandable, rather than a one-time purchase of the hardware. Understanding the service/information content of products is critical, as witnessed by Apple Computer, which has finally awakened late in the game, in comparison to Microsoft, to the fact that what they should have been marketing is the Apple operating system and not the computer hardware.

- *Quality.* No discussion on products and services is complete without referencing quality. Since the late 1970s, quality has become the cornerstone of product and service value. Originally centered around reliability, the concept of quality has been broadened to encompass many dimensions, as detailed in Table 2.1. Today, the position of quality in the marketplace has changed significantly. In the past, quality was an order-winner, a value characteristic that differentiated products and services from those of competitors and which translated into higher levels of customer satisfaction. By the late 1980s, however, high quality had become a requirement of market entry, and its value changed from being an order-winner to an order-qualifier. In fact, customers now expect to receive the same level of satisfaction from mass-produced products and services as they once received from those that were custom-made.

Today, the concept of quality has been significantly expanded to meet the requirements of mass-produced customized products. As pointed out by Goldman, et al. [14], quality has evolved from *reliability*, to a concern with *choosing* between a multiplicity of products and service, to today's requirements for product and service *individualization*. Customers now want to work with suppliers who have the ability to assist them in selecting the right combination of product and/or services offerings, and then configure the purchase to meet their unique requirements. For example, IBM's customer order system contains a sophisticated configuration translator. When the customer orders a specific model and set of features, the system translates the marketing terminology into a buildable product, and then hands the order off to production. The whole process is focused on removing redundancies, reducing cycle times, and speeding a quality product to the customer.

Market leaders realize that the dynamics driving today's products and services require them to reengineer their manufacturing and distribution processes. Briefly, this requires the following activities:

- Continuous development of product design, manufacturing processes, delivery systems, and value-added services that provide for flexibility and configurability to respond to customer requirements while fostering optimization

- Dedication to the creation of core systems that assist the customer in the identification of specific solutions and ensure speedy transaction and delivery

- Fostering of mutually beneficial partnerships that enrich the competitive values of the seller and nurture the customer

- Support for internal operational structures that thrive on change, entrepreneurship, and a propensity for breakthrough creative thinking

- Absolute abhorrence of waste, command and control management styles, and the pursuit of debilitating localized performance measurements

Information and Communication Technology Dynamics

It has been said that we are at the end of the Industrial Revolution and now stand at the dawn of the Information Age. In the past, wealth was measured in terms of labor, materials, production assets, and finished products. Today, as the value of accurate data information grows and the speed of communications accelerates, *information* has increasingly begun to be seen as the fundamental source of wealth. What this means is that today's market leaders must view information and communication technologies not only as critical management tools that shorten cycle times and increase productivity of business functions through automation but also as key enablers providing the enterprise with the opportunity to activate highly competitive cultures and structures both within the organization and outside in the supply channel in the search for marketplace excellence. The convergence of information technology and the implementation of empowerment-based management techniques stimulating quality and productivity dictate the awakening of new business perspectives characterized by the alignment and unification of the operational and strategic activities of separate enterprises along intersecting supply channels.

At the core of the concept of SCM can be found Information and Communication Technology (ICT). As has been previously stated, SCM is a total management philosophy that seeks to blend internal business functions and allied business partners to form a single, unified supply channel system. The success of SCM is measured on how closely the resources and capabilities of those enterprises constituting the supply channel can be merged and focused on customer satisfaction. The thread that draws channel partners together is information, and the tools that they utilize are the information and communication technologies they develop and implement. As information nodes appear across the business landscape, they can be linked together, creating an information network. The more business information nodes that can be networked, the more robust the information sharing and the more effective the decision-making processes across multiple levels of converging supply channels. Leading-edge companies recognize that ICT provides them with enormous potential to leverage ICT to fuse the diversity of the supply channel into unified supply systems focused on creating customer value through the timely communication of marketplace information.

For example, Wal-Mart's private communications satellite systems enables the company to send point-of-sale data to its 4000 suppliers daily, providing them with an instant window into the retailer's requirements. J.C. Penny's quick-response system enables rapid and low-cost purchase and distribution of imported merchandise from suppliers to stores and catalog customers. Hallmark has developed a highly integrated continuous replenishment system with its retailers to ensure the right greeting cards are in stock for coming holidays. Retailers simply scan at purchase time the *universal product codes* (UPC) found on each card. This sales data is then transmitted daily to Hallmark. The replenishment program, which handles more than 5 million stock numbers, evaluates Hallmark's own forecast and daily sales to determine order quantities. Today, nearly 6,000 stores participate in this replenishment system [15].

Forms of ICT

Information and communication technologies can take several forms based on a company's business objectives. Computerized applications, such as Electronic Data Interchange (EDI), Automated Identification (Auto-ID), workstation trans-

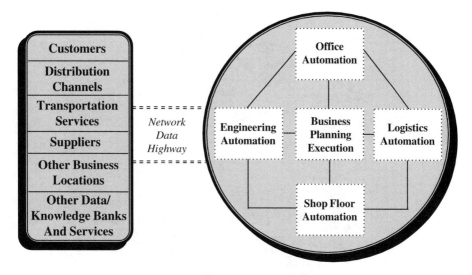

Figure 2.2. ICT network elements.

portation and routing systems, and office automation are technology enablers that focus on increasing productivity by *automating* time-consuming, expensive, manual or repetitive processes. On the other hand, business enterprise information systems, such Enterprise Resource Planning (ERP) and Computer Integrated Distribution (CID), provide business system architectures that enable the convergence of virtual teams both inside the organization and outside in the supply channel to focus on operational excellence and total customer service. A critical component of this process of alignment is the existence of open-system computer architectures that provide for the seamless interlinkage of business systems with other enterprise operational systems, thereby unifying the marketing, production, and distribution activities of the entire supply channel. Figure 2.2 illustrates the architecture of the information and communications structures necessary to sustain competitive leadership to the year 2000 [16].

Today's ICT consists of a divergent group of possible computerized systems as portrayed in Figure 2.3. ICT as it exists today in the spheres of manufacturing and distribution can be divided into the following areas.

- *Business Information Systems.* Computer systems and software applications offer not only individual companies but entire supply channels the ability to leverage information to gain fundamental improvements in productivity and competitive advantage. The functionality and architecture of today's computer systems enable businesses to choose from a wide variety of applications that fit their business processing and networking needs. The range includes fully integrated business systems,

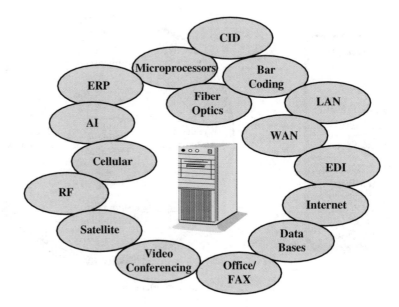

Figure 2.3. Possible ICT products.

such as ERP and CID, on one end of the spectrum, and workstation-specific products, such as freight rating and transportation systems, on the other. Some of the applications available are forecasting and planning systems, automatic replenishment systems, inventory management systems, manufacturing control systems, computerized engineering tools, financial systems, and others.

Effective ICT systems possess the following characteristics [17]:

1. They are based on distributed open systems, more commonly known as "client/server" architecture. This means that business systems and personal computers (PCs) are no longer locked within the boundaries of proprietary operating system architectures but now actually have the ability to "talk" with one another.

2. They are based on distributed relational database technology. Today's computing environment requires database software that supports multiple copies of business databases that are transparent to users both within the enterprise and anywhere around the globe.

3. They span interenterprise functional boundaries and enable the structuring of true global channelwide information networking. Today's systems provide for the integration of internal business functions, such as accounting, marketing, production, materials procurement, and distribution. In addition, through computer networking, they enable companies to share information regarding

customers, production, inventory, and finances with their supply channel partners.

Today's computer systems provide companies and whole channel systems with several key advantages. To begin with, real-time information processing and network integration significantly reduces the *time* necessary to operate planning and execution cycles. This translates into reduced inventories and overhead and asset expense, and facilitates transactions. Second, the timeliness and accuracy of information enables suppliers to be more responsive to the needs of their customers. Third, ICT provides companies with the ability to optimize their operational functions while reducing redundancies and increasing productivity. Fourth, computer automation frees employees to focus on value-added activities, whereas networking capabilities enable them to tap into the knowledge and experience of a virtual team of global experts. Finally, all levels of management will be able to make effective decisions based on timely, global information that adequately reflects the current state of the marketplace.

- *Communication Technologies.* This area is one of the most diverse and possesses significant potential for SCM. Direct communications technologies range from the telephone to teleconferencing and voice mail systems. Today, satellites can beam communications almost instantly across the world, transmitted through cellular telephones that permit easy access, at any time. By linking communications systems, companies like Federal Express can track the real-time status of shipments around the world, fostering better operational efficiency and control.

  One of the fastest growing business communications technologies is EDI. This communications tool has been defined as the computer-to-computer exchange of routine business transactions in a computer-processed standard format, covering traditional applications such as purchase orders, invoices, and shipping notices. The objective of EDI is to eliminate redundancies in data entry and to facilitate the speed and accuracy of the information flow across supply channels by linking computer applications between companies. Possible benefits of implementing EDI include reductions in office staff and supply costs, mailing costs, telephone and facsimile costs, data entry time and costs, accounts receivable days outstanding, customer service errors, traffic management errors, and on-hand inventories. Besides cost savings, EDI enables companies to improve customer responsiveness, tighten channel relationships, and compete better globally by shrinking product time to market and shrinking delays associated with international documentation [18].

- *Material Handling.* Computerized material handling systems targeted at warehouse automation have experienced dramatic growth in recent years. According to an Arthur Andersen report, 54% of the distributors surveyed used some form of automation today, and they expected this percentage to increase to 77% by the year 2000 [19]. Companies implementing material-handling automation must ensure that the proposed technological tools maximize the use of critical resources—space, labor, capital, equipment, and inventory—while providing a high level of customer service.

Material-handling automation can take many forms and consist of many levels of complexity. The acquisition of material-handling vehicles, such as a forklift, could mean automation for some warehouses. Small-parts-handling systems, such as carousel, movable-aisle, and miniload AS/RS systems, constitute automation. Perhaps the technology creating the most impact today can be found in the implementation of automatic identification systems that minimize or eliminate human operator involvement in the collection of information by using optical and radio devices that input information directly into the computer. Bar code readers and printers and magnetic readers are the principle optical tools. The most common radio device is Radio Frequency (RF), a technique that utilizes vehicle-attached or hand-held units that transmit information via radio frequency.

The benefits of implementing physical handling technologies are significant. Generally, the prime reasons driving automation are increasing labor and logistics costs, shrinking the base of skilled labor, and faster required customer order turnaround times. Other advantages can be found in increased output rates, increased information accuracy, reduction in materials wastage, and increased safety levels.

- *The Internet.* Increasingly the use of the *Internet service* available to any PC user stands as the new frontier of information technology. In fact, "communicating in cyberspace" or "surfing the Internet" has not only become the vogue among today's new generation of PC hackers, but it is also offering a totally new ICT arena for competitive advantage. For a small fee, anyone can gain entry to an almost limitless repository of information, from new software to articles and stock market data. For companies, the Internet opens up a whole new medium where products and services can be communicated to current and potential customers at a fraction of traditional costs.

  Leo Electronics, a broker and distributor of integrated circuit chips and other computer components, posts its entire inventory on its website and invites anyone to have a look. In a business characterized by stiff competition, the size and depth of available inventory, and the ability to respond quickly to customer needs is a significant market advantage. By being able to access Leo's inventory status through the company's website, it is hoped customers will regard Leo's inventory almost as their own. To encourage this process, Leo posts the entire inventory file in four formats—DBF, XLS, WKS, and TXT—to encourage customers to download and integrate the file into their purchasing systems. Leo Electronics made the investment in their website for several reasons: to differentiate itself in a crowded field of competitors, to attain a global reach, to do business with companies that currently are not customers, to provide better customer service, and—bottom line—to control and move inventory faster. Doubtlessly, the next step for Leo will be for customers to enter their customer ID and actually place their purchase orders on-line. It would not be out of the realm of imagination to then see a "send it" button appear on the screen with a UPS or FedEx logo on it [20].

Combining ICT Systems

Although the technologies detailed above each will significantly assist companies and supply channels to achieve and sustain marketplace leadership, the real advantage from ICT systems results when these various technologies are combined. Companies are already integrating EDI, bar coding, rating and transportation systems, and point-of-sale systems to make their information more timely and more accurate. Information systems for forecasting, purchasing, customer service, production, distribution center operations, inventory, transportation, and accounting are being linked together, enabling supply channel partners up and down the channel not just to integrate but to merge their separate operations into a single, unified "virtual" supply organization.

For Commercial Aluminum, Inc., producer of Calphalon cookware, the availability of real-time information was a requirement if the company was to satisfy the increasing demand for its cookware. The key was the integration of the various ICT systems the company possessed. A warehouse management system (WMS), which runs on one AS/400, was fully integrated with the company's Manufacturing Resource Planning (MRP II) system, which runs on another AS/400. This link helped to match production plans and delivery requirements so that the warehouse could monitor the status of actual and expected replenishment orders more effectively. In addition to the cookware made in-house, finished goods from more than 60 outside vendors are also received. All goods are received with an Interleaved 2 of 5 bar code identifying each pallet load. The goods are scanned and then transmitted to the WMS by radio frequency (RF). The WMS then assigns the pallet a putaway location and sends the data by RF back to a lift truck driver with a hand-held terminal. Finally, the driver stores the pallet and enters the information via RF back to the WMS.

Commercial Aluminum manually inputs and receives EDI orders from the MRP II system. Orders are downloaded by token ring to the WMS, which then organizes them into order picking waves designed to maximize warehouse resources and shipping carrier requirements. All picking information is communicated by RF terminals. Finally, as orders move toward the shipping dock, the cartons pass by a fixed-position scanner that reads the bar code. Following instructions from the WMS, the sorter then sends the carton down 1 of 15 shipping lanes for final handling. All outbound shipments receive a UCC 128 bar code label identifying the shipment. In addition, order management prepares an advanced ship notice detailing the contents of the shipment and sends it by EDI to the customer. The impact of the new system has been dramatic. Since start-up, the number of orders processed is up 31% and productivity 49% while labor requirements fell 18%. Inventory accuracy, once averaging only 70%, is now virtually 100%. As compared to a year earlier, total shipments have doubled [21].

Channel Dynamics

Several key developments are driving today's marketing and distribution channel environments. One of the most important is the emergence of the global economy. The end of the Cold War, the growth of new markets in eastern Europe and Asia, communication technologies, the speed of transportation, and the integration of the world's economic activities have propelled companies, large and small, into the global marketplace at a dizzying pace. This explosion in international trade is the result of a number of trends. To begin with, the maturing of the economies of the world's industrial nations has forced companies to look to foreign markets not only as a source of basic materials, low-cost labor, and sophisticated technologies but also as a marketing strategy targeted at broadening the competitive arena. The tightening of global competition has generated the second global trend: the emergence of powerful trading blocks in North America, Europe, and Asia. Third, the growth of incomes worldwide, the development of distribution channel infrastructures, and the speed of communications has also increased global demand for products and services and spawned new market opportunities. Finally, strategies focusing on leveraging the core competencies of enterprises across the globe have generated international alliances and joint ventures among companies seeking to expand processes and opportunities that they would be unable to achieve acting on their own.

In addition to the dramatic political changes and the breakthroughs in information and communication technologies drawing the world closer together, companies are also being impacted by worldwide governmental and environmental issues. The influence of national governments can best be seen externally in their stand concerning free trade and the formation of trading blocks, and internally through tariffs, transportation, commerce, and other types of restrictions. Equally important are the growing global concerns relating to environmental issues. Political commitments to clean air and water, safe transport of toxic and hazardous materials, design of effective of transportation infrastructures, and urban congestion and gridlock have spawned new requirements for industries to initiate meaningful reverse logistics programs.

Channel Developments

Although outwardly political issues have grabbed the most headlines in the last few years, the trend toward globalization is really being driven by several key developments relating to products, costs, and competition. The following business strategies are at the core of globalization:

- *Cost-Driven Factors.* The logic of companies engaging in globalization is clear They are seeking to gain marketshare by extending the reach of their products and services to new markets while achieving cost reduction through scale economies in manufacturing and purchasing. These economic advantages include proximity to markets, low-cost wages, foreign government incentives, low-cost materials and components, and leveraging local expertise.

- *Global Collaboration.* The easing of international tensions, the explosion in information and communications systems, and the structuring of global transportation systems have opened new markets in virtually every corner of the world. In addition, today's company has found it increasingly easier to coordinate design, product, marketing, and distribution processes with international partners to form "virtual" companies capable of responding to any customer opportunity. As a result, neither distance nor deficiencies in products or processes pose serious obstacles to companies seeking to enter global markets.

- *Rapid Introduction of Products and Services.* The ability of companies to utilize agile design, manufacturing, and distribution processes to bring products to market simultaneously across regions separated by global time and distance have made international trade a significant source of competitive advantage.

- *Product and Service Customization.* Breakthroughs in design and value-added processing have enabled companies to speedily and efficiently configure products to meet individual customer requirements. Such value-added processes as repackaging, relabeling, and easily configurable assembly facilitate customized features to meet global demand as it occurs.

- *Quick Response Delivery.* The radical developments occurring in communications and transportation have dramatically lessened many of the traditional risks and interruptions characteristic of global trade. The growth of international logistics infrastructures and communications technologies have made it possible for companies to leverage short product life cycles and customization in an effort to respond quickly and decisively to shifting global demand for products and services.

- *Global Strategies.* The points discussed above have enabled companies to break out of the narrow boundaries of national markets. No matter the company size, today's political environment and global information, communications, and transportation systems have virtually opened any market to any product or service. The goal of competitive global production and distribution is the development of a single supply channel system focused on attaining the best cost and service possible while welding enterprise and allied partner resources together to maximize competitive advantage.

As one of today's top international companies, Benetton, a $1.2 billion clothing retailer, underscores the tremendous opportunities that are available to companies possessing the strategic vision and practical know-how to leverage the global marketplace. Founded in 1964 in Treviso, Italy, Benetton began as a small manufacturer of sweaters, utilizing a pioneer technique for softening wool without shrinking it. By 1975, the company had 200 stores in Italy and began its first exports. In the early 1980s, Benetton began an aggressive expansion campaign to build a network of stores worldwide. By the early 1990s, the company had over 6100 stores spanning the globe from Europe and Russia to the United States and Japan.

Besides the excellence of its products and services and aggressive marketing strategy, Benetton owes much of its success to its commitment to information systems and communications technologies and the structuring of "world-class" logistics operations. Based in Ponzano Veneto, Italy, a team of 100 information specialists operates a variety of ICT applications, including its international EDI network. Facilitating worldwide information exchange, the network is at the core of Benetton's ordering cycle which unites global demand and the company's supply sources. The application of global ICT technologies has enabled the development of an effective logistics system. Over 50 million articles of clothing are distributed globally each year through a highly effective, quick-response network linking worldwide manufacturing, warehousing, sales, and retailers. Like other international marketplace leaders, such as Ford Motor Company, Canon, Xerox, and IBM, Benetton utilizes SCM principles and practices in an unceasing search to explore new marketing and opportunities for competitive advantage [22].

Alliances and Partnerships

In addition to the global dynamics of the marketplace, managing today's supply channels has been significantly shaped by the growth of strategic alliances and partnerships. In sharp contrast to the adversarial relationships between buyer and seller characteristic of the past, contemporary management thought emphasizes the development of close working partnerships among channel members. Some of the reason for this development can be traced directly to globalization. Although it is true that whole supply channels are now vulnerable to competition from anywhere on earth, the ability to leverage the intellectual, material, and marketing resources of business partners worldwide is rapidly rendering the entry into far-flung markets easier and more cost-effective. The ability of companies to exploit the peer-to-peer networking among designers, producers, distributors, and marketers provided by today's information and communication technologies will facilitate the creation of new forms of competitive-enriching collaboration among dynamic global multiventuring companies.

The growth of strategic alliances and partnerships can also be traced to the growth of an information infrastructure that fosters cooperation in national as well as global markets. Trends such as the explosion in information and communications technologies, highly skilled human resource shortages, emphasis on product and service customization, restructuring and downsizing of corporations, and increased attention to time-based competition strategies have encouraged the formation of business alliances as a means of gaining and sustaining market advantage by leveraging the core competencies of channel partners. In addition, by integrating businesses, strategic alliances can reduce the risks associated with the development of product, distribution, and marketing capacities by sharing in the benefits as well as the misfortunes. A comparison between companies that approach the marketplace unilaterally and those that engage in business alliances is detailed in Table 2.2. A common form of partnership is one that exists between a product marketer and a firm providing logistics, customized facilities, or equipment services. Another is between two service firms. The recent development of an intermodal freight program between J.B. Hunt and Santa Fe Railways is an example. Other alliances can occur between product marketers within the same vertical marketplace, or horizontally between independents who sell to the same customers [23].

The Bose Corporation, the internationally known manufacturer and distributor

Table 2.2. Transaction-Versus Alliance-Based Management

Transaction based	Alliance based
Short-term relationships	Long-term relationships
Multiple suppliers	Fewer suppliers
Adversarial relationships	Cooperative partnerships
Price dominates	Value-added services dominate
Minimal investment from suppliers	High investment for both buyer and supplier
Minimal information sharing	Extensive product, marketing, and logistics information sharing
Firms are independent	Firms are interdependent with joint decision making
Minimal interaction between respective functional areas	Extensive interaction between buyer and supplier functional areas

of quality audio speakers, has perhaps taken the most radical steps in their effort to utilize supply partner competencies in place of internal salesmen, buyers, and, in some instances, material planners. The concept, called "JIT II," is, according to its architect, Lance Dixon, "an application of the notion of partnering, eliminating the self-protective relationships between vendors and manufacturers by bringing selected supplier representatives into the plant." These buyers are stationed in Bose's facilities at the supplier's expense, on a full-time basis, and are provided with access to customer data and granted the authority to launch purchase orders with their own organizations to replenish impeding shortages in Bose's inventory. The same organization applies to transportation. When the delivery is by truck, Bose uses Roadway Services. Roadway has a representative in-house who attends purchasing/logistics meetings, checks all daily shipments, expedites orders, performs route analysis and coordination with Roadway terminals, and sometimes changes Bose's dock procedures. The result is that Bose averages nearly 98% on-time delivery. JIT II succeeds at Bose because all partners involved have developed mutual trust and long-term commitment, a true partnership in which there is mutual willingness for growth, change, and a common focus on absolute customer service [24].

In the past, the competitive environment was dominated by isolated companies offering standardized products and seeking the lowest unit production cost. Today, the requirements for total customer service, low price, high-speed delivery, and highly customized goods have forced companies to enter into alliances and collaborations of all kinds in an effort to maintain marketplace superiority. Some of these alliances focus on creating economies of scale by utilizing the capabilities of partners, thereby avoiding the cost of adding capacity. Some seek to merge technical, design, and marketing skills to gain specialized competencies and speed time to market. Others will embark on temporary alliances with competitors to realize temporary advantage. For example, witness IBM's collaboration with Apple and Motorola to develop and market the PowerPC chip and operating system designed to end DOS-Mac software exclusivity. Still others attempt to use the complementary capabilities of business partners to achieve channel vertical integration or economies of scope. Whatever the means, channel partnering has a common goal: The capability to assemble quickly focused skills and technologies to design, produce, deliver, and service products that will provide decisive competitive advantage [25].

Logistics Dynamics

The pace of change driven by today's customer, technology, product and service, and supply channel dynamics has significantly altered the structures and objectives of the logistics function. In the past, logistics was seen principally as an operational activity, consisting of a series of independent functions—warehousing, transportation, and finished goods inventory management—and focused around delivery and cost management performance measurements. In contrast, forward-looking companies now perceive logistics as a strategic, cross-functional management activity whose mission is to plan and coordinate all activities necessary not only to achieve delivered service and quality at lowest cost but also to enable today's enterprise to realize new opportunities for competitive advantage. The prime role of logistics must, therefore, be understood as providing the fundamental linkage connecting the marketplace and the operating activities of the business. The proper scope of logistics spans the enterprise, from the acquisition of materials, through manufacturing and channel management, to the delivery of the product to the customer.

Competitive Imperatives

A thoughtful perspective toward logistics has today become absolutely essential for those companies seeking to achieve and maintain competitive superiority. As the competitive context of the global marketplace changes, spawning new challenges and complexities, the capacity of logistics functions to leverage operational resources has become paramount. In order to compete successfully along the dimensions of time, customer service, and cost, today's logistics functions must be guided by the following competitive imperatives.

- *Customer Service.* Perhaps the fundamental role of logistics is to actualize time and place utilities necessary to realize enterprise initiatives for the continuous improvement of customer service across all operations and management dimensions. These dimensions include the flawless performance not only of the basic service elements of product availability, quality, and delivery but also of value-added services such as quick response, delivery to JIT requirements, and transaction processing customization. In addition, today's logistics functions are often responsible for meeting customer expectations regarding the ease of customer order inquiry, order placement, and order transmission, timely and responsive postsales support, and accurate, timely generation and transmission of order information both within the enterprise and with external parties. In addition to the traditional concerns with warehousing, transportation, shipping, and customer order management, some companies have included, among the functions of logistics, production planning, so that virtually the entire chain of order fulfillment has become the responsibility of logistics.

  The ability of logistics to execute the order management cycle is absolutely critical for the survival of some companies. For Orval Kent Food Co. Inc., the nation's largest processor of refrigerated salads, "world-class" logistics is essential in a business where products are extremely perishable and customers demand perfection. Orval Kent will make a product to order with a minimum of 48 hours' notice and promises to deliver within 2 days. Such quick response can only occur with finely integrated order management, production, and delivery functions. In addition, as grocery chains and food suppliers strive to reduce inventories, Orval Kent has been forced to engineer production and shipment to meet smaller lots and deliver at the customer's convenience to stay competitive. To meet these extreme logistics demands, the company has created cross-functional teams that service specific customers. These teams include representatives from customer service, logistics, sales, and other departments dedicated to meeting customers' needs and resolving any problems that might occur. For Orval Kent, logistics is at the crossroads where internal profitability and growing customer service pressures meet [26].

- *Reduced Cycle Times.* In a business environment where product life cycles are short and time to delivery is shrinking, logistics must be able to implement techniques that constantly reduce warehousing, order processing, delivery, and invoicing cycle times. The goal behind this competitive principle is simple: The more companies can accelerate the movement of material and product through the channel, the more product they can sell and the greater the profitability and competitive position. The key to speeding up cycle times lies in integrating the supply channel. Connecting suppliers, manufacturers, service providers, and customers across converging supply channels is fundamental in eliminating redundancies, inefficiencies, and costs while improving customer service. Logistics provides the operational structure for the realization of supply channel integration. By continuously reengineering efficient storage, transport, communications, and transaction functions across internal and external channels, logistics enables the smooth and rapid flow of products and information through the entire supply system.

  The ability of logistics functions to create competitive advantage by accelerating inventory cycle time has been well documented in the electronics industry. According to consulting firm Pittiglio, Rabin, Todd, & McGrath, the implementation of efficient logistics techniques have resulted in a steady 5 year improvement in inventory turns among U.S. electronics firms. Also, according to 1994 statistics, inventory as a percentage of revenue decreased from 14.3% in 1993 to 13.4%. In addition, gross margins increased by 3.4% to 37.6%, and revenue growth increased from 13.4% in 1993 to 14.7% in 1994. Furthermore, cash-to-cash cycle time declined steadily from 142 days in 1990 to an average of 114 days in 1994. Finally, inventory calculated as the number of days on hand has diminished 22 days since 1990 [27].

- *Increased Value-Added Services.* Besides superlative delivery quality, logistics must be able to continuously create new approaches to competitive advantage by developing and enhancing the services they offer to the marketplace. Customer-supplier relationships in the 1990s have increasingly focused on value-added services targeted at reducing customer cost, facilitating customer operations, and increasing customer productivity. Three critical flows can be identified:

  1. Implementing management techniques and technologies that accelerate the physical movement of goods from the supply source to the point of consumption

  2. Reducing redundancy in the transmission of critical information up and down the channel, such as demand schedules, market data, inventory supply levels, warranty and product information, and postsales support

  3. Increasing value-added services that improve productivity and eliminate costs such as automatic replenishment, advanced shipping notices, order status tracking, bar coding, packaging, and delivery

  The kind of innovative logistics services that gain customer orders can be seen in the activities of Flexible Packaging, a division of Union Camp, to solve the needs of their marketplace. A mid-sized manufacturer of bags and other premium packaging materials, Flexible was trying to uncover alternative methods of capturing the business of a well-known dry dog food manufacturer. The company's investment in the latest machinery, putting the best graphics on the bag, and streamlining inbound logistics helped, but the competition was still fierce and discounting was a severe problem. Finally, Flexible struck on the idea of extending their inventory control expertise to their customers and to use it as a value-added service to differentiate the company from the competition. The first step was to create an inventory control system that eventually led to assuming control of the dog food manufacturer's entire inventory of bags. The results were astounding. The customer saved millions of dollars in inventory carrying costs, and Flexible packaging increased its business with this customer 200 times—from $150,000 to $30 million [28].

- *Increased Logistics Service Quality.* Accompanying increased value-added services is the pursuit of ever widening dimensions of logistics quality. It is clear that today's marketplace expects high-quality products and services, and that the demand for quality will continuously increase. As they have in the manufacturing industry, issues relating to *quality* and *productivity* have become the number one focus of the distribution industry. According to a 1992 Arthur Andersen survey among top industry experts, *reliability* was chosen as the top measure of performance followed by *timelines* and *accuracy*. By the year 2000, accuracy is expected to move from third to second. According to the panelists surveyed in the report, it was felt that accuracy, whether it be in delivery, billing, or pricing, would provide significant customer advantage by reducing costs and distinguishing world-class companies who do things right the first time [29].

Logistics functions today must pursue a strong commitment to ensuring zero defect delivery, the development of continuous improvement in services, and the implementation of performance measurements that focus on increasing transaction accuracy and value-add to the customer. One method is to establish a program of critical logistics performance measurements. Fundamental metrics, such as customer satisfaction, asset utilization, operating cost, quality, cycle time, and productivity, will enable managers to match company operations, goals, and opportunities for increased performance with marketplace requirements. Competitive benchmarking is another technique. Benefits derived from benchmarking include the following:

1. Identifying and incorporating industry best practices into enterprise functions

2. Providing a source of stimulation and motivation

3. Breaking down organizational and individual resistance to change by exposure to new ideas not necessarily originating in the same industry

4. Assisting professionals in exposing technological breakthroughs occurring in other industries that are not existent or that have never been consider as possible in the home industry.

- *Reduced Total Logistics Costs.* As operational costs grow, margins shrink, and customers increasingly demand competitive pricing, logistics can significantly add to marketplace advantage through improved control of operational and processing costs. In fact, it can be stated that the effective management of enterprise costs begins with the efficient functioning of logistics and achieving the proper balance between cost and service level. Identifying the impact of logistics decisions on total channel cost is difficult. Logistics functions cut across company organizational structures, and managers must be careful when determining cost policies. A decision, say to cut inventory costs, could result in a decline in customer service levels and an increase in transportation costs. As a result, the effective management of logistics costs can occur only when logistics is viewed as an integrated system.

  Cost-effective logistics functions enable the supply channel to achieve targeted levels of customer service and profitability at the lowest possible operating cost and capital investment. The components of logistics costs can be seen from two perspectives. The first focuses on macromeasurements determined by calculating the proper level of customer service, which will provide sales and sales revenue, and the corresponding profits, which will yield cash flow and access to capital and return on investment, which will measure the extent of return on investment in channel processes. The second component seeks to determine costs arising from the following elements of logistics: maintaining service levels and the cost of lost sales, transportation, operating warehousing facilities, order process and information maintenance, and inventory carrying costs.

- *Organizational Responsiveness.* Fundamental to the achievement of competitive logistics is the continuous reengineering of the logistics organization to meet the challenges of a more demanding marketplace, configurable products and services, and the enabling power of information technology. As customers continually change their expectations regarding the speed and quality of the services demanded of their suppliers, logistics operations must become more flexible and adaptable. Organizationally, this means that logistics functions will continue the recent trend of eliminating hierarchical, restrictive command and control structures in favor of leaner and flatter operating organizations. As the logistics organization becomes more transparent, companies will seek to leverage information technologies and the skills and capacities of their supply chain partners in order to effectively adapt to meet the changing logistics requirements of today's marketplace.

  At Moen Incorporated, the Ohio-based manufacturer of bathroom, kitchen, and other plumbing fixtures, integrating the entire customer fulfillment function has been critical to competitive success. To meet its constantly changing customer requirements, the traditionally separate departments of transportation, distribution, and customer service have been merged into a single organization. One of the major benefits has been that the group's associates understand all of the functions and can talk intelligently with the customer. In addition, by encouraging Moen's logistics and customer service associates to visit both customers and suppliers, alternative avenues to improve distribution and delivery processes have been unearthed. When combined with a state-of-the-art warehouse and customer management computer system, Moen has been able to achieve same-day order shipment, down from the 5 days it previously took to cycle order requirements [30].

The continuing changes driven by the global marketplace, products and services, and information and communication technologies have dramatically changed the traditional operational environment of the logistics function. Rising service expectations, corporate downsizing, automation, outsourcing, and other

Top logistics opportunities ranked in descending order out of a total of 200 points

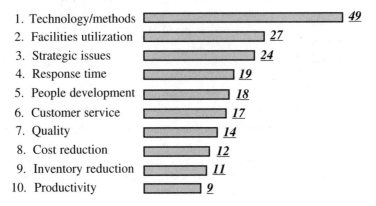

Figure 2.4. Top logistics opportunities.

alternatives to private warehousing and transportation have altered the playing field and changed the nature of how logistics creates competitive advantage. Based on the findings of a team of experts assembled by Tompkins Associates, a well-known logistics consulting firm, the 10 top opportunities that are shaping tomorrow's logistics function have been detailed in Figure 2.4.

Summary and Transition

When business historians of the future discuss the business environment of the 1990s, they will undoubtedly characterize it as one of momentous change in the way companies were run, how they developed, manufactured, and delivered products on a global basis, and how they communicated and processed information. When they turn to investigate the guiding management philosophies that assisted companies to plan and control their businesses during that turbulent decade, they will unquestionably point to SCM. Previously, companies had focused their energies on integrating internal logistics organizations in an effort to reduce costs and support sales and marketing objectives. SCM, on the other hand, will be seen as enabling enterprises to move beyond the narrow operations-oriented confines of integrated logistics to design, develop, and maintain value-enhancing strategies that provided for the activation and unification of the unique competencies of business partners along converging supply channels. Finally, SCM will be seen as marking a fundamental management breakpoint, presenting a whole new set of operational paradigms designed to effectively leverage global supply channels. Among the key organizational attributes of SCM will be found that the ability to manage functions *outside* the business will overshadow the ability to manage functions *within* the enterprise; and the ability to merge dissimilar competencies into a coherent customer-satisfying resource will supersede preoccupation with traditional local functional management performance measurements.

Supply chain management has become today's most exciting management concept because it enables not only individual enterprises but whole supply channels to exploit the five marketplace dynamics discussed in this chapter. Before the existence of these dynamics, SCM was unnecessary and, without today's information and communications technologies, virtually impossible. However, as each of the five dynamics arose, and they arose together because they are fundamentally interdependent and interrelated, the traditional management paradigms associated with mass-marketing approaches, centralized computerized manufacturing resource planning systems, and internal logistics organizations were incapable of blending the resources and competencies necessary to activate the competitive opportunities found in each dynamic to provide fundamental and sustainable competitive advantage. On the other hand, by providing a comprehensive management philosophy capable of unifying the value-enhancing resources within the enterprise and allied business partners along the entire supply channel, SCM has provided for the structuring of virtual organizations capable of breakthrough thinking and radical approaches of realizing continuous total customer satisfaction.

Supply chain management provides the enterprise with the ability to respond effectively to the needs and expectations of today's customer by illuminating the key value attributes companies offer to the marketplace and then engineering production and supply processes to be agile enough to continuously meet the particular requirements of individual customers as they arise. SCM enables whole supply channels to respond to the dynamics shaping today's products and services by providing the environment necessary to transform enterprises from passive producers and distributors of standardized goods and services to active creators of value centered around the quick-response delivery of uniquely configured, variable combinations of products, services, and information that supply customized solutions capable of evolving as the needs of the customer change. As a total management philosophy, the ability of SCM to leverage the networking power of ICT is essential in blending the competitive-enhancing functions within the organization and up and down the channel and keeping them focused on creating customer value. Today's leading-edge companies recognize that SCM enables them to apply ICT to fuse the diversity of the supply channel into a unified supply system capable of responding to any marketplace opportunity or competitive challenge.

The SCM concept provides enterprises with a means of effectively managing the tremendous changes occurring in the global marketplace and the emergence of strategic alliances and partnerships by facilitating the activation of several core business strategies. Among these strategies can be found the leveraging of international cost and economic opportunities, the structuring of the global "virtual" enterprise, the ability to rapidly introduce new products and services, the enabling of product and service customization, and the implementation of quick-response deliveries that allow supply channels to respond decisively to shifting global demand. Finally, SCM has been made possible by the rise of modern integrated logistics. Whereas logistics functions provide the fundamental operational linkages between the enterprise and the marketplace, SCM provides for the development of strategies that enable management of the entire supply channel as a unified whole rather than from the narrow perspective of the operational performance objectives of individual channel constituents.

Notes

1. F. R. Denham, "Making the Physical Distribution Concept Pay Off," *Handling and Shipping* (October, 1967), pp. 54–59.

2. Donald J. Bowersox, *Logistical Management: A Systems Integration of Physical Distribution Management and Materials Management*, 2nd ed. New York: Macmillan, 1978, pp. 9–10.

3. Bernard J. LaLonde, "A Reconfiguration of Logistics Systems in the 80s: Strategies and Challenges," *Journal of Business Logistics* 4 (1) (1983), 1–11, and Jeffrey Karrenbauer, "Distribution: A Historical Perspective," in *The Distribution Handbook* (James F. Robeson and Robert G. House, eds.) New York: The Free Press, 1985, pp. 10–12.

4. Arthur Andersen, *Facing the Forces of Change 2000: The New Realities in Wholesale Distribution.* Washington, DC: Distribution Research and Education Foundation, 1992, pp. 17–20.

5. Michael Treacy and Fred Wiersema, *The Discipline of Market Leaders*. Reading, MA: Addison-Wesley, 1995, p. 4.

6. Steven L. Goldman, Roger N. Nagel, and Kenneth Preiss, *Agile Competitors and Virtual Organizations*. New York: Van Nostrand Reinhold, 1995, p. 47.

7. Treacy and Wiersema, pp. 32–33.

8. Amy Zuckerman, "Inacom Redefines Logistics," *Inbound Logistics* 15 (12) (December, 1995), 18–22.

9. Goldman, et al., pp. 235–266.

10. Treacy and Wiersema, p. 101.

11. Laurie Joan Aron, "Delivery Speed Keeps Electronics Boutique at the Top of Its Game," *Inbound Logistics* 16 (1) (January, 1996), 30–40.

12. William H. Sell, "Competition 1990's Style," *APICS: The Performance Advantage* 5 (11) (November 1995), pp. 29–34, and Goldman et al., pp. 86–87.

13. Anne Hedin, "When AS/400s Build AS/400s," *AS/400* (January/February 1996), 29–32.

14. Goldman, et al., pp. 58–59.

15. These examples can be found in Bernard H. Flickinger and Thomas E. Baker, "Supply Chain Management in the 1990s," *APICS: The Performance Advantage* 5 (2) (February 1995), 24–28 and Rhonda R. Lummus and Karen L. Alber, *Supply Chain Management: Balancing the Supply Chain with Customer Demand.* Falls Church, VA: APICS, 1997, p. 66.

16. David F. Ross, *Distribution: Planning and Control.* New York: Chapman & Hall, 1996, pp. 14–15.

17. Thomas G. Gunn, *In the Age of the Real-Time Enterprise.* Essex Junction, VT: omneo, 1994, pp. 29–32.

18. Margaret A. Emmelhainz, *Electronic Data Interchange: A Total Management Guide.* New York: Van Nostrand Reinhold, 1990, p. 4, and William K. Riffle, "EDI: Let's Look at the Basics." *APICS: The Performance Advantage* 3 (6) (June, 1993), 226–28.

19. Andersen, *Facing the Forces of Change 2000*, p. 206.

20. Keith Biondo, "Sleepless in Cyberspace," *Inbound Logistics* 15 (12) (December 1995), 2.

21. Gary Forger, "Recipe for Short-Order Success," *Modern Materials Handling/Scan Tech News* (January 1996), S–10 to S–11.

22. David L. Anderson and Dennis Colard, "The International Logistics Environment," in *The Logistics Handbook*, pp. 664–665.

23. Christopher Gopal and Harold Cypress, *Integrated Distribution Management.* Homewood, IL: Business One Irwin, 1993, pp. 198–204; Donald J Bowersox, Patricia J. Daugherty, Cornelia L. Droge, Richard N. Germain, and Dale S. Rogers, *Logistics Excellence.* Burlington, MA: Dial Press, 1992, pp. 141–147.

24. Richard H. Green, "JIT II: An Inside Story." *APICS: The Performance Advantage* 2 (10) (October, 1992), 20–23.

25. Goldman, et al., pp. 29–30.

26. Toby B. Gooley, "How Logistics Drives Customer Service," *Traffic Management* (January, 1996), 45–56.

27. Pittiglio, Rabin, Todd, & McGrath, *1995 Inventory Performance in the Electronics Industry.*

28. Keith Biondo, "You Can Be Different," *Inbound Logistics* 16 (2) (February, 1996), 2.

29. Andersen, *Facing the Forces of Change 2000,* p. 139.

30. Toby B. Gooley, "How Logistics Drives Customer Service," *Traffic Management* (January, 1996), 45–56.

3

Evolution of Supply Chain Management

Although supply chain management (SCM) has only recently appeared as one of today's most powerful strategic business concepts, its development can be traced back to the rise of modern logistics. In fact, although SCM represents a radically new approach to leveraging the supply channel in the search for order-of-magnitude breakthroughs in products and markets, it, nevertheless, is closely connected with and in many ways is the product of the significant changes that have occurred in logistics management. Over the past 30 years, logistics has progressed from a purely operational function to become a fundamental strategic component of today's leading manufacturing and distribution companies. As logistics has evolved through time, the basic features of SCM can also be recognized—first in their embryonic states as an extension of integrated logistics management, and then, as a full-fledged business philosophy encompassing and directing the productive efforts of whole supply chain systems. A comprehensive understanding of SCM, therefore, requires a thorough understanding of the evolution of logistics management.

This chapter seeks to explore the various stages that mark the development of modern logistics management, culminating in the emergence of the SCM concept. The chapter begins with a brief overview of the origins of logistics management and the scope of its basic functions and objectives. Following this discussion, the remainder of the chapter focuses on examining the stages marking the growth of modern logistics. Up to this point in time, logistics can be seen as evolving through four distinct phases. In the first management phase, logistics was perceived purely as a tactical function consisting of a decentralized group of enterprise operational activities associated primarily with warehousing and transportation. The second phase of logistics can be characterized as the conscious centralization of logistics functions targeted at optimizing operations costs and customer service. The third phase is composed of two management concepts. The first can be described as the integration of core logistics functions with inventory planning, order processing, production planning, and purchasing. The second concept can be found in the extension of logistics integration outside the firm to embrace the entire supply channel, beginning with the supplier and concluding with delivery to the customer. The fourth, and final, phase of logistics can be found in the emergence of SCM. By being perceived as part of a continuously evolving operational and strategic business process, the chapter attempts to deepen the SCM concept by examining its roots and subsequent maturation as one of today's most critical management paradigms.

The Logistics Function—An Overview

Of the business functions of the modern enterprise, logistics has had the most visible impact on the economic condition of society. For centuries, businesses have been faced with the fundamental problem that demand for goods often extended far beyond the location where they were made and that products were not always available at the time when customers wanted them. It is the role of logistics to not only to fill this gap in the market system by providing for the efficient and speedy movement of goods from the point of manufacture to the point of consumption but also to act as a catalyst facilitating the wealth of nations. Where the capabilities of logistics have been limited, people live close to the source of production and have access to a limited range of products. On the other hand, societies that possess complex and inexpensive logistics systems are marked by elaborate and enriching market systems, a wide spectrum of available products, the rapid exchange of goods, and accelerating standards of living. Robust logistics systems enable the modern enterprise to leverage and focus productive functions while extending the reach of its products to meet national and international demand.

Historical Beginnings

Logistics can be said to be driven by two fundamental forces. The first force can be simply stated as the sum total of what the marketplace, at a given point in time, has determined as constituting the general wants and needs of business and society. The marketplace can be briefly defined as a particular arrangement of potential customers sharing a common need or want and who are prepared to exchange possessed value for those products and services perceived as constituting the solution to that want or need. The second force driving logistics can be found in the ability of distributive functions to bridge the gap between customer requirements and product and service availability. The critical elements of this force consist of time to delivery, cost, and ease of exchange. These two forces have differed over time and place, each, in turn, sculpturing the historical contours of the socioeconomic terrains to which they belonged and the structure of the logistics functions necessary to achieve marketplace goals.

As far back as the turn of the century, economists have considered the activity of *distribution* as the function by which the commodities of production are moved through the supply channel and the exchange process is determined. Writing in 1915, Arch W. Shaw felt that the various activities performed by distribution could be separated into two groups: *demand creation* and *physical supply*. Demand creation consists of the communication of the value found in the product or service that in turn will arouse in the consumer a desire for it and cultivate a willingness to pay the price and expend the effort to acquire it. However, this desire on the part of the customer would possess no economic value if the goods were not available at the time and place wanted. It was distribution's role to solve this basic problem of creating exchange value by ensuring that the output of production was matched to the customer's requirement as efficiently and as quickly as possible [1].

This fundamental interrelationship between the demand-creating nature of the market and the demand-satisfying capabilities of logistics can be seen throughout history. The earliest forms of logistics arouse when early civilizations found that surpluses of grain, raw materials, or manufacture could be traded for scarcer commodities that adjourning neighbors happened to have in abundance. In these early stages of commercial development, market demand occurred when available products or services were offered to the buyer. For the most part, all sales transactions occurred in bulk, with the purchaser personally viewing the entire purchase and arranging for its movement to the next stage of distribution. Because the majority of transactions occurred between producer and consumer directly, logistics was a relatively simple affair; its impact happening after the sale had been completed.

With the dawn of the modern industrial era, the nature of products, the demands of the marketplace, and the role of logistics were significantly altered. During those periods of business history before the Industrial Revolution, products were unique hand-crafted objects where the craftsperson dealt directly with the consumer. However, with the coming of the factory system, product standardization, and improvements in finance, law, and commercial superstructures, producers lost their primary character as hand-crafters and found themselves emerged in the problems of production. At the same time, the mass production, scale economies, and mass consumption of standardized goods necessitated and was nurtured by the growth of national and even international markets and marked the beginnings of mass distribution. As distance began to separate buyer and seller, the ability to bring goods to the customer grew in importance. Specialized middlemen and transportation services arose to satisfy this need. In a very short period of time, the structure of commercial exchange as we know it today had emerged: Producers shoulder the burden of production and distribution, or sometimes dispose of the product through the agency of various wholesalers who, in turn, sell to retailers, who then sell to the final customer.

The separation by time and space of the producer from the consumer and the development of alternate modes of distribution is the fundamental principle of the *supply channel system*. Classically, the nature of the intermediaries used and the structure of the supply channel directly impact all other marketing decisions. Speed of delivery, for example, depends on how widely opportunities for product and service transaction are geographically dispersed and the time it takes for goods to move from a supply channel node to the point of sale. Pricing decisions often depend on whether a mass merchandise, an exclusive, or a selective distribution strategy is chosen. In addition to internal elements such as cost, promotion, and quality, the external structure of the supply channel is equally important in defining channel utility. For the most part, products and information travel through one or more intermediaries, such as company-owned distribution functions, wholesalers, dealers, brokers, and retailers as they make their way through the channel pipeline. The structure of the supply channel that emerges out of these relationships is forged by the manner in which the transfer of ownership and the flow of goods from producer to consumer is determined. When viewed from this perspective, the supply channel can be described as a network of interdependent partners who not only supply products and services at the right time, place, quality, quantity, and price but who also stimulate demand for the whole system through marketing and sales activities.

Beginnings of the Logistics Organization

Although it has often been pointed out that the term *logistics* has been used for centuries to describe the component of warfare associated with the transport of soldiers, arms, and military supplies, it is with the comparatively recent rise of trade, industrialism, and the modern market system that contemporary logistics management can date its inception. Even the term "logistics management" is comparatively new. At the beginning of the century the term *physical distribution* was commonly used. Over the course of the century, it was also called, among other terms, Business Logistics, Logistics of Distribution, and Materials Management, and it was not until the 1980s that the term *logistics* gained popular usage. However, although the logistics function has long been acknowledged as one of the central building blocks of the Industrial Revolution, it has only been during the past 40 years that academics, practitioners, and consultants have begun to recognize it as a critical management science. Why has it taken so long for the logistics concept to develop? What was the early organizational structure of logistics and why did it develop the way it did? How were marketing, production, and logistics linked together? What were the business factors arising in the 1950s and 1960s that provided the background for the rise of modern logistics management?

The "question" of the role of the distribution channel in the economic process has been the topic of critical discussion for over a century. The "question of distribution" can be simply formulated as follows. Companies expend material, labor, and overhead to make products that must be joined with the cost of distribution to arrive at the total cost. With the rise of Scientific Management techniques, product design and process engineers had been able to ascertain with precision the optimal costs for production. However, the same capability to calculate and control the cost of distribution was woefully inadequate. For example, Arch W. Shaw stated at the turn of the century that

> The business man is concerned with the production and distribution of goods. Factory production he finds relatively well organized. The era of the rule of thumb is passing, and the progressive business man can call upon the production expert, technically trained, to assist him in solving his problems of production. But the marketing of the product has received little studious attention. As yet there hardly has been an attempt even to bring together, describe, and correlate the facts concerning commercial distribution. . . . The most pressing problem of the business man today, therefore, is systematically to study distribution, as production is being studied. In this great task he must enlist the trained minds of the economist and the psychologist. He must apply to his problems the methods of investigation that have proven of use in the more highly developed fields of knowledge. He must introduce the laboratory point of view. [2]

Fifty years later, much the same sentiment was being expressed by Peter Drucker. The "problem of production," he stated, had been solved. However, the effective management of distributive functions both within the enterprise and outside in the supply channel was so poorly understood that he termed it "The Economy's Dark Continent" [3] and the "Frontier of Modern Management" [4]. Even during the mid-1960s, when it was noted that the actual cost of logistics accounted for 30 to 50% of total cost and was certainly the largest single item in the cost structure, it is noteworthy that minimal effort had been expended on investigating logistics organizational structures and processes, and it remained largely unexplored territory.

There are several explanations for the neglect and subsequent late development of logistics management. To begin with, most company executives during this period considered the effective performance of logistics activities to be of secondary importance. Companies were much more concerned about marketing, sales, production, and merchandising than about the possible advantages offered by logistics. Second, prior to the availability of modern computers and the application of analytical management methods, executives simply had no way of compiling in a timely and accurate fashion the vast amounts of information necessary to integrate logistics functions. It was generally believed that an overall attack on logistics activities would not provide any meaningful improvements. A third major factor can be found in the prevailing economic climate of the period previous to and shortly after the end of World War II. Up until the prolonged profit squeeze of the early 1950s, culminating in the recession of 1958, companies had not considered the optimization of logistics functions to be a critical factor of corporate survival. By 1960, however, it had become obvious that the enormous costs involved in transportation, warehousing, and supply channel management had awakened executives to the significant advantages to be gathered from the application of new methods of cost reduction to the emerging science of logistics management.

However, of all the explanations as to why logistics remained for so long a company backwater, probably the most important can be found in the inability of the management staffs of the time to perceive logistics as a single process. Business management has been described as a planning and control function [5]. Purposeful management and planning become more effective as the number of variables declines, and become more complex and uncertain as possible variables increase. The standard management method of handling complex operations functions was to break them down into smaller administrative entities that, in turn, could then focus on optimizing productivity and cost elements within their own boundaries. Hopefully, if each department achieved their targeted efficiencies and performance goals, then it could be expected that the entire process, however segmented, would also achieve maximum performance. Up until the 1960s, that is what happened to the logistics function. Because the number of variables involved in planning and controlling logistics as an integrated activity are immense, it was felt that these tasks were handled best by separating them and allocating functional responsibility among several department managers. The concept of a single logistics function had simply been departmentalized away.

The decentralization of logistics remained the prevailing management method of handling logistics activities until the early 1960s when cost pressures and the availability of computerized information tools enabled forward-looking companies to begin the process of integrating logistics into a single enterprise function. During the next 35 years, this development could be said to consist of four distinct management phases. The first phase was characterized by the solidification of logistics departmentalism. In the second phase, logistics began the migration from decentralization to centralization of core functions driven by new paradigms centered around cost optimization and customer service. Phase three witnessed expansion of the integrated logistics concept to embrace business activities beyond core warehousing and transportation functions, including outside suppliers. The fourth, and most current, phase can be described as the era of *supply chain management*. These phases are portrayed in Figure 3.1. The remainder of this chapter will explore each of these phases in detail.

First Management Phase—Logistics Decentralization

Assigning historical events or processes to neatly defined periods is a fiction of the historian, designed to facilitate understanding by connecting together what appears to be homogeneous series of actions or a particular set of collective

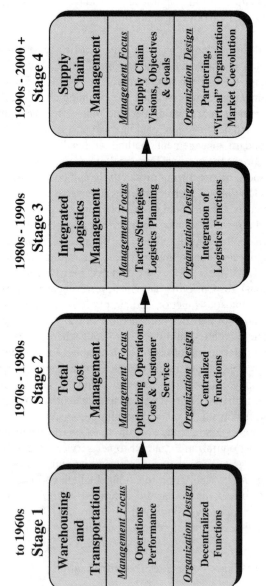

Figure 3.1. Four management stages.

attitudes. In this chapter, the attempt to place the logistics activities of the last 35 years into distinct periods suffers the same propensity for artificiality. Still, the changing views of logistics management occurring during this time do lend themselves to be separated into four unmistakable eras. The first management stage of logistics was the culmination of a century of business development and provides a clear link between the earliest enterprise organizations and today's modern logistics and SCM paradigms.

Overview

Historically, the first phase of modern logistics management occurred in the period extending from the late 19th century to the early 1960s. During this era, enterprise executives did not perceive logistics as a source of significant competitive advantage to the firm. For the most part, logistics was understood purely as those operations activities associated with the physical distribution functions of warehousing and transportation. In fact, logistics was considered of secondary importance next to market, sales, and production, accorded little management status, and assigned less qualified staff. Viewed essentially as a tactical function concerned with inventory management and delivery, it was felt that logistics could not make much of a contribution to profitability, nor was it worthy of much capital investment. Management attached little nor no strategic importance to such critical issues as the integration of cross-functional logistics planning systems, incremental process improvement, or work force policies. For the most part, firms segmented logistics activities, dividing responsibility for logistics management among separate business functional areas. Logistics had little strategic impact outside of day-to-day activities. Logistics management was regarded as a tedious concern with rate calculations and inventory chasing: There was a "minding the store" attitude among all enterprise management groups when it came to logistics functions.

Organizational Fragmentation

Much of the reason for the inability of logistics to be anything other than an operational function driven by endless expediting and daily crises was the fact that the various areas constituting today's logistics organization were fragmented and attached to several different enterprise departments. Not only were activities that naturally flowed into one another, such as purchasing and inventory management, separated from one another, but departmental performance measurements actually pitted the efficiency of one against another. The result was a rather disjointed and relatively uncoordinated management of goods and information as they moved within the organization and outside in the supply channel.

Doubtlessly, much of the basis for the early decentralization of logistics can be attributed to the particular nature of logistics functions. To begin with, logistics systems management posed several perplexing organizational problems that made it difficult for managers to fit it into the classical departmentalized business areas characteristic of most organizations at the time. Perhaps the most important was the fact that logistics is inherently a function that spans the organizational boundaries of most companies. Logistics is not a function of marketing and sales; it is not synonymous with traffic management; it is, assuredly, not the responsibility of manufacturing. Yet, logistics is a critical part of these and other company departments such as purchasing, customer order processing, and inventory planning. Fitting the pieces together appeared to be an impossible task. Once logistics was departmentalized, the situation was compounded by the conundrum of reconciling apparent conflicts between what were, in reality, related organizational units. And, although there was difficulty linking internal functions, there was also the complications involved in bridging legal and proprietary boundaries that existed outside of the company.

In addition to these factors, the whole field of logistics was woefully ill-defined as a management science. There was a distinct lack of professionalism, training, and skills among logistics managers, most of whom had either evolved out of warehousing or purchasing or were chosen from the ranks of managers who lacked the aptitude or skills to make it in marketing, sales, or finance. Furthermore, there was wanting a set of suitable performance metrics that could be used to measure the success of logistics as an integrated whole. When these conditions were linked to the fact that the sheer size of the number of transactions and decisions performed on a daily basis were beyond the managerial skills and information processing resources of the time, logistics decentralization appeared to be the most reasonable solution. Companies handled the great variety of logistics tasks the best they could: They followed the orthodox management method of divide and conquer, and disconnected logistics functions, assigning them to several separate company departments.

The departmentalization of logistics forced companies to organize their business structures to meet the particular market demands placed collectively on logistics functions. Although actual corporate structures varied by company, for the most part, manufacturing and distribution enterprises were organized as illustrated in Figure 3.2. Inventory control could be given to purchasing, manufacturing, sales, or even accounting, each claiming, with equally valid grounds, the inclusion of the function into their departments. Transportation often was split between purchasing or manufacturing, which was responsible for *inbound* inventories, and market or sales which was responsible for finished goods on the *outbound* side of warehouse management. In reality, organizational patterns were structured to meet marketplace requirements. In one company, the primacy of managing critical production processes required that logistics be subordinate to manufacturing. In another, the high value of finished goods required a marketing, sales, or even an accounting orientation. In a third, the cost of transport placed dominance on transportation management functions.

By decentralizing logistics functions, executives hoped to convert the complex and often bewildering array of logistics activities into an isolated collection of

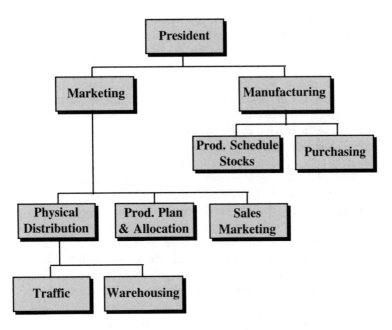

Figure 3.2. Decentralized logistics.

administrative operations. It was reasoned that the departmentalization of logistics functions would provide an effective substitute for what at the time was a disjointed collection of interfunctional logistics planning and operational management systems. Although there were often fundamental conflicts between departmental managers over performance targets, corporate goals focused on integrating and improving the productivity of departmental functions seemed to infer general improvement of the entire logistics process. In fact, it was thought that the maximization of logistics fragmentation would serve to maximize enterprisewide performance. As long as each of the departments with a segment of logistics responsibility managed somehow to get goods and information through their portion of the delivery system, no one questioned the fragmented administration of the system. The fact that many companies budgeted for logistics functions on a departmental basis strengthened the prevailing performance measurements and organizational structures and stymied any initiatives targeted at integrating the process. As a result, logistics functions, such as managing suppliers, transportation, and warehousing, were assigned roles of secondary importance to the primary missions of their parent departments.

Disintegration of the Decentralized Logistics System

In an era when process and delivery cycle times were long, global competition was in the early stages of retooling, and the marketplace was driven by mass production and mass distribution, the decentralization of logistics functions was, at best, a minor issue for most companies in the United States. Although a rare executive sometimes spotted the underlying contradictions between efficiency and separation of logistics functions, for the most part, no one was eager to upset the prevailing system. By convention, the costs for logistics functions were normally absorbed in the operating budgets of their respective departments. The impact of separate marketing and production decisions on collective logistics costs was unknown. What is more, accounting conventions at the time prevented detailed measurement of total logistics efficiency and, more seriously, obscured the important cost trade-offs among these functions that could be used to run logistics as a whole in a more economical and effective manner [6].

By the early 1960s, it had become apparent that logistics system decentralization was clearly an ineffective management paradigm. Three critical problems had begun to surface. First, changing marketplace demands were slowly forcing executives to recognize the need for the development of an integrated logistics system planning approach. Responding to expanding product lines, shorter cycle times, growing competition, and other factors revealed the waste and inefficiency of logistics decentralization. Second, the ability to manage changing business requirements exposed the glaring lack of unified logistics planning and control responsibilities. Although logistics activities were somehow successfully performed, responsibility for them was scattered throughout the organization. No one individual was responsible for distribution, and no one had the authority to coordinate the various logistics tasks. Finally, the ability to pursue effective cost trade-off decisions was impossible. Total logistics performance was often caught in a performance measurement paradox. Although all company departments dutifully sought to meet corporate profitability goals, each interpreted these objectives differently. For example, traffic sought to reduce transportation costs by requiring a higher payload-to-cost ratio, even if this decision resulted in higher inventories. Purchasing's target was to minimize the cost of ordering, frequently causing excessive inventories and handling costs. Inventory management wanted to dramatically cut finished goods, a strategy feared by sales who demanded increased inventory availability and costly short-cycle delivery to win customers.

Some companies attempted to solve these organizational problems by creating a department for one or more of those logistics functions found most critical to marketplace leadership. For example, a company that depended on transportation for marketplace advantage would rename it physical distribution and specific distribution activities found in marketing, manufacturing, and finance were grouped under the new department. The new department, however, continued to operate the same as before, with the logistics manager still reporting to the responsible senior executives in marketing, manufacturing, or finance. Although an improvement, such a limited organization was ill-equipped to tackle the difficult problems of expenditure control, cost trade-off analysis, and performance measurement confronting the logistics functions of the enterprise. The quasi-physical distribution department still remained interested primarily in the functions it performed for its parent organization. Such an structure also inhibited the new department from being influential in corporate strategy formulation [7].

Still, despite the petrifaction of the decentralized logistics system, by the mid-1960s several economic and marketplace forces were converging that would provide logistics managers the opportunity to break out of the traditional departmental approach and begin the process of reformulating the structure and purpose of the logistics organization. Much work needed to be done. As late as 1969, Bowersox lamented that the management science of logistics was still in its infancy. Logistics suffered from a lack of standardized definitions and vocabulary. Questions regarding organizational structure needed to be answered. The true aggregate costs of logistics in the United States had never been accurately assessed, and there was little known about likely future trends. Should logistics be part of a firm's marketing function? Should it be associated with manufacturing? Should it be a department on its own? How would the implementation of new information technologies impact logistics? What was the relationship between the research on logistics being done in academic circles and practical application in the field [8].

Second Management Phase—Total Cost Management

The abandonment of the decentralized logistics organizational paradigm and movement toward functional centralization were driven by several economic and organizational issues. As Karrenbauer [9] points out, the collective research of the period around 1970 yielded the following principal factors responsible for the movement toward logistics centralization: (1) severe economic pressures, centered around spiraling inflation, high interest rates, shortages of materials and energy, high levels of unemployment, and synchronization of world economies; (2) development of computerized information technologies and the application of quantitative analysis tools to logistics management; (3) adoption of a total logistics systems paradigm; (4) increased competition throughout the world's distribution system; and (5) increased competitive pressures in the domestic marketplace. Together, these and other forces signaled the end of the first stage of logistics management.

Overview

The second management phase of logistics coalesced around two critical focal points. The first can be described as the concerted effort of manufacturing and distribution companies to centralize the functions of logistics into a single management system. By merging logistics activities that had been previously fragmented into different autonomous enterprise departments, it was reasoned that a more centralized approach would enable management to examine more closely the interactions among warehousing, transportation, inventory procurement, and customer service, and then to define how each function could be coordinated to optimize the system as a whole. Companies began to remove control of materials management and physical distribution from marketing, sales, and production and to reorganize them into a central logistics department under a single management staff.

The second focal point of phase 2 logistics centered around the development of the *total cost concept*. The basis of this management philosophy resides in the assumption that because logistics costs and a targeted level of customer service were reciprocal, effective management of logistics could be attained by calculating the cost trade-offs necessary to balance total logistics systems costs with the firm's marketing and sales objectives. Under the decentralized system, the goals of maximizing customer service while minimizing logistics costs were separated from one another. Management did not, for example, explore the fact that a possible reduction in the cost of delivery could be achieved without impacting service levels and could actually provide for an increase in overall profitability.

Furthermore, the total cost concept assumed that the operating costs of each of the functions of logistics were inversely proportional and that management should strive to minimize the total cost of logistics, rather than focusing on reducing the costs of one or two specific logistics functions. In fact, it was observed that efforts to reduce the costs of a single logistics function in isolation could actually result in an increase in total logistics system costs. For example, a decision to lower inventory carrying costs by reducing the level of stocked quantities would most likely lead to an increase in transportation costs necessary to maintain targeted service levels and balance channel inventories. It was reasoned that only through effective cost trade-off management based on a centralized view of logistics could executives effectively align logistics system costs with sales and revenue objectives [10].

Growing Functional Centralization

The movement toward logistics functional centralization was driven by three converging factors. Although, as will shortly be explored, some of the impetus came from new thinking about the dynamics of logistics itself, several external economic and marketplace developments were also drawing management attention to the need to reengineer the entire organizational structure and operations processes of traditional logistics. To begin with, as inventory carrying costs skyrocketed during the period, the marketplace increasingly demanded smaller order quantities and more frequent delivery from their suppliers. Decreasing cycle times sounded the death knell of the old decentralized system. Second, a virtual explosion in product lines during the period meant that the responsibility of distribution to deliver products on time, avert obsolescence, and prevent channel inventory imbalances had dramatically increased. Third, new concepts of marketing, pricing, and promotion necessitated changes in the cumbersome, fragmented operations of traditional distribution functions. Finally, for the first time, executives were beginning to perceive the logistics function as a potential source of competitive advantage. By cutting distribution channel costs and reducing customers' freight bills, logistics could serve as a powerful resource assisting the enterprise to achieve key marketing objectives.

These factors occurring outside of the organization were also meshing with new ideas growing simultaneously within the profession. Three critical factors can be said to have shaped the development of logistics thinking during this period. The first centered around a new view of the potential for competitive advantage realized when the physical distribution functions of logistics were integrated. In an article written in 1965, Stewart felt that physical distribution was the key linkage between manufacturing and demand satisfaction and was, therefore, critical to the basic profitability of the enterprise. He conceived physical distribution as a system of interrelated "cogs," composed of such functions as distribution planning and accounting, order processing, receiving, warehousing, shipping, and customer service, all centered around the inventory management cog [11]. His point was simple but powerful: Instead of a set of dysfunctional activities, logistics was, in actuality, the management of a contiguous, continuous flow of materials and information that traversed the enterprise, beginning with inventory planning and receiving, continuing on through the firm's distribution channels, and concluding with timely delivery to customer.

During the period of the late 1960s and early 1970s, this radically different view of logistics as an *integrated system* began to dominate discussions among academics, consultants, and practitioners. The concept could be distilled down to a few simple principles:

1. It is the performance of the total system which is the key to logistics advantage, and not the performance of decentralized departmentalized functions.

2. Expenditures on material handling, field warehousing, transportation, and inventory control are of importance only as they relate to total cost and performance of the entire integrated system. This point will be discussed later as the *total cost concept*.

3. Instead of isolated, autonomous functions, there is a direct relationship among logistics activities which may increase or retard total system performance. This point is the essence of cost/benefit trade-off.

4. The close integration of logistics functions can produce a synergy resulting in a level of performance unattainable by those same functions acting in isolation. This heightened performance can take the form of increased customer service or lower total system cost [12].

This new view of logistics as an integrated management system was facilitated by the second critical factor shaping phase 2 logistics—the advent of new information and materials management technologies. During this period, computing technology became much more sophisticated, less costly, and more accessible. The computer's ability to process large amounts of data made it a perfect tool to record and maintain accurate inventory records, forecast demand, provide the statistical data for inventory replenishment, create sales and purchase orders, and drive shop-floor functions. For the first time logistics managers had the ability to make planning decisions and then have the information component available to provide for the continuous monitoring of the physical and information flows within the system. Likewise, the mechanization of materials handling permitted companies to speed up delivery functions and engineer more flexible distribution system response. In addition to its positive impact on productivity, the use of technology had the effect of stimulating logistics functional integration. Because technology-based information flows across the organization, bridging artificial departmentalisms and localized priorities, it rendered the traditional logistics decentralization paradigm increasingly obsolete.

The third factor providing for logistics centralization during this period was the growing realization that effective delivery of logistics functions was critical for increasing customer service. In the past, companies had developed marketplace strategies that focused the organization around the existence of a strong product development capability and a capacity to mass produce products or deliver services at volume and cost levels that provided them competitive superiority. By the early 1970s, it had become apparent that a strong marketing campaign required superlative logistics support if success was to be realized. In an era of mass production and mass distribution, often the effectiveness of logistics functions was the difference between companies who were market leaders and those who were not. By centralizing logistics functions, companies could actually increase marketshare and profitability by such activities as minimizing stock outs and facilitating channel inventory planning and control, reducing customer inventories by slashing cycle times, solidifying customer relationships through integration of supplier and customer replenishment functions, expanding market coverage to new or previously marginal customers, and permitting marketing and sales people to focus more on the task of demand generation instead of spending their time on logistics administration.

Total Cost Concept

While revolutionary changes were transforming the organization and traditional goals and objectives of logistics, a strong concern with analyzing and controlling total logistics system costs was contributing to a growing move toward organizational centralization. Although the interest of manufacturing and distribution executives in total logistics cost began to crystallize in the late 1960s and early 1970s, economists had expressed a deep apprehension about the neglect of logistics costs as far back as the turn of the century. In 1915, Arch W. Shaw felt that the entire business function of distribution was so chaotic and wanting in "scientific" planning that it threatened the social and economic fabric. "Society," he said, "can no more afford an ill-adjusted system of distribution than it can inefficient and wasteful methods of production. The social cost is no less real [13]." In 1929, Ralph Borsodi wrote in a book specifically targeted at addressing the problem of rising distribution costs that, while costs for goods between 1870 and 1920 had nearly trebled, production costs had gone down by one-fifth: What was being saved in production was being lost in distribution. This disturbing fact, he felt, had been caused by three developments: (1) costs for both distribution and marketing had risen out of proportion to production cost; (2) high transportation rates and unnecessary handling was one of the principle culprits of rising costs; and, (3) manufacturers had abused marketing and distribution by persisting in wasteful selling practices that had needlessly increased logistics costs [14].

The concerns of Shaw and Borsodi are just a sampling of the historical record of academics and practitioners who, until 1970, warned about the debilitating effect of excessive logistics costs. Much of the reason for the lack of methods to monitor logistics costs can be attributed to the prevailing practice of logistics decentralization. Cost problems often arose as a result of the discontinuities in measuring the impact of direct and indirect decisions on the efficiency of the logistics system as a whole. This dynamic becomes more understandable when it is considered that the ramifications of logistics policies on the total system are potentially immense. For example, changes in delivery frequency may influence customer ordering decisions, which, in turn, will impact replenishment policies, stocking levels, and the size of the channel, not to mention hidden administrative costs involved in accounting and staffing.

A great deal of the blame, however, can also be leveled at the accounting practices of the time. Because logistics bridges traditional organizational departments, measuring the impact of logistics costs becomes a difficult matter. Whereas demand-creating costs, such as advertising, selling, merchandising, sales promotion, and market research are relatively easy to quantify and analyze, the flow of costs associated with servicing demand as it accumulated through the activities of warehousing, transportation, and order processing and customer service are, however, much harder to identify. Accounting for distribution had always been hampered by the difficulty of developing standards for cost control and analysis and determining bases for cost allocation. For the most part, companies were content to set budgets and standards by department rather than recognizing cost elements against the background of the entire logistics system.

By the late 1960s, executives were becoming painfully aware of the cost of their decentralized logistics functions. Several facts were evident [15]:

1. The impact of logistics functions on enterprise profits were greater than most managers thought. When calculated, logistics cost accounted for as much as 50% of the selling price of the product.

2. Areas for improvements in logistics costs were largely unexplored because they resided in a managerial no-man's-land, outside the scope of responsibility or control of any operating executive.

3. Applying standard cost-cutting measures were woefully inadequate in reducing logistics costs. Although departmental measures could reduce portions of total logistics costs, they were piecemeal and had relatively little impact on total costs. No one manager had the responsibility or power to tackle logistics system costs.

4. Because corporate interest had traditionally focused on the costs arising from manufacturing and sales, management normally had critical financial operating reports in these areas. Because logistics had been splintered, there were no corresponding reports demonstrating how logistics affects the total costs and total profits of the business, and what management action is necessary to tap this significant area for profit opportunity.

It is difficult to determine exactly when manufacturers and distributors began to abandon the traditional view of logistics cost in favor of a more integrated approach. However, it can be stated with certainty that with the publication of Lambert's *The Development of an Inventory Costing Methodology* [16] in 1976, the *total cost concept* had been fully delineated. The fundamental principle of the total cost concept is simple and can be stated as a management planning and execution activity that strives to minimize the *total* cost of logistics rather than the cost of each activity. As pointed out above, simply reducing costs in one area may actually lead to increased total cost. Effective logistics cost management occurs only when cost improvement initiatives are formulated based on an integrated view of logistics.

Despite the simplicity of the total cost concept, effectively managing the firm to achieve aggregate cost savings while contributing to competitive advantage involves a matrix of management issues. To begin with, total cost management requires the convergence and focusing of operational functions around pursuing the best cost opportunities. Among the areas of operational cost can be found transportation, warehousing, inventory, and production. A second cost area revolves around marketing considerations. Opportunities for cost improvement can be found in distribution channel management, improved customer service, and improved inventory turnover and reduced obsolescence. A third area for potential logistics cost management is communications and data processing. Included in this area are costs for order processing, inventory control, payables, receivables, and shipping documentation. These costs tend to rise as the complexity of the logistics system increases. The final piece of the total cost concept involves aligning enterprise management objectives with logistics. This involves reengineering the organization to leverage productivity opportunities, charting the impact of logistics cost decisions on sales volume and profits, and determining the investment required and the probable return on investment.

Birth of Modern Logistics

The convergence of efforts to transform the organizational structure of logistics from a decentralized group of uncoordinated functions to a single integrated department and the rise of the total cost concept can be legitimately described as the origin of today's modern logistics organization. One of the most obvious indications was the significant changes occurring in organizational structure. By the 1970s, many companies had begun the process of moving the functions of purchasing, production planning, and distribution, formally under the control of separate management authorities, under a single logistics manager who was now accountable for all of these functions. To ensure that logistics was part of the firm's critical decisions and strategies, the logistics manager has been elevated to the same level as the managers of marketing and manufacturing. In a 1978 study of the 1000 largest manufacturing firms in the United States, it was found that the most common form of logistics organization (60% of respondents) was a combination of a central staff with line-operating responsibilities handled on a divisional basis. A significant minority (22%) of companies reported that all logistics activities were handled on a divisional basis only, with no centralized staff [17].

The new structure indicated that logistics managers of the mid-1970s had moved far from the days of decentralized functions. In the 1978 study referenced above, 78% of the companies responding had recognized and adopted an integrated logistics management approach, although only 50% felt that they had successfully implemented the concept. In any case, the implications were clear. Logistics managers were now expected to possess the ability to make decisions in terms of the whole logistics system and not just local departmental optimization. Logistics managers were now becoming aware that to assist in sustaining competitive advantage, they needed not only to think of the flow of goods occurring within the boundaries of the company but also of the flow of goods to the customer's receiving dock. Sometimes, this meant thinking about the flow of materials backward to the source of supply. Such thinking also cuts across company departmental boundaries, even to include competitors and potential markets.

Third Management Phase—Integrated Functions

During the 1980s, enterprise executives became increasingly aware that focusing solely on total cost management, although critical in aligning logistics costs with targeted service and profit objectives, represented a passive competitive approach. Gradually, managers began to move away from a preoccupation with minimizing logistics costs and toward a perspective that attempted to expand enterprise functional integration to maximize profits. Operational factors, such as speed of delivery, value-added services, and product availability realized when the entire organization worked closely together, could, in themselves, provide a powerful source of internal support that could significantly assist marketing and sales in winning marketshare. Phase 1 and phase 2 logistics organizations neither sought to activate the potential of their logistics functions nor to leverage the natural links connecting logistics with other customer-oriented functions. As firms began to search their organizations for sources of marketplace advantage to meet the onslaught of the global competition of the 1980s, it became apparent that an integrated logistics system could be employed as a dynamic force capable of winning customers beyond the execution of product and price objectives.

Overview

The third phase of modern logistics management is composed of two central concepts. The first consists in the identification of a new organizational role for logistics. Phase 2 organizations focused on centralizing warehousing, transportation, and customer service to leverage short-term tactical advantage and cost economies. In contrast, in phase 3 organizations, the sphere of enterprise logistics has been greatly expanded both in the number of functions that have been integrated and in the creation of a strategic logistics content. To the traditional core logistics activities were added inventory management, order processing, production planning, and purchasing. By considering all of these business functions as part of a single logistical system, the coordinated management of the combined resources of the demand-creating and demand-satisfying functions of the enterprise could be focused not just on cost management but also on how they could be engineered to provide a continuous source of unassailable strategic competency that would underscore the firm's leadership in a range of products or services and provide vistas for the creation of new competitive space.

Second, the heightened role of logistics provided managers with the opportunity to develop strategic logistics plans that could then be integrated with the strategic plans of other enterprise departments. In this sense, logistics could function as an active catalyst derived from and in support of the firm's overall competitive strategy. In the past, logistics was seen in a supporting role to the more important areas of marketing, sales, production, and finance. In contrast, phase 3 organizations recognize the strategic value of their logistics functions and seek to draft plant charters and mission statements to guide logistics development and ensure alignment with other company functions. This new view afforded logistics an equal position alongside marketing, sales and manufacturing in the formulation of strategic plans, the determination of enterprise resource allocation, and the definition of the scope of marketplace objectives. By closely aligning logistics and manufacturing capabilities and marketing and sales objectives, the enterprise could present customers with a unified approach, guaranteeing product, price, quality, and delivery competitiveness [18].

New Marketplace Values

The decade of the 1980s was an era of significant change in logistics as well as in American business. If the period could be compressed into two quintessential business catchwords, they would be *competition* and *quality improvement*. Competition came in the form of tremendous pressure from overseas manufacturers bringing new forms of organizational structure and dazzlingly simple process methods providing for unheard of levels of productivity and quality that challenged the supremacy of U.S. products and management styles. The challenge came in the form of business philosophies centered around the JIT and Total Quality Management (TQM) management paradigms. It was not just that foreign competitors were offering American consumers higher-quality products at less price—the threat also came from their ability to compress time out of development cycles, engineer more flexible and "lean" manufacturing processes, tap into the creative powers of their work forces, and create entirely new forms of competitive advantage. In this new environment of global competition, companies turned with increasing interest to their logistics delivery systems to assist in reviving their once unassailable market positions.

Companies in the 1980s faced a variety of challenges that increasingly pointed to the need for the close integration of their logistics functions. There were several key pressures that had arisen that were *external* to the enterprise. To begin with, of primary concern during the early years of the decade was the availability and cost of energy which was expected to escalate, particularly for logistics activities such as transportation. A second major challenge was the anticipated increase in the cost and availability of capital. During the early 1980s, interest rates had climbed so high that they were finally capped by government mandate at a staggering 21%. The third major challenge came from changes due to the continued deregulation of transportation. Deregulation fundamentally altered traditional logistics by providing lower costs, shrinking the number of carriers used, and accelerating the outsourcing of functions to third-party logistics firms. The fourth external factor affecting logistics management was the virtual explosion in international trade resulting from a number of trends such as the maturing of the economies of the world's industrialized nations, growth in the practice of global outsourcing, and formation of strategic alliances and partnerships. Finally, the last pressure could be found in the continued demand for increased customer service. This factor took the form of marketplace demand for increased supplier flexibility, provision for computerized tools such as EDI, and the continuous growth of value-added services.

Besides these pressures from outside the enterprise, there were also several *internal* developments reshaping the contours of the logistics organization. One of the most important was the realization that only those companies that provided superlative quality with ever-declining cycle times would be tomorrow's market leaders. For logistics, increased quality meant improved customer service, increased value-added services, and streamlined and flawlessly executed delivery processes. Reduced cycle times meant decreasing product time to market, increasing productivity, shrinking pipeline inventory investment, providing more flexible and timely service without increasing costs, and reducing supply channel risk. Together, attention to logistics quality and time could provide the enterprise with unique market-winning competencies that could not be easily copied by competitors.

Beyond these customer-facing objectives, logistics functions were also under severe organizational and operational pressure. Writing in 1985 of the challenges before logistics managers, LaLonde, et al. [19] singled out rising logistics costs, the cost and availability of capital, the effective use of information technology, and the design and implementation of effective customer service systems as the foremost challenges before the logistics organization. The focus on cost and structure concentrated around such issues as work-force retraining, development of corporate environments conducive to employee productivity, application of technology to improve productivities, careful attention to the use of fixed and variable assets, and identification of customers, products, and markets possessing the greatest potential for competitive advantage. During the 1980s, companies also turned their attention to corporate restructuring. This restructuring took the form of the divestment of unsuccessful businesses, delayering management, matching cost generation with debt service, and focusing on creating "lean" organizations.

Organizational Changes

The primary thrust of phase 2 logistics was to centralize the physical distribution functions of the enterprise in an effort to reduce costs and improve throughput. In this stage, executives sought to integrate finished goods planning, warehousing, transportation, customer service, and other functions directly related to product delivery. The reason for the focus on finished goods management was simple. Finished goods represented the largest and most visible segment of enterprise costs, sometimes as high as 40% of total inventory in a manufacturing company. Second, the effective management of finished goods had a direct impact on the performance of customer service. Finally, engineering organizational changes in the way finished goods were managed was a relatively low-impact, high-gain exercise. Managers could experiment with new service structure configurations without venturing into production processes or other powerful enterprise areas such as finance and sales [20].

Complementary to phase 2 physical distribution centralization, the objective of phase 3 logistics was to integrate the entire material pipeline flow beginning with procurement, progressing through manufacturing and customer order management, and concluding with physical distribution and customer service management. By integrating the entire internal materials flow, companies had the capacity to more effectively manage 80–100% of their inventories and develop targeted strategic plans that leveraged and focused enterprise resources on exploiting the best marketplace opportunities at the lowest cost. The change in managerial emphasis was dramatic. In a 1973 study (Fig. 3.3), for example, only a small proportion of logistics activities were actually under the control of physical distribution departments [21]. The highest percentage activity was outbound transportation of finished goods, and in this, only 37% of the firms surveyed assigned line control responsibility to the logistics department.

By 1990, the picture had changed dramatically, as logistics responsibilities had considerably expanded. According to the results of two Ohio State University surveys, logistics responsibilities had broadened as firms continued the integration of their warehousing, traffic, customer service, and inventory control functions. Figure 3.4 exhibits the changing pattern of logistics responsibilities for the years 1985 and 1991. By 1990, traditional areas of responsibility, such as warehousing, transportation, and inventory management are almost 100% under the control of logistics managers; other areas not even mentioned on the 1973 study and, at the time, performed by marketing and manufacturing, such as order entry and purchasing, were 50% of the time under the control of logistics departments.

The increasing power of logistics was also mirrored in the changing organizational structure of the typical manufacturing and distribution enterprise. During the period 1975 through 1991, companies increasingly sought to centralize logistics functions and to move away from divisional (where each division operates its

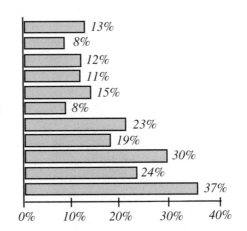

Figure 3.3. Responsibilities assigned physical distribution (1973).

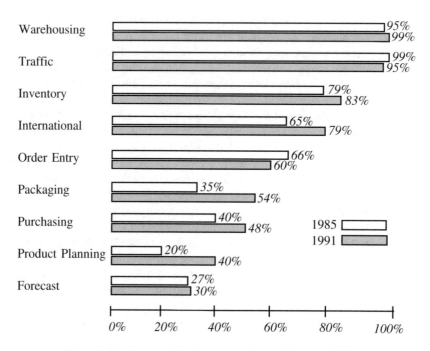

Figure 3.4. Percent of logistics executives with responsibility.

own separate logistics function) and separate division (where a logistics division provides services to other product divisions) structures. According to the Ohio State University surveys mentioned above, about 35% of the 1991 respondents categorized their logistics organizations as centralized, up from around 18% in 1975. Separate logistics divisions declined from around 20% to about 10%, with divisional logistics structures remaining about even. The predominant organizational type during the period, a combination of some centralized and some divisional managed logistics processes, experienced a slight declined from a high of around 41% in 1975 to 38% at the end of the survey period [22].

Logistics as a Competitive Weapon

The final significant characteristic of phase 3 logistics management was the rise of the logistics organization as a competitive weapon. Up until the 1980s, most executives did not view their logistics organizations as a source of competitive advantage. At best, they considered them as playing a neutral role in corporate strategic planning. Competitive advantage was seen as emanating from marketing, sales, product, and financial strategic plans. Logistics functions were common regarded as expense-intensive nuisances possessing little or no marketplace value other than the efficient storage and timely delivery of inventory. It was felt that the logistics organization simply could not make a significant contribution in winning new markets nor was it worthy of much investment in capital or the time to develop long-term operational plans. Its role was to remain flexible and reactive to the demands of marketing and sales.

By the mid-1980s, however, companies began to perceive their logistics organizations as an essential component of their total ability to provide the marketplace with unassailable quality and value. It was argued that the effective management of logistics functions could assist in the realization of the firm's strategic goals by providing both a cost/operational advantage and a service/value advantage. By enabling companies to pursue opportunities in cost reductions, continuous incremental improvement of logistics processes, and closer integration with suppliers, logistics could build and leverage core capacities differentiating the enterprise from its competitors. On the other hand, a focus on logistics efficiency and effectiveness assisted in the determination of essential components of logistics strategic value. When effectively applied to creating customer service value by ensuring product availability, order cycle time, operational responsiveness, logistics system information, and postsale customer support, the elements constituting logistics value were recognized as providing a key source of marketplace advantage.

This new view of the importance of logistics strategy can be seen in the dramatic change in the components of the strategic plan. Perhaps the most noteworthy aspect was the elevation of logistics to equal partnership with marketing and manufacturing. As is illustrated in Figure 3.5, logistics was now conceived as standing at the critical juncture between manufacturing, which is internally oriented, and marketing, which is externally oriented. The mission of logistics in this strategic model is to develop coherent long-term goals focused around customer service, establish detailed operational metrics, validate the overall supply chain charter of the firm, and ensure that logistics objectives are in alignment with the corresponding corporate, marketing, and manufacturing strategies of the firm. Logistics is to be considered as the link connecting the marketplace and the operating capabilities and competencies of the enterprise.

The growth of logistics as a competitive weapon was the culmination of phase 3 logistics management. The integration of logistics functions that spanned the supply channel from product procurement through delivery had enabled a level of competitive advantage not realized by enterprises of the past. Despite the significant operational breakthrough, however, what was wanting was the ability of firms located along the supply channel network not only to integrate their logistics functions but also to leverage the convergence of the marketing, product development, procurement, inventory planning, manufacturing, quality management, and core competencies of whole channel systems to provide reservoirs of "virtual" resources and competitive advantage unattainable by even the largest corporations acting on their own. The realization of this objective is the subject matter of phase 4 logistics management.

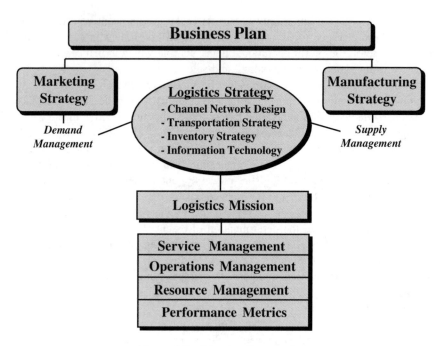

Figure 3.5. Planning framework.

Fourth Management Phase—Supply Chain Management

During the early 1990s, companies began the process of responding to new marketplace challenges by expanding on the functions of integrated logistics management to realize radically new opportunities by tapping into the wide range of productive capabilities to be found in the matrix of allied companies gathered along the supply channel. The acceleration in the globalization of the marketplace, increasing expectations concerning service quality, third-party outsourcing, alliances and partnerships, the reengineering of the organization, and the impact of information and communications technologies had forced companies to look beyond the integrated logistics model in search of new strategic possibilities. Whereas decades of solid management and organizational development had elevated logistics to a critical position in the typical enterprise, it had become obvious, however, that the requirements involved in shaping and integrating whole supply channel systems were beyond the strategic capabilities of a function that was focused primarily on the execution of operations activities. The goal of effective channel management was not only to provide for "world-class" customer service at the lowest possible cost but also to continuously explore new avenues for leveraging the productive and innovative resources to be found when channel partners converge in the common pursuit of unique sources of competitive advantage and whole new regions of competitive space.

These conceptual elements of phase-four SCM slowly emerged out of several management initiatives designed to link supply chain strategy to overall business strategy. The first clear movement toward SCM can be found in the growth of the concept of Quick Response (QR) arising out of the U.S. textile and apparel industry and the Crafted With Pride council founded in 1984. By the early 1990s, the QR concept had been expanded to embrace integrated partnerships of retailers and suppliers focused on utilizing information technologies to foster quick response to sales demand by distributors and manufactures along the channel pipeline. From another direction, the grocery industry was working on their own version of quick response. In 1992, a group of industry leaders created the Efficient Consumer Response (ECR) Working Group. The objective of the group was to investigate the grocery supply chain to identify new technologies and best practices that would provide significant new opportunities to make the grocery supply pipeline more competitive.

Recently, companies from a wide variety of industries have transformed many of the concepts of ECR into a new demand and supply management paradigm called Continuous Replenishment (CR). The central objective of CR revolves around the ability of supply chain members to utilize information technology to migrate from supply systems centered around pushing product down the supply network to *pulling* inventory onto retailers shelves driven by actual customer demand. Through point-of-purchase systems, demand information is forwarded by computer to supplying elements within the channel, permitting them to provide the retailer with the necessary inventory replenishment just-in-time. Companies such as Hewlett-Packard, Whirlpool, Wal-Mart, Baxter, and Georgia-Pacific have each in their own way to achieve new areas of competitive advantage by reducing cycle times, removing redundancies within the channel, and providing the right product at the right time to their customers. Taken together, these movements mark a distinct break with phase three logistics concepts and have inaugurated a whole new era of channel management based on exploiting the advantages to be found in closely integrated partnerships of companies linked together along the supply chain network.

Overview

The competitive pressures of today's global marketplace have compelled companies to implement what only can be called a dramatic paradigm shift, migrating them from phase 3 logistics to phase 4 SCM. As was discussed above, the fundamental feature of phase 3 logistics management was the integration of internal logistics activities and strategies with those of business partners targeted at improving customer service and total cost across whole channels. At the core of phase 4 organizations is a new view of the potential for competitive advantage, of which logistics integration is but a part, that can be found by building close relationships with channel partners. Prior phases regarded logistics as a source of internal strategic advantage at best, providing the resources necessary to achieve marketing and sales objectives. In contrast, phase 4 SCM builds on the basics of logistics operational integration to create "virtual" organizations that span channel boundaries and link together critical competencies, enabling whole channels of supply to explore uncharted areas of competitive advantage. This new management paradigm consists of three distinct elements:

1. *Expanded View of Logistics Operations Management.* Phase 4 SCM requires companies to move beyond simply focusing on optimizing logistics activities only, to one where all enterprise functions—demand management, marketing and sales, manufacturing, finance, and logistics—are closely integrated to realize order-of-magnitude breakthroughs in product design, manufacturing, delivery, customer service, cost management, and value-added services before the competition. Whereas cost control will remain critical to marketplace success, logistics performance will increasingly be measured against total enterprise JIT and quick-response objectives. This internal orientation requires top management to focus the firm's strategic planning and organizational structures around the strength of their logistics functions.

2. *Extension of Integrated Logistics Management to Encompass Opportunities for Competitive Advantage Occurring Outside Company Boundaries.* In its most basic form, a focus on external integration enables business functions within and across enterprises to search for productivities and new competitive space by leveraging innovative relationships with their vendors, customers, and third-party alliances. Through the use of information and communications technologies that network whole supply channels, companies have the capability of viewing themselves and their channel partners as extended enterprises possessed of the ability to realize radically new ways to create marketplace value.

3. *New Strategic View of Logistics and Channel Management.* The tactical elements of phase 4 SCM focus on traditional logistics operations tasks such as accelerating the flow of channel inventories, optimizing internal business functions with those of outside partners, and providing the mechanism to facilitate continuous channelwide cost-reduction efforts and increased productivity. Although key elements, the real strength of phase 4 SCM is to be found in its *strategic* dimensions. The external orientation and networking capabilities of phase 4 SCM organizations enable companies to establish a shared marketplace and competitive vision connecting channel partners, to structure coevolutionary channel alliances providing for order-of-magnitude breakthroughs in products and services, and to manage complex channel relationships enabling

companies to lead market direction, generate new associated businesses, and explore radically new opportunities.

New Operational Paradigms

The impact of phase 4 SCM on the traditional manufacturing and distribution enterprise is dramatic. Besides providing for new avenues to periodically reinvent their organizations to ensure internal productivities, companies will also have the opportunity to continuously redefine the boundaries of their organizations and align them with the competitive possibilities found in their channel partners. To be successful, such a strategic imperative requires today's enterprise to implement new values and operational cultures. The radical changes brought about by global competition have radically challenged traditional thinking not only about organizational structure and design but also about the use of channel alliances, the participation of the customer in product design, manufacturing process, and logistics issues, the nature, diversity, and motivations of the work force, and the application of information and communications technologies.

Figure 3.6 is a possible model that can be referenced by enterprise executives in rethinking their organizations to leverage the opportunities provided by SCM [23]. As can be seen, the neat organizational charts found in most firms have

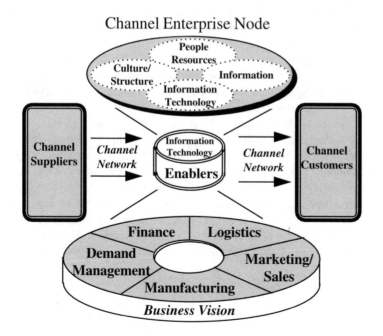

Figure 3.6. Structures of SCM.

been swept away in favor of an integrative approach that conceives critical business functions as unified around a common enterprise-competitive vision and closely interlinked with suppliers and customers. The purpose of the model is to emphasize the convergence of supply channel resources and core competencies networked internally and externally together through information and communications technologies.

As the intersecting circles at the top of the diagram indicate, the enterprise possesses four strategic resources that, together, compose the nucleus of the organization's core competencies. Perhaps the key to leveraging these strategic resources is understanding that they cannot hope to produce the level of competitive advantage required by today's fast-moving global marketplace by functioning in isolation. Industry leaders are continually in the process of realigning and reintegrating these four resources. In reality, the four strategic resources are so dynamically linked that a change in the contents of any one resource acts as a catalyst for simultaneous change in the remaining three. For example, the implementation of an integrated enterprise information system will require reengineering of the existing organization, enable the closer linkage of the entire channel network, provide for interactive systems of information previously nonexistent, and change the educational, technical skills, and work values of employees.

The interdependence and alignment of the four strategic resources also has a direct impact on the business functions of the enterprise. Traditionally, the five departmental groupings, detailed in the bottom portion of Figure 3.6, often possessed their own internal cultures and structures, information flows, people resources, and performance measurements. Usually, there was little continuity between them; often the needs and resources of one function were in misalignment with or even in opposition to those of another. In contrast, phase 4 organizations have eliminated the isolated departmental boxes on the tradition organizational chart and have substituted integrative structures that provide for the flow of business processes across internal departmental boundaries and network the resources and competencies of businesses spanning the supply channel. The integrative nature of such a business environment provides for a common sharing rather than a decoupling and segmentation of channel processes, with a common business vision permeating each function and each channel node and providing focus and direction for the entire supply network. The path to leveraging company and channel competencies lies in a critical understanding of the endless competitive possibilities found in the people, business cultures, and information and communications technologies of each participant comprising the supply channel network.

Understanding Phase 4 Strategic Resources

The first and most important of the SCM strategic resources is the people who define the company's work culture and plan and execute the firm's business functions. In the past, organizations were characterized as rigid hierarchies where process flows and information were the responsibility of narrowly defined departments and functional categories. Phase 4 management requires that companies abandon this narrow and debilitating view of their people resources. During the past decade, redefining the human organization has been a common theme of professionals and practitioners seeking to leverage the wealth of skills and resources within the enterprise's people organization. Increasingly, companies seeking competitive advantage have ended hierarchical control and narrowly defined and fixed divisions of labor and departmental performance, and have implemented work force values centered around employee commitment, self-management, continuous learning, and team work.

The integrative aspects of today's information and communications technologies (ICT) have facilitated the networking of people resources and is a key enabler in the activation of intraenterprise and interenterprise virtual teams. The use of ICT tools permits peer-to-peer networking by linking each information node so that communication can occur with every other node without having to pass serially through departmental organizations. Networking assumes that people resources have ready access to information regardless of location, and that people have full access to the knowledge of others wherever it may be within the enterprise or out in the channel system. The command and compliance management paradigms of the past had placed barriers that isolated people resources. Networking removes these barriers by enabling the competencies of the work force existing in different functions and different companies to be welded together to form "virtual" teams directed at pursuing the best marketplace opportunities or devising breakthroughs that generate whole new competitive space. Peer-to-peer networking is focused on achieving cross-functional leveraging of the ideas, skills, knowledge, and visions of whole channels of supply in the quest for marketplace leadership [24].

The second strategic SCM resource can be found in the culture and organizational structures found within each enterprise. These strategic resources contain the firm's attitudes toward customer service, quality, innovation, job and task flexibility, and competitive values. For the most part, phase 2 and 3 organizations were characterized by hierarchical "command and control" operating cultures, fragmentation of effort and authority, and narrowly defined performance measurements. To successfully compete in the global marketplace and leverage the full potential of their internal people resources and the competencies of partners out in the supply channel, companies must effect the following four broad changes:

1. *Continuous Organizational Reengineering.* Today's marketplace leader must be able to quickly adapt to new organizational designs, new channels of information transfer, and different distributions of authority. Organizationally, this management value means stripping away all impediments that add cost and redundancies to production, distribution, and administrative processes and that block the effective application of

current core competencies as well as the development of new competencies that provide gateways to tomorrow's opportunities. The goal of phase 4 management is the nurturing of an organization that enables growth and diversification, seeks to create fundamental interlinkages among corporate, business units, and channels partners, and perceives planning and performance as centering around collective rather than centralized or decentralized patterns of authority.

2. *Change in Workforce Values.* The interplay of different levels of organizational goals cannot be aligned and integrated without real changes to work force cultural values and attitudes. Phase 4 organizations require a new operating culture concerned with teamwork, the elimination of all forms of waste, value-added services, ongoing employee education and training, commitment to total quality, removal of operational barriers, and an endless search for an environment characterized by open communications and information sharing between people and functions. In the past, the work force had been regarded as basically reluctant participators, problem-solvers, and team members that had to be led. Today's most successful enterprises seek to foster purposeful employee initiative and empowerment centered around a defined companywide sense of shared direction that reconciles the needs of individual objectives and concerted, coordinated effort [25].

3. *Reevaluation of Performance Measurements.* Phase 4 organizations require managers to radically change the methods by which their companies measure performance. In past management phases, companies were splintered into narrowly defined departments and measured success by tracking departmental variances and detailed costs. In contrast, phase 4 organizations are characterized by metrics that seek to measure the progress of the entire enterprise as well as the success of the channel network in meeting customer expectations and creating new competitive space. These global measurements coalesce around key elements such as customer satisfaction, asset utilization, operating costs, quality, and cycle time.

4. *Abandonment of the Traditional Cost/Service Level Trade-off Approach.* Past managers tried to maintain a direct ratio between cost and customer service to assist in decision making. For example, if marketing required a higher level of product quality, the goal was to determine the cost of attaining that quality level and then to analyze whether or not the anticipated increase in marketshare justified the expense. Phase 4 organizations have found that a single-minded dependence on cost trade-off thinking stands as an impediment to marketplace leadership. Although no company can long stand on a faulty understanding of the cost of business, the real goal is to continually delight customers by creating

products and services that anticipate and fulfill their unarticulated needs. Using artificial cost/service standards to brake a company's collective creativity is a guarantee for marketplace failure.

The robustness and availability of intrachannel and interchannel information comprise the third strategic resource. Information unifies the firm's business functions into a single competitive unit and links each node in the supply channel, providing data concerning customers, market opportunities, and costs essential for decisive decision making. The ability of (ICT) to *integrate* and *network* both enterprise business functions and the supply chain requires companies to rethink not only traditional information flows but also the nature of information itself. In earlier logistics management phases, the flow of information was restricted by organizations that required it to flow serially from one department to another, with each business function abstracting and summarizing the data they needed to execute isolated-area decisions.

The radical breakthroughs in computerized technologies, such as computer-to-computer networking and client server architectures, have accelerated the use of integrative information techniques, thereby enabling companies to create a structure of networked business and channel functions referencing and being coordinated through a common data architecture. Local forecasting systems feed global marketing data to inventory and logistics capacity planning systems; quick-response order processing systems are linked electronically to internal and external customers throughout the supply channel; data concerning customer product designs and item specifications are interactively input into the databases of their suppliers; logistics systems receive their input from warehousing and shipping and provide feedback to marketing and accounting. The result is that the entire supply chain network is directed toward achieving the goals of a clearly articulated business vision as well as positioned to continuously create new products and market space that keeps the channel one step ahead of the competition.

As new technologies change the way business functions, customers, and suppliers are networked, commonly accepted definitions of information will have to be reformulated. Table 3.1 provides an example of shifts in commonly used concepts of information. Companies seeking marketplace leadership must be prepared not only to continually rethink information but also to provide for a channelwide open dialogue of information. The proper understanding and effective utilization of information is a fundamental resource in attaining tomorrow's business, organizational, and technological strategies [26].

The final strategic resource can be found in the application of ever-new forms of ICT systems that enable the entire channel network to facilitate service and cost objectives while exploiting fresh business opportunities as new forms of technology emerge. The application of ICT systems to solve business problems can take several forms. It can consist of the implementation of a single enterprise resource planning system that seeks to integrate the entirety of the enterprise on

Table 3.1. Characteristics of Integrative Information

Type	Traditional Use	Integrative Information
Activity	Action-based	Intellective based
Timeliness	Important only to local department systems	Critical for global technologies
Accuracy	Decayed as it moved away from departmental systems	Perpetual 99–100% global accuracy
Measurement	Rarely audited, never used to measure other departments	Perpetually audited, provides criteria for global performance
Knowledge	Sentient, physical/manual knowledge and skills	Symbolic, abstract manipulation of data
Integrative	Serial interfacing	Interactive
Decisions	Rarely provides data for decision making outside department	Provides data for global interactive decision rules
Relational	Parochial, departmental dialects and meanings	Provides for a common reference context
Availability	Departments closely guard information from others	On-line, real-time, and interactive globally
Feedback	Physically observed conformance to procedures and processes	Computer-generated data matched to programmed decision rules

a common database. It can also take the form of a stand-alone or a group of computerized applications whose purpose is to automate a tactical management function or to provide decision support for a specific department or function. In any case, the real value of ICT systems occurs when business functions are linked to provide an integrated information node. As information nodes appear across the enterprise and across the supply channel landscape they can be linked together, creating an information network. The more business nodes in the system that can be networked, the more robust the information sharing and the more effective the decision-making process. In this sense, ICT can never be perceived as an enterprise strategic resource independent of the other three strategic business resources portrayed in Figure 3.6. Leading-edge companies understand that superior performance requires the close combination of technology and organizational values and infrastructure.

Perhaps the most important requirement for the successful implementation of ICT systems is aligning ICT architectures with business processes. Several key elements come to mind. To begin with, in order to achieve the results expected by implementers, the ICT systems selected must be focused on the type of business problems to be addressed, such as managing globalization, more effective logistics, superior customer service, or combinations of these goals. Establishing the right information and communications systems is critical. Technically, the

ICT systems implemented should provide companies with the ability not only to network with their global facilities but also to exploit computerized linkages backward to suppliers and forward to customers. In addition, ICT systems should possess the capability to utilize relational database management systems, modern applications that promote functional integration within the enterprise and outside in the supply channel, sufficient bandwidth to cover all anticipated network volume and ensure speedy communications, and flexibility and configurability so that the right modules can be leveraged to meet the needs of each business function [27].

Information and communications systems provide for the convergence of the four strategic enterprise resources while enabling and directing business functions toward the achievement of targeted business strategies. Many companies make the mistake of employing ICT systems purely as tools meant to automate business functions. Technology is falsely perceived as a "quick fix" to the range of marketplace challenges that lie before them. Phase 4 organizations, on the other hand, understand ICT systems as constituting an integral part of the integrative process rather than solely as a support function. In reality, the benefits to be gathered from today's ICT systems can only be realized when implementers recognize the interdependency between ICT and the other three strategic resources. The real challenge of phase 4 management, therefore, is to be found in directing the enterprise's people resources, structure and culture, and channel network, activated by ICT systems, in the pursuit of whole new regions of unassailable competitive leadership.

Summary and Transition

Supply chain management represents a radically new approach to leveraging the market-creating competencies existing both within the each individual organization and outside in the supply channel. SCM challenges companies not only to continually search for new logistics processes that eliminate costs, remove wastes and redundancies, and reduce cycle times, but also to leverage the tremendous opportunities to be found in supply channel partnerships to create dramatically new areas of competitive leadership. Yet, despite the revolutionary nature of SCM, it is closely connected with and in many ways is the product of the various stages of development logistics management has progressed through over the past 30 years. During that period, logistics has evolved from a purely operational function to become a fundamental strategic component of today's leading manufacturing and distribution companies. Because the kernel of SCM can be found in the logistics functions of the past, a comprehensive understanding of SCM requires a through understanding of the development of logistics management.

Historically, logistics management has served a central position, linking enterprise market and sales strategies on the one hand, and manufacturing and inventory execution on the other. As far back as the turn of the century, economists considered the activity of *distribution* to be the function by which the commodities of production are moved through the supply channel and the exchange process is determined. However, despite its importance, the logistics concept was slow to develop. Most executives considered logistics to be only of tactical importance and because of the scope and huge volume of transactions occurring in physical distribution, virtually impossible to manage as a integrated function. In fact, it was not until around 1970 that companies began to perceive of their logistics functions as adding a critical dimension to their ability to create and sustain competitive advantage.

The development of logistics management can be said to have occurred in four distinct phases. The first phase spans the period from the late 19th century to the early 1960s. During this era, logistics was considered of secondary importance, next to marketing, sales, and production, and had little strategic impact outside of day-to-day operations. To improve performance, logistics functions were decentralized, with operational responsibilities split among marketing, manufacturing, and finance. During the 1970s, companies began to recognize the operational and cost inefficiencies of the decentralized system. The second phase of logistics management attempted to solve these problems by, first, centralizing logistics functions into a single management system, and then constructing logistics activities to achieve optimal total cost for the whole system. The transformation of the organizational structure of logistics from a decentralized group of uncoordinated functions to a single, cost-focused integrated department marks the origins of today's modern logistics organization.

The third logistics management phase began in the 1980s and was characterized by a concerted attempt by companies to move away from a passive approach to logistics to one where logistics was conceived as providing internal enterprise strategic advantage. This was accomplished by expanding the number of functions integrated and exploring how they could be engineered to provide a continuous source of unassailable strategic competency. The rise of SCM in the early 1990s inaugurated the fourth phase of logistics management. As the competitive pressures of the global marketplace grew, companies found that even the best "world-class" logistics operations barely kept them in the competitive race. To capture and sustain market leadership, executives found that they had to transform previous concepts of logistics, which concentrated primarily on performing internal and external distribution related operations, from being only a source of internal competitive advantage to a source of *external* advantage as well. This radical change in the scope of logistics is at the heart of the SCM concept. SCM can be defined as having two dimensions. The first consists of an operational strategy centered around accelerating the cycle times of inventory and information within the channel pipeline and optimizing the linkages between internal functions and supply partners. The second dimension consists of the continuous networking of the competencies of intersecting supply channels focused around the creation of shared marketplace and competitive visions, coevolutionary alliances providing for order-of-magnitude breakthroughs in products and services, and radically new marketplace opportunities that enable supply systems to remain one step ahead of the competition.

Although optimizing logistics operations and extending integrated logistics management to complimentary activities performed by business partners out in the supply channel are critical, it is the new strategic dimension added by SCM that defines the essential nature of phase 4 logistics. To uncover the new sources of innovation and customer enrichment promised by SCM, enterprises must radically change the way they develop and implement corporate strategic direction. This is the topic of the next chapter.

Notes

1. Arch W. Shaw, *Some Problems In Market Distribution*. Cambridge, MA: Harvard University Press, 1915, pp. 7–12.

2. *Ibid.,* pp. 41–44.

3. Peter F. Drucker, "The Economy's Dark Continent," *Fortune* (April 1962), 103–104.

4. Peter F. Drucker, "Physical Distribution: The Frontier of Modern Management," speech reprinted in Donald J. Bowersox, Bernard J. LaLonde, and Edward W. Smykay, eds., *Readings in Physical Distribution Management: The Logistics of Marketing*. London: Macmillan, 1969, pp. 3–8.

5. David F. Ross, *Distribution: Planning and Control*. New York: Chapman & Hall, 1996, pp. xi-xix, and Robert N. Anthony, *The Management Control Function*. Boston, MA: Harvard Business School Press, 1988, pp. 3–24.

6. Alan C. McKinnon, *Physical Distribution Systems*. New York: Routledge, 1989, p. 4.

7. Donald J. Bowersox, "Emerging Patterns of Physical Distribution," in *Readings in Physical Distribution Management,* pp. 276–277.

8. Donald J. Bowersox, "Physical Distribution Development, Current Status, and Potential," in *Readings in Physical Distribution Management,* pp. 375–376.

9. Jeffrey Karrenbauer, "Distribution: A Historical Perspective," in James F. Robeson and Robert G. House, eds., *The Distribution Handbook*. New York: The Free Press, 1985, pp. 4–5.

10. David F. Ross, *Distribution: Planning And Control*, pp. 17–19.

11. Wendell M. Stewart, "Physical Distribution: Key to Improved Volume and Profits," in *Readings in Physical Distribution Management,* pp. 375–376.

12. Donald J. Bowersox, "Distribution Logistics: The Forgotten Marketing Tool," in *Readings in Physical Distribution Management,* pp. 375–376.

13. Shaw, pp. 44.

14. Bernard J. LaLonde and Leslie M. Dawson, "Early Development of Physical Distribution Thought," in *Readings in Physical Distribution Management,* pp. 375–376.

15. Raymond LeKashman and John F. Stolle, "The Total Cost Approach to Distribution," in *Readings in Physical Distribution Management,* pp. 207–208.

16. Douglas M. Lambert, *The Development of an Inventory Costing Methodology: A Study of the Costs Associated with Holding Inventory.* Chicago, IL: National Council of Physical Distribution Management, 1976.

17. D. M. Lambert, J. F. Robeson, and J.R. Stock, "An Appraisal of the Integrated Physical Distribution Management Concept," *International Journal of Physical Distribution and Materials Management* 9 (1) (1978), 74–88

18. Ross, *Distribution: Planning and Control,* pp. 20–21.

19. Bernard LaLonde, "Evolution of the Integrated Logistics Concept," in James F. Robeson, and William C. Copacino, eds. *The Logistics Handbook.* New York: The Free Press, 1994, pp. 9–10.

20. Wendell M. Stewart and William J. Markham, "The Role of the Physical Distribution Manager," in *The Distribution Handbook*, pp. 35–37.

21. James M. Masters and Terrance L. Pohlen, "Evolution of the Logistics Profession," in *The Logistics Handbook*, pp. 26–27.

22. *Ibid.,* pp. 27–28.

23. This section is based on David F. Ross, "Aligning the Organization for World-Class Manufacturing," *Production and Inventory Management Journal* 32 (2) (Second Quarter 1991), 22–26 and *Distribution: Planning and Control*, pp. 26–36.

24. Charles M. Savage, *Fifth Generation Management.* Burlingtom, MA: Digital Press, 1990, pp. 152–156.

25. Gary Hamel and C.K. Prahalad, *Competing for the Future.* Boston, MA: Harvard Business School Press, 1994, pp. 317–323.

26. Ross, *Distribution: Planning and Control,* pp. 33–34.

27. Thomas G. Gunn, *In the Age of the Real-Time Enterprise.* Essex Junction, VT: omneo, 1994, pp. 104–106.

4

Developing SCM Strategies

At the heart of supply chain management (SCM) can be found the continuous unfolding of dynamic organizational, marketplace, and product strategies that enable today's enterprise to leverage both the core competencies within its own boundaries as well as the almost limitless capabilities of supply channel partners in the search for new sources of competitive advantage. Unlike conventional strategic planning, which focuses around determining budgets and detailed metrics concerning the expected performance of existing products and processes, companies focused around the SCM paradigm perceive the formulation of business strategies as an opportunity to explore radically new and innovative approaches to the marketplace. Whereas SCM business strategies are concerned with the creation of new visions of logistics that transcend conventional techniques of purchasing, producing, moving, storing, and selling products and services, their real importance resides in their ability to assist executives in designing a clear blueprint for the development of new organizational architectures that will prepare their companies to build uncontested marketplace advantage in the future. SCM strategies can accomplish this objective by deploying new management models, productive processes, and technologies to achieve consistent and continuous breakthroughs in the creation of the products and services customers really want, the acquisition of new productive competencies or the migration of existing competencies to new functional areas, and the targeted utilization of the capabilities of supply channel partners to engineer an interenterprise vertical channel of many diverse competencies drawn together as a seamless, coherent customer-satisfying resource.

This chapter seeks to define and then explore how companies can develop market-wining channel business strategies by leveraging the strategic and operations planning components of SCM. The chapter begins by exploring the rise, fall, and rebirth of strategic planning. As will be seen, fundamental to the renewal of competitive strategy is the changing nature of competition that has required companies to move from an enterprise-centric view to one that utilizes SCM concepts of channel partnership for the development of new forms of product/ service combinations and new competitive space and the reengineering of competitive enterprises. Following this overview, the chapter then proceeds to explore the elements of the strategic planning process. The discussion focuses around defining the fundamentals of business planning, determining enterprise goals, developing competitive business strategies, and creating detailed operations strategies. The chapter concludes with a detailed analysis of the elements necessary to develop effective logistics strategies. A seven-step process is outlined to guide planners, and practical examples of logistics strategic activities are provided.

Changing Views of Business Strategy

The formulation of effective enterprise goals and their translation into operating strategies that will guide the business functions of the firm in the search for competitive leadership is perhaps the most important role performed by a company's management staff. The success of the business plan requires the development of product and service strategies that bring bold innovation to the marketplace, the continuous regeneration of flexible organizations positioned to create radically new processes and integrate different combinations of competencies, the implementation of information, communications, and automation technologies designed to network the entire supply channel, and the ability to leverage strategic alliances with channel partners and outside services providers that facilitate the synthesizing of new capabilities and techniques within or among enterprises.

Overview

The development and implementation of enterprise strategic plans is a comparatively new business activity. Although it can be said that no company in the past could succeed without identifying basic product, service, and customer goals and the strategies necessary to achieve those goals, business planning was mostly concerned with the narrow calculation of enterprise physical and financial assets and how they could be optimized to exploit existing products and processes. Broadly speaking, during the 25 years after the conclusion of the Second World War, U.S. companies could achieve marketplace, profitability, and performance objectives without much attention to the strategic planning needs of their firms. It was an era of mass production and mass distribution in which companies were organized to exploit economies of scale and scope made possible by vertical integration, centralized hierarchical organizations, standardized products with long life cycles, and stabilized manufacturing processes that focused on cost efficiencies and long production runs. The prevailing business culture of most manufacturers and distributors was to capitalize on existing equipment, facilities, and processes to push mass-produced goods into the marketplace. Customers either had to adapt their requirements to match mass-market products or pay high prices for customized goods and services that satisfied their individual needs.

The decade of the 1970s, however, brought a series of upheavals fueled by spiraling energy costs, economic inflation and uncertainty, and growing overseas competition that required companies to move beyond their tradition focus on operations optimization. No longer could U.S. firms take marketplace dominance for granted. The result was a growing interest in preparing the company for change through a total enterprise planning process that not only focused on business strategic elements, such as the identification of corporate goals and the effective allocation of resources, but would also guide the organization through the labyrinth of shrinking margins, shortened product life cycles, synchronization of world economies, increasing labor and materials costs, and struggle for competitive advantage that characterized the years that followed. The ability of manufacturing and logistics functions to respond rapidly to an environment of perpetual change had become a necessity for corporate survival.

During the 1980s, the requirement that firms seriously pursue effective strategic planning was heightened by subtle, yet devastating, challenges aimed at the very heart of the mass-production and mass-distribution systems that had dominated the operational and functional core of most U.S. businesses for almost a century. Suddenly, corporate planners were to discover that focusing on ways to further penetrate markets, more effectively allocate resources, and cut process and overhead costs were simply not enough to keep their companies on the cutting edge of marketplace leadership. Traditional global giants like Xerox, RCA, Westinghouse, IBM, and Sears were beginning to lose market dominance, and to lose badly, to more agile and responsive companies like Canon, Sony, Hitachi, Compaq, and Wal-Mart. What had happened was advances in information, communications, and automation systems and the rise of new management methods clustering around JIT, TQM, and *lean manufacturing* concepts had enabled bolder, more innovative companies to capture market advantage through shortening production runs, leveraging the core competencies of allied partners, building to customer order, and achieving dramatic reductions in new product to market development time.

By the early 1990s, the industrial world was frantically searching for new competitive strategies that would provide for more flexible manufacturing processes and faster change over times, the creation of empowered work teams, the engineering in of superior quality without increase in cost to the customer, the proliferation of products to give customers more choice at no increase in cost, and the aggressive pursuit of development technologies that would leverage existing capabilities to rapidly create new products or transform existing ones. The goals of the new paradigm were straightforward. To be able to meet customers' requirements for high-quality, customized products and services, quick-response deliveries, and low prices, companies needed to increase productive flexibility in the face of decreasing product development cycles; that is, the cost of flexibility in production and distribution had to continuously decrease as new products were more rapidly introduced in order to lower the investment hurdle and fund the next generation of new product development [1].

Today's Business Planning Environment

The last 10 years have been devastating for many companies once thought to be bulwarks, not only of American but also of global business. Once proud and profitable companies such as Philips, TWA, Texas Instruments, Boeing, Commodore Computers, and Digital Equipment Company (DEC) have seen their positions of market leadership erode or disappear altogether in an avalanche of new technologies, management paradigms, and marketplace changes. Perhaps even more menacing is the seemingly unstoppable success of new competitors who magically appear to have mastered nontraditional methods of rapidly restructuring their basic goals and processes, achieving dramatic breakthroughs in quality and productivity, leveraging the innovative power of new technologies, identifying whole new markets, and actualizing the creative powers of their work forces. Clearly, the organizational, process and procedural, and strategic assumptions that had long guided the functioning of U.S. business had become outdated; many companies were squarely facing the threat of extinction.

The response of companies in the early 1990s to the dramatic decline in productivity and profitability was, and often continues to be, a single-minded focus on fine-tuning the structures, values, and skills of their organizations to meet the realities of the new marketplace. Utilizing the concepts behind JIT, TQM, *lean manufacturing*, and *continuous improvement* management methods, executives single-mindedly are attempting to rebuild their companies from the inside out in an effort to close the gap between themselves and emerging industry leaders. Long standing business strategies have been jettisoned in favor of policies centered around corporate restructuring. Downsizing, process redesign, elimination of unprofitable products and underperforming businesses, employee empowerment programs, and other strategies are being implemented in a frantic effort to stem the loss of competitiveness, trim corporate fat, and raise asset productivity. As Hamel and Prahalad have pointed out [2], "denominator management" has swept the field as executives attempt to increase efficiency and productivity by cutting, often brutally, headcounts and assets.

The corporate reengineering frenzy of the past half-decade was a relevant response to the radically new marketplace challenges confronting every enterprise. However, a strategy of "rightsizing" and continuous incremental improvement eventually meets a law of diminishing returns. At what point should restructuring end? How efficient should processes be before the cost of maintaining them does not bear the gain? When should the outsourcing of functions stop and which enterprise competencies should be protected? The reality of restructuring and process reengineering is that they are, for the most part, ways of catching up to the competition. Whereas developing methods to be more responsive to customers, more purposefully focused on quality and productivity, and more capable of slashing cycle times are, indeed, critical elements necessary for marketplace survival, they have today become the bare minimum for competitiveness. Achieving these objectives will allow companies to qualify to play in today's marketplace; they do not, however, constitute by themselves the path to marketplace leadership.

Already, the evidence pointing to the strategic failure of "incrementalism" is beginning to accumulate. Although the concerted efforts of enterprise executives to make their companies leaner, meaner, and more productive is important in repositioning organizations in their effort to "catch up" with the competition, such a strategy is narrowly focused on short-term objectives and is incapable of providing the basis for continuous innovation and corporate growth required of tomorrow's marketplace leaders. More to the point, becoming more flexible, more focused, and better able to leverage assets are important goals. Today's best companies, however, must also be capable of fundamentally reconstructing themselves, rapidly regenerating their core marketplace strategies, and developing products and processes innovative enough to reinvent their industries.

For example, in 1993 mega retailer Herman's World of Sporting Goods undertook a massive restructuring and quality improvement project designed to shed waste and poor management and to recapture its share of the retail recreational products marketplace. Its strategy was to close dozens of stores outside the core U.S. northeast market, make operations more flexible, and streamline logistics processes. Besides critical operational improvements, the company's centralized mainframe decision support system was dismantled in favor of a client/server computer environment targeted at empowering its management and staff. Vendor direct-to-store shipments were discontinued, with all inbound receipts temporarily consolidated into one giant warehouse where value-added processing techniques and planned branch store shipments were used to leverage economies and provide for pinpoint deliveries. The supplier base was halved; quick-response and automatic replenishment techniques were developed. By stabilizing inventories and rigorous cost cutting the management team expected significant results [3]. In June 1996, despite its frenzied attempt at business process reengineering, Herman's went out of business.

The prevailing strategies for dealing with competition in many companies have often been limited to cutting fat (quality management) and overhauling sloppy processes (reengineering). When the success of these initiatives eventually diminishes (and they will diminish in the long run despite the initial gains in productivities and profits), companies are faced with two alternatives: either they go out of business like Herman's or they begin the arduous task of fundamentally reconstructing their business strategies and reinventing themselves, their products and services, manufacturing processes, channels of distribution and performance measurements, and redefining who their customers are. Today's market leaders have the ability to look beyond the fascination with incrementalist strategies to the formulation of new competitive visions that bring bold innovation to customers, integrate the competencies of entire channel ecosystems to create capabilities beyond the competition, and enable them to be the destroyers of old industries and the creators of whole new marketplaces.

Rebuilding Strategic Advantage

The ability to create winning strategies that focus on developing new markets and innovative products and services rather than searching for ways of continually slashing and shrinking process costs is a radical departure from the management styles of the past half-decade. Formerly, enterprise executives oriented strategies around organizational issues such as restructuring, reengineering, and merger and divestment; operational issues such as asset management, resource allocation, and performance benchmarking; and marketing and product issues such as minimizing cycle times, maximizing market share, competing for new product leadership, and pursuing avenues for competing as individual players within existing industry structures. In contrast to these traditional strategic planning objectives, building the competitive strategies of tomorrow requires executives to revolutionize both the content and structure of the planning process itself. Overall, questions of a much larger nature must be answered: What will be the nature of competition tomorrow? How are enterprises to gain insight into the marketplace challenges and opportunities of tomorrow? How can the organization be continuously reinvented to capture the dynamic competencies of its work force? How can the competencies and resources of business channel partners be accessed to provide radically new sources of product development, manufacturing, and distribution advantages unavailable to the competition? Finding answers to such questions requires companies to do more than focus on streamlining the work force and process flows: They must devise strategies and visions and create paths that only those who aspire to conceive of new opportunities and new marketplace vistas dare to tread.

The Decline of Corporate Planning

As firms look to business restructuring and reengineering techniques as the hoped-for panacea for their competitive problems, executives have been correspondingly abandoning their interest in strategic planning. In the past, companies spent large amounts of time and money developing corporate business plans. Reams of paper detailed, from every angle, how to position the enterprise in the marketplace and protect current competitive advantage. Today, many companies are disbanding or downsizing their corporate planning functions. Many executives now spend little time thinking about competitive strategy and designing the strategic plans necessary to take their companies into the next century. Is it that a preoccupation with incrementalist management provides a more immediate way to increase the bottom line than plotting the future? Is it a feeling that all that concern in the past over strategic planning never really seemed to pay off; that competitors and markets really function in a totally random fashion and cannot be predicted with any kind of accuracy?

What is being argued is that the decline of corporate planning is the direct result of the total inability of the traditional planning process to provide distinct and thoughtful strategies about how a company plans to compete in the future. Normally, corporate plans are a wearisome compendium of charts and statistics where debts and credits must balance and projected profits must be justified to the penny. Corporate planners, following the precepts found in countless business textbooks and MBA programs, focus the bulk of their energies around developing highly detailed, telephone-book-sized analyses of cost-to-profit statistics, market segmentation analysis, diagnostics of product/service potential to labor and over-head ratios, sales targets, and so on. In reality, traditional business plans are more often actually operations and tactical plans and not really about strategic direction at all. Most plans seek to develop individual strategies for each business or functional unit, complete with metrics and objectives that are independent of one another and the enterprise as a whole. Corporate planners, then, painstakingly integrate these separate unit plans together to provide the appearance of a unified enterprise approach. Instead of providing clarity, consistency between functions, and recognizable objectives common to the entire enterprise, such strategies are usually dominated by a single business function, such as sales, which biases the allocation of corporate resources and diminishes the perspectives and contributions of other functions [4].

Whereas these operations strategies are most assuredly necessary for the successful execution of the everyday functioning of the enterprise, it is important to note, however, that they are concerned with the manipulation and calculation of existing and quantifiable facts. For example, take the creation of the business forecast. In this plan executives must balance expected growth and aggregate revenues arising from sales and other sources with incurred costs and other liabilities. The goal is the creation of detailed targets and performance benchmarks that can be used to guide ongoing decision making. Core operational strategies work best when product, industry structure, and the marketplace have been clearly defined. They are about charting the progress of the existing business. They are focused on positioning the firm's current competitive capabilities in an attempt to determine the optimal combination of markets, channels, product differentiators, and value chain configurations.

Companies spend most of their time developing core operations strategies for several reasons. To begin with, it is what executives have been traditionally taught to do, and what Wall Street expects from companies when devising investment portfolios. In the era of mass production, when industries, markets, and products appeared to be immutable and changed very slowly, calculating profits against costs provided key indicators of a company's competitive position. Over time, business schools and accepted business practice reinforced this approach. Sophisticated formulas, reporting procedures, accounting methodologies, and even computerized software were created to assist corporate planners to broaden and deepen core operations planning. Furthermore, it is also a relatively simple process once the rules and conventions are known. It is always easier to generate information based on concrete, observable data than it is to develop strategies designed to uncover new sources of competitive advantage. Charts and graphs exhibiting last year's sales juxtaposed to this year's projected sales and then extrapolating what the company will have to do to make the target, what assets will have to be acquired, what performance goals will have to be reached, and similar metrics are based on measurable "facts" that seem to provide clear direction and can be translated into recognizable goals shareholders, as well as the whole enterprise, can understand.

Firms concentrate on their core operations strategies because, frankly, that is what they are best at doing. Executive planners normally have a very close understanding of the workings of their organizational structures, productive processes, suite of products/service offerings, marketplaces, and where they stand in comparison to industry rivals. Nevertheless, it has become painfully apparent to managers that this region of business planning is insufficient by itself in gaining and sustaining industry leadership. Strategies focused around operations management techniques such as restructuring, reengineering, continuous improvement, and total quality management will help a company catch up to marketplace leaders. They will never, however, provide companies with the vision, foresight, and innovate capabilities to realize new competitive space and catapult them into positions of industry leadership. Core operations strategies are not about regenerating competencies and products, leveraging technology and channel coalitions to rethink what tomorrow's competitive leadership will look like, shaping the future of industry, or preempting global markets.

Changing Nature of Competition

Although focusing on core operations strategies is critical for business survival, they can severely limit the perspective of executive planners. Perhaps the most critical fault of this perspective is that it is based on a view of competition that has all but vanished from today's hyperactive and dynamic global economy. In the age of mass production, competitiveness was defined as the ability of one company to offer an array of standardized products and services at cost-effective prices to a specific market segment that could beat the similar offerings of rival companies who sought to capture the same market. Market winners then strove to solidify their positions of leadership by improving products and services by listening to their customers, continuously reducing process and overhead costs, and investing in new development and new processes to keep the engines of competitive advantage moving forward. Because neither products nor markets changed very little, companies that got off to an early and very profitable start could leverage considerable resources, global distribution channels, advertising and marketing clout, and the newest technologies to maintain that lead.

The problem with this introspective, enterprise-centric view of competition is that marketplace rivals, sometimes with unorthodox methods, sometimes from the other side of the earth, have today been quick to challenge, for marketplace supremacy, successful industry leaders who continue to base their business fortunes solely on core operations strategic planning. As a result, some firms that once had exciting, solid products with loyal customers have found, as have Westinghouse, Honeywell, IBM, Digital and others, their competitive positions shrinking and their industry leadership evaporating in the onslaught of more innovative companies. Some of these competitors will attempt to force their way into a once-stable business ecosystem by cloning successful products, processes, and organizational structures. Others will leverage special competencies to violate traditional market boundaries, raiding targeted customer segments once considered the exclusive preserve of more mature, established companies. Still others will utilize radically new technologies or management styles that not only steal business away but also increasingly render obsolete the intellectual and physical assets of their more senior rivals. Such innovative companies have found new ways to leverage cooperating business communities to undermine traditional notions of vertical and horizontal integration and to anticipate, if not shape, the rapidly evolving structures of industries and markets [5].

The traditional view of operations strategic planning suffers from several core defects. To begin with, it has become obvious that economies of scale and scope no longer guarantee market leadership. The prize of customer loyalty today goes to suppliers that possess flexible processes that can produce customized products in any lot size at the lowest cost. Also, traditional strategies labor under the constraints of being too narrow in scope. Three elements come to mind. Normally the time frame of the typical business plan spans no more than a year or so, instead of looking at competitive opportunities maturing many years, perhaps decades into the future. Furthermore, companies often develop their plans around the product or business unit rather than the entire firm or even the whole supply channel. Finally, executive planners all too frequently focus their attention on competing solely through their existing goods and services. Competitive leadership today requires enterprises to move beyond this limiting idea of what traditionally constitutes competitiveness and focus their efforts on competing through new visions of the future, the development of new competencies, and the blending of new arrangements of supply chain partnerships that provide for the evolution of new industry ecosystems [6].

The ability to look to the future potential of products and markets beyond today's capabilities is perhaps the key element of competitive strategy. Assuming that the pace of demand and the needs of the marketplace will remain fairly constant can cause companies to overlook enormous possibilities. Take, for instance, the well-known story of IBM's failure to acquire the rights of copying machine technology from Xerox in the late 1950s. Pressed for cash, Xerox offered its patents to IBM. After intense market study, it was determined that even if the new copier replaced 100% of the market for carbon paper, dittograph, and hectograp (the leading methods of copy reproduction at the time), the return would not sufficiently justify the investment. The offer, accordingly, was turned down. Obviously, IBM's error had been one of strategic foresight. The competitive power of the copying machine lay not in replacing existing reproduction techniques but in its ability to perform tasks beyond the reach of those technologies [7].

Learning to compete in the next millennium will mean that executives must understand and create business plans on two levels. First, they must continue to create the traditional operations plans providing for the day-to-day objectives and performance measurements that will benchmark their companies' progress. However, corporate planners must also spend a proportionately larger share of their time developing those competitive strategies that will provide a window for determining who the firm's customers will be and what markets will be profitable in the future, what competencies will be required, what will be the opportunity environment, and what supply channel networks stretching across several different industries will be necessary to develop the business ecosystems of tomorrow. Developing overall successful business plans is similar to taking a journey. The map shows the routes, the cities, and the checkpoints on the way to the destination. Once the trip begins, however, travelers must constantly measure the progress of each mile, replenish provisions, and note how far they are ahead or behind schedule. Competitive strategies are the map providing the sense of direction and illuminating the opportunities to be found at the end of the journey; core operations strategies, on the other hand, provide the detail yardstick where costs and profits and the daily struggle for competitive survival can be measured and informed decisions concerning the next step can be made intelligently.

Rebuilding Strategic Advantage

Developing competitive strategies is about forming new ideas and creating new enterprise structures and values that permit companies to unearth and exploit emerging, radically new marketplace opportunities—to claim new competitive space that rivals are either unprepared or too timid to explore. Thinking about the future and what it will take to get there is hard work, much harder than cranking out detailed operations and process reengineering strategies designed to demonstrate what it will take to "catch up" to industry leaders. It is one thing to research competitors, benchmark competitive gaps, and draft plans on how performance and processes deficiencies are to remedied, and another to grapple with the risky task of gaining foresight into future market needs. As was mentioned earlier, executives have been trained to perform incrementalist analysis very well. What they often do not have the capability to do is to realistically locate the whereabouts of tomorrow's markets, challenge their employees to unleash their capacities for innovation and entrepreneurialism, locate new competencies, form coalitions that broaden their firm's portfolio of industry experience and capabilities, and engineer new processes and products.

Today's industry leaders create competitive strategies that permit them to explore new possibilities that leverage their strengths and build new possibilities for capturing the marketplace. Reference Microsoft's attempts to capture Internet access software leadership from pioneer Netscape. Ever searching for new avenues of competitive advantage surrounding their immense lead in PC operating systems, Microsoft quickly got into the Internet access business shortly after Netscape and other companies introduced their web access software. By the late summer of 1996, an all-out war, dubbed Web War I, was underway. Hostilities erupted when Microsoft Corp. promised to give away the new Microsoft Internet Explorer 3.0 browser as part of an initiative to challenge Netscape for on-line supremacy. To sweeten the pot, Microsoft threw in free Internet access until the end of 1996 to the on-line, full-text version of the *Wall Street Journal* and ESPNET Sportszone. Although Netscape is currently used by 84% of all people using the Web, the company was quick to announce only a week later their challenge to Microsoft in the form of Navigator 3.0. How deeply Microsoft will cut into this lead remains to be seen; the important fact of the example is to show Microsoft's ever-strategic vigilance to search for new markets and new competitive horizons in the world of personal computing.

Building the kind of competitive strategies characteristic of a Microsoft, a Sony, or a Hewlett-Packard, requires a significant broadening of the objectives of enterprise planners. Comparisons of means and goals are revealing. Instead of concentrating solely on incremental improvements in costs and marketshare, competitive strategies call upon executives to search the future to determine who customers will be and how company and channel partner competencies and resources can be shaped to preemptively seize market leadership. Instead of focusing on the ritual of the annual corporate plan and its preoccupation with existing industry structures, business unit budgets and costs, competitor benchmarking, and segmentation analysis, the creation of competitive strategies requires an open-ended, continuous process centered around elucidating the core competencies, new functionalities, channel coalitions, opportunity horizons, and availability of migration strategies that will provide insight into the technologies, competencies, and products necessary to create leadership in tomorrow's marketplace. Finally, vibrant competitive strategies require the participation of the entire enterprise, including business partners, rather than a narrow set of staff MBAs and business executives.

The ability to leverage competitive strategies resides around four basic premises:

1. *Executing Effective Strategies.* In today's global marketplace, only those companies that can develop energetic and marketing-creating competitive strategies will be able to seize market leadership. Although core operations strategies will remain as essential barometers of short-term and medium-term performance, competitive strategies are designed to provide the foresight, structure, and competitive will to shape tomorrow's best companies.

2. *Building Effective Business Channels.* Time-based competition and the disintegration of the vertical corporation have made it impossible for companies to compete on a stand-alone basis. In fact, it has become apparent that the marketplace belongs to those who are better than their rivals at creating and competing through evolving networks of channel partners, rather than competing based solely on their own product and market efforts. Seizing tomorrow's mega-opportunities belongs to those companies who can work with others, despite potential rivalry. On any given day, for example, Microsoft might find Intel to be a supplier, a buyer, a competitor, and a partner. The goal is not to selfishly exploit surrounding supply channel partners but rather to develop unbeatable alliances that provides sources of collective competitive capacities to successfully invade rival industries or invent totally new ones.

3. *Creating Market-Winning Innovation.* As product life cycles dwindle and requirements for configurable goods and services increase, it is imperative that companies move beyond trying to serve the market based on a purely product/cost basis. And, although it is critical that companies continue to produce better products with processes aimed at ever-increasing quality levels and shrinking costs, what networked enterprises must continuously refocus their efforts on pursuing bold, transenterprise innovation that is targeted at providing customers superior solutions unattainable elsewhere.

4. *Engineering Competitive Enterprises.* According to Hamel and Prahalad [8], there are several paths to building a competitive winner. First, companies may seek to change the rules of engagement in a mature industry (as Wal-Mart did in the retail industry). Second, companies may seek to redraw traditional industry boundaries (as Disney did in its acquisition of ABC in an endeavor to reshape the face of entertainment). Finally, really innovative companies may seek to invent whole new industries and marketplaces (as Microsoft did in computer software). Instead of a preoccupation with shrinking labor costs, assets, and process cycle times, tomorrow's market leaders have found that a strategy that is focused on creating whole new markets and reinventing old ones is the winning strategy.

At the core of strategic success today stands an informed organization possessed of the capability to seize marketplace initiative swiftly, decisively, and without notice to the competition. Such organizations create havoc with their competitors, generating confusion and chaos, producing paralysis, shattering cohesion, and effecting the collapse of rival organizations far richer in skills, assets, and initial marketplace advantage. Writing in 400 B.C., the Chinese philosopher of war Sun Tzu stated that the best structured organization for war is the one that understands

> that its essential factor is speed, taking advantage of other's failure to catch up, going by routes they do not expect, attacking where they are not on guard. [9]

Such is the strength of today's world-class companies in their ability to see opportunities for markets that do not as yet exist, rapidly focusing resources when they are needed, and driving and sustaining innovation before their competitors.

The Business Planning Process

The corporate business plan seeks to validate the broad business goals and to define the strategies that will be communicated to the firm's operational business units and supply channel partners. In developing effective business plans and strategies, the enterprise's top management team must determine answers to such questions as the following:

- What are the long-range goals of the business?
- What is the strength of current relations with supply channel partners, and what new partners are needed to gain unique sources of competitive strategy?
- How radically can the company change the rules of engagement in the industry to which it belongs?
- Can the company expand its competitive boundaries to secure new markets?
- Is it possible for the firm to create whole new products and define new competitive space?
- What are the organization's competitive strengths and weaknesses?
- What is the competitive structure of the organization, and how can core competencies be built to improve competitive strength?
- What changes will be necessary to gain and maintain market competitiveness today and in the future?
- How best should the firm manage its productive assets, and what return on investment is expected in order to meet profit targets and shareholder expectation?

In answering these and other questions, enterprise executives must seek to align company goals, strategies, and ambitions with organizational, economic, and marketplace realities. To remain at the forefront of competitive leadership, the enterprise must be constantly repositioned to exploit marketplace changes and breakthroughs in technologies and to identify shifts in economic conditions, products and services, customer wants and needs, and laws and governmental regulations.

Fundamentals of Strategic Planning

The business strategies of today's leading-edge companies can be said to consist of three distinct but interlocking pieces, as illustrated in Figure 4.1. The strategic business planning process begins with the definition of enterprise goals. Goals are broad, long-range objectives and attitudes that guide the efforts of the firm as an organizational entity as well as the individual actions of all of its employees. According to Drucker, enterprise goals formulate answers to the questions concerning the nature of the business, its broad objectives, its customer base, and targets the enterprise should be moving toward [10]. Defining business goals is an interactive process whereby the executive team of the enterprise periodically raises fundamental questions concerning the well-being of the enterprise and redirects the operating strategies of the firm necessary to meet the challenges that emerge.

The second piece of a successful business plan can be found in the articulation of the strategic competitive mission of the enterprise. The strategic mission consists of several dynamics. To begin with, the focus of the strategic mission is concentrated on future rather than existing capabilities of the organization and the marketplace. Although the strategic mission normally can be said to start with the firm's current marketing plan, operational structures, products and pro-

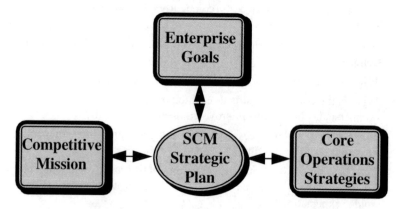

Figure 4.1. Elements of SCM strategic planning.

cesses, and customer service objectives, it is really about identifying emerging opportunities for new competitive space and reinventing the enterprise to activate the innovative competencies located both within company boundaries and business partners located outside in the supply channel. The strategic mission seeks to identify the big questions facing the firm: Who will be tomorrow's industry leaders? What new technologies will be available that will revolutionize the marketplace? What combinations of products and services will prove to be the winning combination? Which companies will form the key partnerships and alliances? How can the skills and innovative spirit within the enterprise be leveraged to proactively shape the nascent markets of tomorrow?

The final piece of an effective business plan can be found in the formulation of the core operations strategies of the company concerned with how to position and measure the success of existing products, markets, and businesses within the current industry structure. Of critical importance is determining which product strategies, pricing and promotional schemes, distribution channels, and supply chain partners will provide the best marketshare and the highest profits. Among the critical activities in core operations planning are the creation of the forecast of the company's expected growth, assets, return on investment, and total net income objectives for a specific time period, determination of the current assets and competencies necessary to support financial and marketing targets detailed in the business forecast, and the disaggregation of the forecast and asset plans down to the firm's business units.

Determining Enterprise Goals

Building effective business strategies begins with the formulation and articulation of a broad mix of values, beliefs, and institutional attitudes that identifies the nature of the enterprise, its aspirations, expectations, and vision of the future. The maturation and gradual unfolding of these organizational *goals* evolve over long periods of time, slowly forging an indelible personality on the firm and its employees that becomes very difficult to change once in place. Hayes and Wheelwright [11] describe these vague but powerful attitudes and values as an enterprise's *business philosophy* and define it as "the set of guiding principles, driving forces, and ingrained attitudes that help communicate goals, plans, and policies to all employees and that are reinforced through conscious and subconscious behavior at all levels of the organization." The business strategy provides the framework for purposeful action and the grounds upon which competitive, marketplace, governmental, and environmental norms are developed.

An enterprise's business philosophy usually consists of several levels of often diverse goals existing in differing degrees of development which converge to shape the firm's approach to the marketplace and its employees, products, and relations with allied channel partners and governmental agencies that together constitute its business ecosystem. Some of the goals are obviously financial, such as profitability, corporate growth targets, and return on investment. Others concentrate on fashioning unassailable market leadership spanning industries and continents. Still others focus on providing quality of work life, service commitment, furthering of community and societal objectives, minimization of risks to promote orderly growth, and so on. Slogans such as Ford's "Quality One" and UPS's "we are the best ship in the shipping business" attempt to communicate basic enterprise values, and product and service commitments to the marketplace.

Corporate goals serve a multitude of purposes. They help executive planners focus their efforts on identifying tomorrow's opportunities and reshaping the enterprise and its channel relationships, to proactively configure company resources to shape nascent industries, or to fundamentally reconfigure existing business resources and processes to reinvent current industries. They enable corporate, business unit, and functional business area strategies to be integrated around a common game plan. They also provide the basis for operational decisions and establish the boundaries of the competitive options available. Finally, corporate goals assist managers in the everyday execution of decisions in making the best trade-offs among performance measures such as cost, inventory investment, delivery, and serviceability, and between short-term tactics and long-term strategies.

It can be said that an enterprise's business goals are determined by five interwoven elements that give a firm its distinct character. The first element can be found in the company's *history*. Every company has a record of larger-than-life founders, past achievements, traditions, and policies that provide the company's workplace community with a sense of continuity and identity. Think of the Ford Motor Company, of the genius of Henry Ford, its century-long importance and commitment to the economic life of the nation and of the world. When executives begin the process of redefining the business goals of a company, they must ensure that the new directions sought for the enterprise are a logical extension of the firm's past and are supportive of long-agreed-upon objectives and organizational intent. The second element defining a business's character can be found in the *current preferences* of the owners, managing directors, and management staff of the firm. Their ambitions, visions, and tangible objectives are designed to reshape existing corporate goals and direct the enterprise in the pursuit of new competitive regions and unexplored marketplaces. Third, *governmental, environmental,* and *social factors* can have a dramatic affect on business goals. For example, transportation deregulation has had a major impact on the trucking industry and has spawned a whole new industry of firms providing various ranges of third-party logistics support. The availability of *financial* and *physical resources* form the fourth element determining a firm's corporate goals. For example, a small retailer could not hope to compete head to head with Wal-Mart in terms of overall price and product availability. Finally, companies pursue marketplace objectives based on their *distinctive competence.* Today's automakers have significant competencies to tackle the problem of developing viable electrically powered cars. On the other hand, Sears's venture into the realm of insurance and finance deviated from core business goals and was an ill-conceived diversification [12].

The ongoing rethinking of an enterprise's business goals is an interactive, iterative exercise designed to renew the soul and substance of the business. Business goals contain a company's long-term market vision and serve as the foundation for the development of the firm's competitive and core operations strategies. They provide the justification for the decisions the company makes and the groundwork for the policies to be followed by all levels of the organization relating to customers, products, services, and business partners, the structure of the supply channel, and the sense of community of purpose and enthusiasm guiding everyone in the organization in their endeavors to realize personal and, by extension, company goals.

Developing Competitive Business Strategies

The articulation of comprehensive business goals defines the overall objectives the enterprise would like to achieve in the long run. Translating these broad, long-term mission statements into a defined set of particular actions to be pursued necessary to accomplish these objectives is the function of business strategy. Today's strategic planners know that effective strategic plans consist of two regions. In the first can be found the core operations strategies that define the served market, the range of existing products, the estimated margins and value-added outputs of operational processes, and the resources and organizational structure necessary to sustain growth, profit, shareholder, and return-on-investment targets. The second region consists of the competitive strategies that constitute the firm's vision of the future, how it will create the industries of tomorrow, how it will preemptively build critical process competencies, develop alternative product and service offerings, and reshape the point of customer contact, what channel alliances will be necessary, and how to continuously reposition the enterprise for competitive advantage.

Although developing competitive business strategies is a critical requirement, defining a detailed planning process that can be used by all types of enterprise is clearly an impossibility. The extreme divergence in the business environment and the bandwidth of enterprise possibilities ranging from manufacturers and distributors to service companies and public institutions renders such a formula meaningless. Still, a broad outline that all types of enterprise can use when formulating their competitive business strategies can be suggested. There are essentially five major steps to be considered.

In the first step, strategic planners must match the capabilities of the firm against the objectives stated in the business mission. Competitive strategies can only be made relative to the firm's long-term business goals. One of the best methods for examining a business's competitive potential is to perform a WOTS-up (Weakness, Opportunities, Threats, Strengths) analysis. In this performance evaluation, the strategic planning team would dispassionately evaluate the company by detailing the following:

Weaknesses. This area consists of the competitive shortcomings found in the firm's products, product development processes, customer service, distribution channels, technologies, human resources, and other key areas. The goal is to assess each weakness in light of competitor strengths and the changing requirements of the marketplace.

Opportunities. At this point in the evaluation process, planners would begin identifying the major opportunities to create radically new competitive space, reshape existing industries, develop breakthrough coalitions with exciting partners that could drive the business ecosystem in an entirely new direction, and other possibilities presented by changes in technology or the global market business environment. The ideas that surface during this activity will be used later when planners begin formulating the strategic vision that will serve as the fulcrum of the entire planning process.

Threats. The success of just about every company is threatened by competitors who openly contend for market supremacy or who silently seek to gain marketshare by obsoleting the core strengths and competencies of their rivals. Just as deadly can be dramatic changes occurring in the marketplace that can quickly destroy competitive advantage and nullify competencies once thought unassailable.

Strengths. In this area of the analysis, the core competencies, product and process, work force, and other strengths of the company are detailed. A list of an organization's strengths serve as the starting point for cataloging the capabilities available to build tomorrow's marketplace leadership.

This first step is meant to be a comprehensive activity in which all aspects of the enterprise from equipment and plant to business partners are reviewed. The goal is to gain a basic measurement of how well the company measures up to competitors and markets in the search for tomorrow's most innovative products and services.

The objective of the second step of the competitive strategy planning process is to critically assess the future opportunities for industry leadership that lay before the enterprise and the competencies and resources needed to proactively shape the direction of marketplace evolution. The goal of this step is to illuminate the route by which a company can get to the future first and engineer a position of unassailable leadership before the competition. This process of generating marketplace foresight is a great deal more than just developing the traditional business forecast. That activity is concerned with charting the marketability of the existing product/service mix, leveraging established processes, and assessing the sales potential of current markets. Marketplace foresight, on the other hand, is a creative, thoughtful process whereby the convergence of future trends in the global business arena, breakthroughs in communications and information technologies, the availability of new materials, the creation of new supply chain structures, and a host of other possibilities can serve to point the company toward exciting new prospects for creating and guiding the evolution of whole new industries and markets.

The task of formulating the content of a firm's market foresight should be the responsibility not purely of top management but of a large, eclectic cross section of individuals within the company. Executive teams have normally fulfilled the role of charting the course of the business, ensuring consistency and purposefulness in investment programs, negotiating strategic alliances and acquisitions, and authorizing budgets and resource acquisition. In contrast, market foresight tasks teams need to have a much more future-oriented view. Such teams are driven by intersecting streams of intracompany experience and debate. The more divergent the perception and the closer those differing perceptions can be welded into a single comprehensive strategic vision, the more effective will be the outcome of the market foresight process. Imagine a company where product engineers understand customer trends, marketing people have a deep understanding of the newest process technologies, and service functions are acutely aware of the service requirements of customers regionally as well as on the other side of the globe. Take, for example, Nabisco's drive to develop a family of reduced-fat cookies and crackers that did not taste like tree bark. Armed with some breakthrough food technology and a total company effort, the result was a new product called SnackWell's, which in 3 years has evolved into a $500 million brand [13]. That kind of innovation requires teams with a creative view of tomorrow's opportunities and a capacity for preemptive competency-building to produce marketplace winners.

The agenda of market foresight teams is straight forward and consists of the following points:

- *Customers.* In this element, marketing foresight teams will attempt to envision what types of product/service solutions customers will be expecting from the marketplace 5 or even 10 years or more in the future. In addition, teams will also try to uncover whole new areas of customer needs that the current product/service mix does not currently address. Think of the software companies that invested early in the development of Internet access software, thereby gaining a jump on even the largest PC software competitors.

- *Products and Services.* A critical element in market foresight planning is investigating new products and services and emerging market niches. Part of this plan will focus on the ability of companies to engineer critical breakthrough upgrades to existing products and services. Other opportunities may be found in the development of radically new product lines made possible by advancing technologies.

- *Core Competencies.* Creating new product/service mixes and new markets naturally will require the acquisition of new skills and competencies not found in the existing organization. Will these resources by acquired or developed? What will be the role of outsourcing, partnering, or the formation of virtual company arrangements?

- *New Markets.* In this element planners will need to assess what markets the company expects to be available in the future. Are these newly formed markets a part of mainstream changes (such as in the computer industry) or special-niche markets (such as in the chemical and pharmaceutical industries) uniquely suited to the company's special expertise? Are these new markets to be regional or global, and how should they be controlled and serviced?

The third step to be taken by market foresight teams is the critical evaluation, elimination, and acceptance of the many new product/service, organizational, and market ideas that have been generated up to this point in the planning process. This review process can take several avenues. First, planners will have to begin selecting those existing products and services they feel can be engineered to provide outstanding sources of future competitive advantage. A corollary of this activity is the search for hidden reservoirs of innovation and entrepreneurship among the firm's work force. Second, the capital, staffing, physical plant, and resource requirements necessary to actualize the product/service plan will have to be allocated. Third, the bundles of new products and services and their accompanying development groups and funding requirements deemed to possess exciting future potential should be selected. Such initiatives can take a variety of forms, from skunkworks, spinoffs, and grassroots entrepreneuralism to acquisition. Finally, foresight teams must identify the markets and where in the value chain they expect to introduce their new product/service offerings. The critical part of this analysis is understanding the nature of the needs of the entire value chain, beginning with customers and how they will use the company's products and services to provide value to their customers, and ending with how direct and indirect supply chain partners can be integrated into the process. Actual rollout, ongoing maintenance, and performance measurement of these plans are the preserve of the firm's operations strategies.

The fourth step in defining the strategic competitive plan is the construction of the strategic architecture that will provide the foundation for the continuous search for new products and services, the deployment of new functionalities, the reinvention of current capabilities or the acquisition of new competencies, and the reconfiguration of customer interfaces. An enterprise's strategic architecture is not a detailed blueprint. As a long-range planning tool, it cannot be used to design the details of prices, operating budgets, staffing requirements, distribution channels, production rates, inventory targets, and so forth. Planning teams that find themselves immersed in such detail are destined for incrementalism and eventual inertia. In contrast, a strategic architecture attempts to detail, for the enterprise, what product/service mixes it must start building now before the competition does. It maps out what markets must be explored now if the company is to dominate them in the future. It also illustrates what resources and competencies the firm must begin acquiring now if the company is to achieve the position of leadership it envisions for itself tomorrow, 5 years, even 10 years from now. Reference the relentless unfolding of Microsoft's strategy to dominate every aspect of the PC software market touched by its operating system. Market foresight planners at Microsoft are not concerned with optimizing today's resources or revenues: They are concerned with preparing the organization to realize tomorrow's emerging opportunities and profit potential [14].

Besides establishing the basic framework for the competitive plan, the strategic architecture also enables planners to activate and direct the intellectual resources and desire to succeed found within the enterprise's work force. Considering the gap normally existing between the firm's current resources and capabilities and the future marketplace opportunities identified by market foresight teams, the competitive plan must be able to communicate to the work force a sense of overall competitive direction. Clear objectives provide them with the expectation of exploring new skills and achieving fresh objectives as they are challenged each day to uncover the path to tomorrow's competitive opportunities. At the core of this strategic attitude is the ability to locate, acquire, activate, and continually renew the core competencies of the enterprise. Companies with winning strategies, such as Motorola, Wal-Mart, Saturn, and others, succeed because they have found ways to continuously converge strategic foresight with the ability to build employee consensus on the efficacy of future goals, construct work around high-value activities, configure capabilities in new and exciting ways, rapidly realize the time from investment to payback, leverage the competencies of business partners, and shield company resources from the competition.

The final step in competitive strategic planning is establishing the measurement methods detailing how well the enterprise is realizing targeted marketplace opportunities. Companies must be careful not to fall back on the performance metrics characteristic of operations planning. Graphing sales revenues to costs or calculating resource productivities are insufficient for providing key the indicators of the ongoing success of the firm's competitive strategies. The obvious measurement is determining how closely the enterprise is consistently achieving strategic goals. This can be measured by examining a company's relative competitive position to competitors: Is it indeed creating new competitive markets and venturing into unknown opportunities, or simply playing the "catch-up" game to market leaders? Once opportunities have "taken off," has the firm been able to shape the resultant emerging industries, thereby blocking competitors and building an unassailable position? Has the company learned when new market initiatives have failed, or lapsed into incrimination and inertia? Have global brand and distribution positions been correctly anticipated to meet new product rollout?

Second, have the core competencies necessary to sustain marketplace value in the new opportunity arena been assembled? Are competencies growing in the proper direction or are they stagnating? Have competencies existing within the company been properly deployed, or are they still jealously guarded behind corporate business units? Has a process been established to nurture company competencies, set competency acquisition goals, and protect competencies from atrophy or being raided by competitors? Have key business partners been identified and key channel resources harmonized with company resources? Third, has the company's ability to continuously identify new opportunities in emerging markets grown or retreated? Are managers and staff more conscious of opportunities to reinvent themselves and the company, or concerned only with incrementalist measurements? Has the company been able to develop a community of people on all levels of the organization who perceive themselves as creative innovators ceaselessly on the lookout for new sources of competitive advantage, or are they taskmasters concerned only with departmental performance?

In summary, winning competitive strategic planning processes are based on the following principles:

- *Observation.* Today's organization must be able to discern and absorb market, product, technological, and operational information wherever it occurs. Accurate information enables enterprises to see the playing field clearly, to identify potential problems as well as opportunities, and to block the movements of competitors before they have become a threat.

- *Orientation.* Based on the information available, executives must be able to constantly reposition their companies to meet changing competitive circumstances. The effectiveness of the principle depends on the existence of agile organizations and technologies that can be repositioned to blunt attacks and pursue openings in competitive space. There is a direct corollary between the speed by which companies can reposition their productive processes and reallocate resources and the level of success necessary to exploit competitive opportunities.

- *Decision.* Accurate observation, focused objectives and resources, and agile processes and organizations are essential to executing effective decision making. Cloudy objectives and indefinite information make decision making difficult and outcomes uncertain. Winning decisions are those that assist companies in gaining foresight into locating tomorrow's markets and what it will take to achieve success.

- *Execution.* An effective strategic plan is not a detailed road map to guide daily activities. The strategic plan identifies the resources and competencies necessary to achieve the hoped-for breakthroughs in products and markets necessary to seize competitive leadership. As firms begin the real process of turning strategic vision into products and services, exploring alliances with leading-edge customers and suppliers, undertaking joint development with possible competitors, and applying new technologies, strategies must be altered and stretched to meet new opportunities.

Working with Operations Strategies

In contrast to competitive business strategies, core operations strategies are concerned with defining the detailed budgets, capital and asset targets, sales and profit objectives, and departmental performance measurements necessary to the smooth functioning of the enterprise. The goal of the operations planning process is the drafting of specific benchmarks designed to ensure optimum performance of the firm's resources to meet the unfolding of the firm's competitive strategies. In this way competitive and operations strategies work hand-in-hand. The former provides the broad topography of potential business opportunities, whereas the latter enables company executives to see the details, the roads and signposts as well as rivers, cliffs, and streams, as they traverse the landscape. The resulting operations plans should link the enterprise together through commonly determined sales, inventory, profit, and return-on-investment targets and confirm the expectation that the firm is on the right track. Today's market leaders know that they must have both plans in place to ensure continuous creation of and search for new opportunity horizons, as well as the detailed metrics to provide for purposeful organizational reengineering and process improvement.

In developing effective operations plans, the firm's top management team must determine answers to such questions as the following:

- How best should the firm manage its fixed and current assets?
- What return on investment is expected in order to meet profit targets and shareholder expectations?
- What current markets should the firm target, and what are its products and services?
- What is the structure of the organization, and how can its available strengths be leveraged to achieve competitive advantage?
- What operational budgets will be necessary to ensure the smooth functioning of the whole company?
- What are the customer, inventory, delivery, production, and warehousing performance measurements?
- What competency and process enhancements should the company be pursuing today to guarantee marketplace leadership in the future?

In answering these and other questions, enterprise planners must seek to align competitive strategies with economic and marketplace realities. Operations planning seeks to reposition the enterprise within the context of competitive strategies to exploit marketplace changes and competitive advantages and to identify critical shifts in economic conditions, products and services, customer wants and needs, and laws and government regulations. As illustrated in Figure 4.2, the operations planning process begins with a definition of the firm's financial plans. These plans can be separated into the *investment* plan, which determines the net earnings required to achieve expected corporate growth, return on investment, and dividend payout objectives, and the *profit* plan which determines the revenue required to generate the earnings needed to actualize the investment goals.

The second step in the operations planning process is the creation of the business forecast. The business forecast seeks to detail the firm's expected growth and total projected net income targets for a specified time period, usually for 1 year, but often for several years into the future. In developing this portion of the business plan, executives must balance expected aggregate revenues arising from sales and other sources with incurred costs and other liabilities, and the impact the revenue plan will have on company resources and stockholder investment. In addition, business planners must review qualitative factors, such as economic, political, social, environmental, technological, and competitive forces in drafting the corporate forecast. This portion of the business plan takes the form of a detailed statement of sales volumes, cost of goods, margins, selling and operating expenses, and taxes. Once these estimates have been determined, they will be

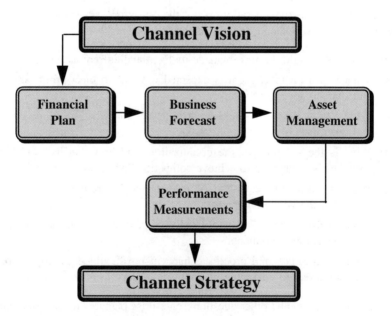

Figure 4.2. Channel operations planning.

passed to the firm's operating departments, who will then translate them into market segment, product and service, and sales and distribution channel strategies.

The third step in the operations planning process is the management of the current assets necessary to support the revenue plans detailed in the firm's business forecast. A typical company's assets can be divided into two general categories: *current assets* normally described as available cash, open accounts receivable, and merchandise and production inventories, and *fixed* or *capital assets* normally consisting of land, buildings, equipment, investment, and other tangible and intangible resources. During this portion of the operations planning process, corporate planners must ascertain whether the enterprise possesses resources sufficient to realize the operating budgets necessary to satisfy revenue objectives and to acquire those capacities that are required for forecast plan execution, but which are currently not available. A critical part of this step is determining the debt financing required, in addition to equity, to support the assets required throughout the business plan.

The final portion of the operations planning process is to determine the detailed departmental performance measurements. Some of these performance targets are oriented around product and service delivery. Critical metrics in this area center around process issues such as design, quality, reliability, level of customer satisfaction, and level of customer expectation met. Other measurements are customer oriented and focus on order management issues, such as product availability, courtesy, speed of cycle time, security, and understanding of customer problems, and postsales issues, such as support, value-added services, response, access, and competence. Inventory measurements attempt to maximize the ratio of investment to the level of customer serviceability. Critical metrics are inventory turns, inventory to current assets, inventory to total assets, and inventory to net working capital. The final set of performance measurements are concerned with the productivity of assets. Manufacturing process efficiency and utilization, flexibility, and operations costs, and distribution channel and transportation utilizations and efficiencies fit into this category.

SCM and Competitive Strategy

The concept of competitive strategy can be an elusive business paradigm without a firm structure to anchor issues relating to core competencies, business processes, and supply channel alliances. In Chapter 1, it was stated that the real power of supply chain management consisted not in the traditional operational aspects associated with logistics management but in the ability of SCM to provide today's enterprise with radically new opportunities to create marketplace advantage by leveraging supply channel partnerships, information and communication technologies, and the knowledge and innovative capabilities of the entire channel's work force. The SCM framework enables supply channel constituents to converge their individual creative resources and strategic visions. It provides for the activation of strategies for the effective management of cross-functional and cross-enterprise business processes whereby each member of the supply channel network can ground the vision of tomorrow's competitive environment and their own desires for innovative self-realization on the bedrock of closely integrated planning and operational functions managed by channel-level process teams.

What is being suggested is that the concept of SCM represents such a revolutionary and powerful competitive force that it should be considered as a fundamentally new source of strategic planning. As was pointed out earlier, the traditional focus of business strategy centered around forecasting, financial analysis, and market and product positioning. The goal was to determine the best opportunities for marketplace leadership and then to allocate capital and productive resources among business units and projects necessary to achieve opportunity realization. The idea, introduced by Michael Porter in the 1980s, that corporate planners should also include competitive factors found within the enterprise and outside in the marketplace, and the concept, developed by Gary Hamel and C. K. Prahalad, that strategic focus should be placed on strengthening and leveraging an enterprise's core competitive competencies have significantly broadened the boundaries of traditional strategic planning. Finally, Hammer has argued that the real locus of strategic planning is not about *positioning* at all, rather it is an exercise in defining a business in terms of "how it works" in determining strategic objectives around the essential operational processes that provide a business with its unique productive character. "What a company *does*," states Hammer, "is central in deciding what it is, and where and how it should compete" [15].

Although constituting critical elements of strategic planning, all of these concepts have a common drawback: They are all focused primarily around the enterprise. Although providing for the creation of strong corporate plans, they begin to weaken as discussion approaches company boundaries. On the other hand, the SCM concept requires strategic planners not only to be enterprise focused but also supply channel focused. SCM strategies require firms to identify themselves and the bases of competitive advantage less by the products and services they offer to their customers and more by the processes that they use to create marketplace value as it moves through the organization and out into the supply channel system. This concept is critical. No matter how good company-level products and services, organizational components, resources, and productive processes may be, they cannot attain the same range of competitive force as when those same attributes are linked together with allied partners to form a comprehensive supply chain network strategy. Real growth and expansion in the 1990s, the kind that consistently preempts the competition and increases marketshare, occurs when companies move beyond a preoccupation with internal capabilities and focus around strategies that continuously search to converge their own strengths with the productive resources and innovative knowledge of their channel partners.

One of the best examples of the strategic SCM concept can be found in the third-party logistics services industry. Originally, companies like Ryder, Schneider, Leaseway, and J. B. Hunt began as purely contract transportation companies. However, as requirements for JIT delivery, expanded value-added services, and heavy cuts in company transportation staffs increased during the 1980s, transportation companies began concentrating on how they could leverage what they did best—provide inexpensive, quality logistics services—to capture greater marketshare. By focusing on their basic service-oriented processes, these companies began to evolve into full-service logistics providers handling a wide spectrum of services ranging from warehousing, shipment consolidation, and logistics information systems, to rate negotiation, value-added processing, order fulfillment, and inventory replenishment. Expansion into such areas meant that they were more than simply offering contract services: They had become long-term logistics partners with the companies they serviced.

As the demand for third-party services continues to grow in the 1990s, third-party service providers have often found that they cannot provide the type and quantity of necessary service acting alone. According to a recent survey [16], 16 of the top 21 third-party service companies had formed alliances to broaden their service offerings. These alliances took several forms. Eight contract logistics company CEOs said they were involved in marketing alliances, one in a joint venture, and one that was both a joint venture and a marketing alliance. An example of such an alliance can be found in the group of third-party firms servicing Case Corporation, an international manufacturer of heavy equipment. Managing Case's expansive logistics environment required a closely knit alliance consisting of the Fritz Companies Inc., GATAX Logistics, and Schneider Logistics. Although each logistics partner is responsible for some aspect of Case's logistics activities, the objective is not just to achieve cost economies. According to Larry Trumbore, director of integrated logistics for Case, the alliance is about "a big opportunity to leapfrog to a logistics competitive advantage" through integrated logistics management and supply chain reengineering. George A. Gecowets, executive vice president of the Council of Logistics Management, feels that the "Case-Fritz-GATAX-Schneider logistics service alliance provides a classic example of the benefits of working in a virtual response business environment, where all four companies link electronically so they can operate as one. Alliances like these allow each partner to concentrate people and resources on those things that it does best. In doing so, each partner will improve its own operations and, thus, evolve into a formidable competitor in its own markets. [17].

Crafting SCM strategic architectures, such as the one at Case Corp., requires companies to utilize strategic planning techniques that seek to expand the vistas of traditional financial, internal resource, and marketing positioning. Such questions as the following need to addressed: Does the firm have a broad enough view of the future and of the industry forces shaping it? Is the product/service vision competitively unique? What is the level of supporting consensus inside the firm and outside among supply channel partners? Have the proper strategies necessary to acquire the required channel competencies and clarify opportunity approaches been formulated? Is the channel strategy resilient enough to cover potential discontinuities, the impact of the competition, and evolving customer needs? Failure to provide satisfactory answers to these strategic, channel-spanning questions can have a devastating impact on the channel strategy, resulting in truncated opportunities for growth, a dangerous reliance on internal capabilities, and inability to reposition channel resources to respond to new opportunities.

Once company strategists expand their perception of strategy beyond the narrow confines of their own organizations, they will be able to leverage SCM as a practical method to realize competitive strategic objectives through the spectrum of the following initiatives:

- Mobilizing the work force around a channel network strategic architecture. Competitive strategies provide direction as to where the firm would like to go; SCM identifies the necessary organizational competencies and values and how they should be gathered and effectively structured.

- Leveraging resources inside and outside organizational boundaries. The key to meeting the decreasing cycle times and increasing requirements for marketplace value can only be attained by a continuous reinventing of the organization to search out and realize innovative breakthroughs in products and processes. This extends to the use of "virtual" companies where strategic partnerships can assemble breakthrough capacities before competitors who depend solely on their own resources.

- Supporting joint marketing, product design, development, and rollout. The SCM philosophy enables companies and their channel partners to jointly converge their innovative and productive capabilities on the creation of products and processes that preempt the competition and create whole new areas of competitive space.

- Supporting distribution and channel management to create new sources of customer value and reach new marketplaces. SCM can significantly assist companies to streamline their logistics functions and supply channels to leverage new products and new ways of providing customer value. In addition, SCM provides the channel structure whereby market planners can develop new markets and competitive alternatives.

- Fostering organizational changes. Repositioning the organization cannot be successfully performed without a reference base. SCM sets the parameters around which new process interlinkages can be engineered, a collective view of the supply channel can be forged, core competencies identified and renewed, virtual organizations assembled, and a basis for continuous improvement in customer service, logistics, production processes, and channel management formed and communicated throughout the enterprise.

- Providing a basis for clarity and direction for top management. SCM enables executives to develop realistic and collective agendas for restructuring and transforming the competitive values of the corporation by exploring the limitless potential to be found in the supply channel networks to which they belong. SCM reveals the extent to which the company is driving industry evolution and permits executives to operationalize strategic aspirations into a clear set of corporate challenges.

SCM is perhaps the most powerful force in the development of today's enterprise because it provides realistic structures and measurements that assist market leaders to overturn the current marketplace, challenge the "business as usual" mentality, redraw industry boundaries, set new price-performance expectations, and offer radically new products and service offerings.

Logistics Strategy

Because of its growing strategic importance, no discussion relating to the development of business planning would be complete without examining the impact of effective logistics plans. In Chapter 1 it was stated that SCM consists of three dynamics, two of which were operations or tactically oriented and the third was focused on strategic aspects. The first two dynamics are what today is classically termed *logistics*. The first dynamic views logistics as an operations management function designed to integrate and optimize the operations activities of *inbound logistics* (sales forecasting, inventory planning, sourcing and planning, and inbound transportation), *manufacturing processing* (stores management, production, value-added processing, and work-in-process inventory management), *outbound logistics* (finished goods warehousing, customer order management, and outbound and intracompany transportation), and, *logistics support* (logistics systems planning, logistics engineering, and logistics control). The second dynamic of logistics management can be seen in the extension of logistics integration and functions to complimentary activities occurring outside company boundaries. This dynamic involves developing and maintaining supply chain partnerships and the synchronization of channel operations.

The evolution of strategic SCM has elevated the development of effective logistics operations plans to a new level of critical importance to today's enterprise. If it is the role of strategic SCM to assist in the generation of new marketplaces, seek out new forms of product and service innovation, and build unique core competencies, it is the role of logistics operations strategies to activate the actual capacities and capabilities found within the firm and outside in the supply channel system in the pursuit of operations excellence. If strategic SCM provides the overall goals and competitive possibilities open to the enterprise, logistics management enables the company's work force to purposefully construct the groundwork where channel alliances will be germinated and the ongoing cycle of market analysis, product/service design, production, distribution, and customer sales and service will be played out. In the past, corporate executives considered logistics as a mundane activity that provided little competitive advantage to the company. Today, the development of an effective logistics operations strategy has become the keystone to the success of SCM competitive strategies by creating real value for customers, closely integrating the supply channel, driving down operations costs, focusing the marketing and sales effort, and facilitating operations flexibility.

Leveraging Logistics for Competitive Advantage

The effective and innovate use of logistics functions has provided today's global enterprise with a fundamental core competency and a source of strategic value that cannot easily be replicated by competitors. The following examples illustrate how two successful companies have leveraged the power of logistics. Without a doubt, the key to L.L. Bean's success can be found in its solid, yet innovative use of logistics operations. L.L. Bean is a specialty-merchandise direct marketer that sends out over 130 million catalogs yearly to more than 15 million people. The company's domestic mail order plus retail sales totaled $743 million in 1993 and has been increasing over 17% annually. L.L. Bean's reputation has been made by guaranteeing 100% customer satisfaction for whatever reason. Just as important has been the firm's superlative order processing and delivery systems. Some of the features of L.L. Bean's logistics functions are the following:

- To maintain its superlative customer service, the company has established several Key Result Areas (KRAs), each of which has an objective and a measurement strategy. Examples of customer service KRAs include product guarantee, in-stock availability, fulfillment time, and convenience.

- The company has also enhanced its service reputation through a TQM initiative that empowers people to change processes. Major improvements have been made in such areas as reduced returns for quality, reduced work-in-progress cycle times, and increased return on sales.

- In 1993, L.L. Bean signed a multiyear contract with FedEx, giving the carrier 80% of the 10 million packages it ships annually. Benefits have included faster and more precise delivery times, as well as a better ability to track shipments. This strategy was targeted at modernizing the company's fulfillment and distribution systems [18].

In order to capture the benefits of process reengineering and Efficient Consumer Response (ECR) in the grocery industry, Proctor & Gamble, a $33.4 billion consumer products manufacturer, has been working tirelessly to effect real internal supply chain management improvements for years. In 1994, P&G extended its

SCM initiatives to its customer base. Termed "Streamlined Logistics I," the goal of the program was to simplify pricing, ordering, and payment processes to cut billing errors, reduce paperwork, and replenish inventories more quickly. One of the key moves was to uncomplicate customer ordering. In the past, the company had operated five separate business units with five distinct selling organizations and distribution systems. After the changes, customer now only had to place one order with P&G and receive a single invoice instead of a separate invoice from each selling organization.. In little over a year, the program succeeded in reducing customers' handling costs by $14 million. Buoyed by its success, P&G launched a more ambitious program late in 1995 (called "Streamlined Logistics II") targeted at customers buying truckload quantities direct from the company's 35 plants, which represents about 90% of P&G's freight. The program is straightforward. P&G offers products at a lower list price; in return, customers must adopt P&G's efficient replenishment techniques, closely scheduled pick up and delivery of loads, and EDI transmission of shipping notices. The initiative is expected to save customers about $50 million overall. P&G believes that everyone will benefit from a more efficient supply chain—the company itself, its customers, and the end user [19].

Similar stories of today's best companies looking toward their logistics functions for unique ways of providing competitive advantage and marketplace differentiation have become commonplace. As markets change and product life cycles accelerate and new technologies and management styles provide new methods of buying, producing, moving, storing, and selling products, the importance of logistics operations planning can be expected to grow.

Developing Winning Logistics Operations Strategies

In developing an effective logistics strategy, managers should focus on a seven-step process. The overall objective is to end with the creation of a strategy that is in alignment with business plan objectives, matches marketplace requirements with logistics capabilities, and activates continuous improvement and quality-oriented competitive values. The strategic process can be described as follows.

1. *Defining Corporate Objectives.* The orientation of a company's competitive and operations strategies determine the direction the logistics organization must be headed in order to successfully respond to targeted marketplace challenges and opportunities. These strategies have a direct impact on decisions regarding the complexity of the distribution channel, the degree of channel partner integration, and transportation and delivery requirements necessary to meet marketing objectives. In addition, the strategic orientation determines the extent to which logistics strategies focus on a narrow or wide range of activities in support of products, markets, or technologies. Wal-Mart, for example, conceives of its chan-

nel distribution function as a closely integrated matrix of suppliers, manufacturers, distribution centers, and retail outlets. Close integration between logistics and competitive and operations strategies provides a clear strategic direction for logistics operations and defines the boundaries that mark the parameters against which consistent performance can be measured.

2. *Aligning Competitive and Logistics Values.* The logistics strategy must be carefully integrated with the overall competitive values pursued by all areas of the enterprise. Some of the most important of these values are the following:

- *Products and prices.* This value is broad based and covers activities such as product positioning, advertising, promotions, and pricing. Internally, logistics actualizes these initiatives by lowering inventory carrying costs, implementing communications technologies to speed the order fulfillment process, and activating team-based management styles and cultures focused on a market segment of one. Externally, logistics must foster values that enable the enterprise to meet the rising expectations of customers, achieve shorter product life cycles, and quickly respond to requirements for competitive pricing and value-added services.

- *Cost.* By following this value, enterprises seek to gain and hold customers by offering the lowest cost in the industry. The elements providing for cost advantage are varied and differ according to company structure. Some of the key cost drivers are economies of scale and scope, cultural experience resulting in shortened learning curves, flexibility of capacity utilization permitting business units to expand or contract to meet demand, internal and external linkages providing for the unencumbered flow of goods and services, sharing of expertise, integration, timing of product or service introduction, location, and institutional factors such as governmental regulation, financial incentives, and unions. Efficient logistics functions enable firms to leverage productive processes to continually achieve the lowest market cost.

- *Delivery.* The ability to deliver goods and services at the time and place customers want them is a significant competitive value. Delivery time is such an important value to customers that some will willingly pay premium prices to attain goods on the date desired. Gauging the impact of delivery requirements on a company's information systems and logistics capabilities is critical in the attainment of this value. Delivery objectives may also require the deepening of the distribution pipeline all the way out to the retail level in pursuit of margin and volume opportunities.

- *Differentiation.* The characteristics that differentiate a successful logistics channel from competitors can be used to create and sustain a competitive advantage. Some of the attributes providing for differentiation are policy choices emphasizing quality, technology, service, or other elements, as well as operational objectives such as linkages to suppliers and customers fostering responsiveness or optimization, timing in the form of being a market pioneer, location, speed of organizational learning, integration, economies of scale focusing on largeness or smallness of volume, and institutional factors such as unions.

- *Inventory positioning.* Logistics channels can follow one or several attributes of this value in the search for market leadership. The most obvious attribute is *product availability.* Coca-Cola competes in a low-margin, high-volume global environment through complex, multiechelon bottling and marketing channels that enable the company to offer its products within easy reach of the customer. Another attribute is *quality,* achieved by offering goods of superior product reliability or performance. Department stores like Bloomingdale's and Nordstrom are examples. Still another attribute emphasizes *flexibility.* Marketing channels pursuing this attribute may offer products unattainable elsewhere or possess the ability to respond to unexpected large-volume orders within competitive delivery times.

The logistics strategy must articulate clearly the priority attached to each of these competitive values and how logistics functions will assist in value realization. Because it is virtually impossible for a company to excel in all five of the above values simultaneously, managers must define the extent of their commitment to each, and from that decision, position their logistics organizations accordingly.

3. *Logistics Strategic Analysis.* Now that the parameters of the firm's competitive and operations strategies and competitive values have been defined, logistics planners can turn their attention to performing an analysis of the logistics organization's ability to purposefully respond to these challenges. As illustrated in Figure 4.3, the analysis begins with a review of the market channel structure. This first element on this level of analysis is *channel design.* The configuration of the channel structure is influenced by a number of factors, including customer demands, channel economics, channel power, market size, and channel players' roles. Channel design is also shaped by the intensity of activities and functions that must be performed to satisfy a targeted level of customer satisfaction. Channel design will significantly influence the second element of channel structure: the *physical network strategy.* The

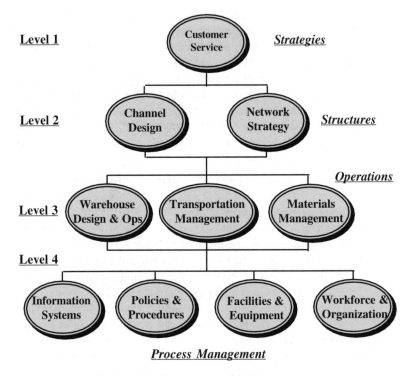

Figure 4.3. Components of logistics strategy.

physical network is concerned with determining how many facilities will be needed, where they should be located, what product lines will be carried, and what transportation services will be needed. The physical network must maximize value to the customer while minimizing costs and optimizing tangential functions such as manufacturing and transportation.

On the second level of the logistics strategic analysis, planners are concerned with logistics operations. On this level can be found the functions of materials management, warehousing, and transportation. On the strategic dimension, several critical decisions impacting these key operations areas must be answered: To what extent should these functions be owned, leased, or contracted? How many of these functions should be outsourced and how many should remain in-house? To what degree should third-party service providers be used? On the tactical level, the analysis must include forecasting, inventory management, production scheduling, purchasing, facility layout, materials-handling technology, productivity, carrier selection, load planning, routing and scheduling, and other issues. Of critical importance are diagnostics detailing how well continuous improvement, quality management, and benchmarking are creating the kind of performance necessary to achieve higher-level competitive strategies.

On the final level of the logistics strategic analysis can be found the logistics support functions including information and communications systems, operational policies and procedures, installation and mainte- nance of facilities and equipment, and the firm's work force. These functions are the backbone of logistics and their level of productivity, quality, accuracy, and dedication will have a direct bearing on the success of the logistics strategy. Information and communications sys- tems, for example, not only facilitate and improve the accuracy of work, they enable the close integration of the entire supply channel. A tightly integrated and high-performing work force is also key to successful logistics performance. A logistics organization characterized by high functional efficiency and excellence, concern with quality, and a passion for innovation is a necessity if the dynamic environment required for the success of corporate competitive strategies is to be established [20].

The internal and external environment of a firm's logistics functions can dramatically influence competitive strategy. Decisions made on the corporate and business unit level must be kept in close alignment with logistics capabilities. The goal of the logistics strategic analysis is to detail and possibly reengineer the elements of the firm's logistics struc- ture to produce a congruent focus centered on the enterprise's corporate mission. Unless a logistics strategy is developed that facilitates the attainment of the corporate strategy, the chances of the enterprise realiz- ing the competitive advantage desired will be seriously diminished.

4. *Enabling the Supply Channel.* When applied to a multiechelon supply channel network, a competitive logistics strategy can also assist in optimizing and coordinating channel linkages by permitting managers to operate the entire channel as an integrated alliance of suppliers, manufacturers, wholesalers, retailers, and customers functioning as a single competitive unit. Closely integrated supply channel partners en- able companies to collectively pursue the opportunities latent in the competitive SCM strategy. By removing the barriers that inhibit the natural connectivity between network supply point nodes, the velocity of the flow of inventory and information can be significantly increased.

Many managers have realized that improving internal functions such as warehousing or transportation have resulted in small gains if the rest of the business units in the channel network have not also improved their processes. By emphasizing channel integration, logistics functions can facilitate the alignment of channel-level goals and strategies by streamlining the interaction of the separate business units that constitute the distribution pipeline, capitalizing on core business strengths, and optimizing distribution network functions. In addition, competitive logistics directly attacks cost management issues by targeting the elimination of waste in the form of excess inventory and lead times and reducing costs in the form of supplier negotiations, transportation, order processing, poor quality, and facilities management. Finally, competitive logistics requires that companies embark on a never-ending crusade centered around customer service and continuous improvement.

5. *Establishing a Philosophy of Continuous Improvement.* The fifth step involved in leveraging logistics as a competitive weapon is the pursuit of continuous improvement along all internal logistics organizational dimensions and vertically throughout the supply channel. Several key directions come to mind. The first involves continually reengineering logistics processes. This objective involves an iterative management process that seeks to continuously align supply channel requirements and logistics resources in order to achieve order-of-magnitude improvements in time, cost, and quality while increasing the velocity of the supply-demand process by enabling quick and flexible response to meet ever-increasing levels of customer service. Logistics reengineering consists of five critical areas. First, reengineering demands the constant reinvention of technologies, products, and services that create sources of superlative value for customers and markets. Second, reengineering requires the integrative networking of the entire distribution chain. Third, the development of the skills and knowledge of the work force must take first place ahead of technology enablers. Fourth, successful logistics reengineering cannot take place without the open commitment, continuous reculturalization, ownership, and reward of the employee base. Finally, organizations must be serious about the continuous cultural and structural changes that characterize any successful reengineering initiative.

6. *Implementing Information Technology.* In this step, managers seek to foster competitive logistics by the implementation of information and communication technologies designed to link the supply channel together, as well as to facilitate internal information flows and cost reductions. Information technologies will assist planners to better plan and control global supply and demand processes, balance logistics resources, deploy inventory, optimize routing and delivery, and continuously shrink costs and turnaround time throughout the channel.

7. *Developing Channel Alliances.* In this step, firms seek to enhance logistics advantage through the continuous unfolding of strategic alliances with suppliers, customers, carriers, and third-party service providers. The goal is to leverage core logistics competencies found in channel partners. Through tools such as EDI, vendor and customer scheduling,

and service outsourcing, logistics functions can better match distribution activities and resources, thereby increasing economies and market coverage.

A comprehensive logistics strategy provides the firm with a mechanism to synchronize enterprise competitive strategies and logistics capabilities. As the firm's business functions begin to work closer together as virtual teams that coalesce and dissolve in response to specific logistics issues, the ability of logistics to refocus operational capacities to meet changing marketplace realities increases. In addition, the integration of SCM competitive and logistics operations strategies enables companies to achieve continuous improvements in customer service accompanied by simultaneous declines in operating expenses. Finally, an effective logistics strategy enables the enterprise to pursue competitive advantage through logistics functions.

Although the agenda for leveraging competitive logistics advantage may seem complex, the following are practical examples by which logistics functions can further enterprise competitive advantage.

- Locating supply points close to customer sites in order to facilitate quick response. In some cases distributors have the ability to "lock in" customers by providing daily, if not several times a day, inventory replenishment by positioning a warehouse near a customer's plant.

- Linking a firm's channel inventory availability with customer inventory systems to facilitate planning and quick response. By positioning terminals in the customer's plants or through EDI, customers can review supplier inventory statuses and directly create replenishment orders, thereby reducing purchase order costs and lead times.

- Utilizing automated warehousing tools, such as automated storage/retrieval systems (AS/AR), materials-handling equipment, and autoidentification/data collection equipment. Combined with skilled operators, warehouse automation can speed up material and information turnaround.

- Exploring the use of third-party service providers for the performance of functions not within company core competencies. Examples include the use of contract carriers, warehouses, and logistics services.

- Integrating distribution as part of the entire order fulfillment cycle and not purely an execution function. This organization reorientation is fundamental in pursuing quick response and continuous improvement philosophies. Companies such as Xerox, Apple Computer, and Georgia-Pacific have consciously focused on elevating the status of logistics in their organizations and in integrating the supply chain, resulting in the placement of logistics as the cornerstone in the firm's search for competitive advantage.

Summary and Transition

The ability to create competitive advantage by developing new markets and innovative products and services rather than simply searching for ways to continuously cut costs and shrink the existing business has become the fundamental challenge before today's enterprise. Questions relating to what will be the nature of competition tomorrow, how are companies going to gain insight into the marketplace challenges and opportunities of tomorrow, and how can the organization be continuously reinvented to capture the dynamic competencies of its work force requires companies to do more than focus on organizational incrementalism. They must devise strategies and visions and create paths that only those who aspire to conceive of new opportunities and new marketplace visions dare to tread.

Based on the JIT and continuous improvement methodologies of the past decade, most firms have turned toward business restructuring and reengineering techniques as the solution to their competitive problems. Unfortunately, this concern with "incrementalism" has executives focusing too narrowly on plans that are more often concerned with operational and tactical objectives than on the creation of thoughtful goals about how a company plans to compete in the future. This traditional approach to business planning contains several critical defects. To begin with, it depends on an industrial system revolving around mass production and economies of scale and scope. Also, the time span of the typical business plan is too short, being expressed in years instead of decades into the future. Furthermore, companies often develop their plans around the product or business unit rather than the entire firm or even the whole supply channel. Finally, executive planners all too frequently center their attention on competing solely with their existing suite of goods and services rather than competing through new visions of the future, development of new competencies, and blending of new arrangements of supply chain partnerships that will provide for the evolution of new industry ecosystems.

The business strategies of today's leading-edge companies can be said to consist of three distinct but interlocking pieces. The first piece begins with the definition of the enterprise's core business goals. Business goals can be defined as broad, long-range objectives and attitudes that guide the efforts of the firm as an organizational entity as well as the individual actions of all of its employees. The second piece of a successful business plan can be found in the articulation of the strategic competitive mission of the enterprise. The competitive mission can be characterized as being focused on actualizing the future possibilities of the organization, rather than on existing capabilities. It is about identifying emerging opportunities for new competitive space and reinventing the enterprise to activate the innovative competencies located within company boundaries and the coalitions found outside in the supply channel. The final piece of an effective business plan can be found in the formulation of the core operations strategies of the company centered around how to position existing products, markets, and businesses within the current industry structure. Among the critical activities in operations planning are the creation of the forecast of the firm's expected financial viability, determination of the current assets and competencies necessary to support financial and operations targets, creation of meaningful benchmarks and performance metrics, and disaggregation of the operations plan down to the firm's business units.

When creating competitive strategies, planners must be careful not to slip into "incrementalism" and the inertia that inevitably arises when future objectives are subjected to detail measurement and trade-off thinking. Inventing winning competitive strategies are based on the following principles:

Observation. Today's competitive organization must be able to discern and absorb market, product, technological, and operational information wherever it occurs.

Orientation. Based on the information available, executives must be able to constantly reposition their companies to meet changing competitive circumstances.

Decision. Accurate observation, focused objectives and resources, and agile processes and organizations are essential in the effective execution of business plans.

Execution. An effective strategic plan is not a detailed road map to guide daily activities.

The strategic plan identifies the resources and competencies necessary to achieve the hoped for breakthroughs in products and markets necessary to seize market dominance. It is the responsibility of operations plans to constantly search for ways to optimize and align company resources with tomorrow's marketplace opportunities in the search for marketplace leadership.

Because of its growing strategic importance, no discussion relating to the development of business planning would be complete without examining the impact of effective logistics planning. If it is the role of strategic SCM to assist in the generation of new marketplaces, seek out new forms of product and service innovation, and build unique core competencies, it is the role of logistics operations strategies to activate the actual capacities and capabilities found within the firm and outside in the channel system in the pursuit of operations excellence. If strategic SCM provides the competitive possibilities open to the enterprise, logistics management enables the company's work force to purposefully construct the groundwork where channel alliances will be germinated and the ongoing cycle of market analysis, product/service design, production, distribution, and customer sales and service will be played out. Today, the development of an effective operations strategy has become the keystone to the success of SCM competitive strategies by creating real value for customers, closely integrating the supply channel, driving down operations costs, focusing the marketing and sales effort, and facilitating operations flexibility.

Notes

1. Steven L. Goldman, Roger N. Nagel, and Kenneth Preiss, *Agile Competitors and Virtual Organizations.* New York: Van Nostrand Reinhold, 1995, pp. 57–59.

2. Gary Hamel and C.K. Prahalad, *Competing for the Future.* Boston, MA: Harvard Business School Press, 1994, pp. 9–10.

3. Laurie Joan Aron, "Logistics on the Rebound," *Inbound Logistics* 15 (2) (February 1995), 34–36.

4. Terry Hill, *Manufacturing Strategy: Text and Cases*, 2nd ed. Burr Ridge, IL: Irwin, 1994, pp. 18–20.

5. James F. Moore, *The Death of Competition.* New York: HarperBusiness, 1996, pp. 62–63.

6. Hamel and Prahalad, p. 300.

7. Michael Hammer and James Champy, *Reengineering the Corporation.* New York: HarperBusiness, 1993, pp. 86–87.

8. Hamel and Prahalad, p. 21.

9. Sun Tzu, *The Art of War.* Boston: Shambhala, 1988.

10. Peter F. Drucker, *Management: Tasks, Responsibilities, and Practice.* New York: Harper & Row, 1973, Chapt. 7.

11. Robert H. Hayes and Steven C. Wheelwright, *Restoring Our Competitive Edge.* New York: John Wiley & Sons, 1984, p. 25.

12. Philip Kotler, *Marketing Management*, 6th ed. Englewood Cliffs, NJ: Prentice-Hall, 1988, p. 37.

13. J. Robert Hall, "Top Dog's View of Supply Channel Management," *Inbound Logistics* 16 (6) (June 1996), 23–30.

14. Hamel and Prahalad, pp. 117–138.

15. Michael Hammer, *Beyond Reeingineering.* New York: HarperBusiness, 1996, p. 195.

16. This results of this report can be found in James Aaron Cooke, "On the Up and UP (and Up)!," *Traffic Management* (January 1996), 49–51.

17. Leslie Hansen Harps, "Case Corp. Constructs Logistics Model of the Future," *Inbound Logistics* 16 (10) (October 1996), 25–32.

18. Robert A. Novack, C. John Langley, and Lloyd M. Rinehart, *Creating Logistics Value.* Oak Brook, IL: Council of Logistics Management, 1995, p. 221.

19. James A. Cooke, "P&G's Mega Gamble," *Traffic Management* 34 (12) (December 1995), 33–36.

20. Kevin O'Laughlin and William Copacino, "Logistics Strategy," in *The Logistics Handbook.* (James F. Robinson and William C. Copacino, eds.). New York: The Free Press, 1994, pp. 57–75.

5

Supply Channel Management

The ability to realize the operations and strategic opportunities latent in the supply chain management (SCM) concept can be found in the approach by which companies leverage their supply channels. Historically, academics and practitioners have narrowly perceived the supply channel as acting primarily as the main artery in the business system through which the flow of goods and services is regulated. In addition, the supply channel is usually defined as a collection of internal organizations and external channel partners linked together in order to effectively perform the functions necessary to move products and marketing information as they make their way through the production, marketing, and sales systems. Finally, the many decisions companies make in shaping the supply channel are seen as directly impacting the nature and driving the dynamics of business competition and are instrumental in determining the success or failure of enterprises and individual initiatives.

Although there can be little debate that this "functional" understanding of the supply channel remains at the very core of today's global market system, it has become evident that it illuminates only a fraction of the capabilities for competitive advantage that can be found when the full potential of the supply channel is energized. Instead of a purely operational function, today's market-winning firms have found that competing for the future also requires a strategic understanding of the supply channel. Companies in the forefront of exploring new opportunities in areas such as the Internet, interactive television, digitization of communications, use of computerized modules in products, alternatives to the internal combustion engine, and others have found in their supply channels an almost limitless reservoir of resources, processes, and competencies far beyond what they could possibly possess within the boundaries of their own organizations. As product life cycles continue to shrink, the cost of new product development and distribution soars, and the ability of competitors to preempt product and marketing innovation intensifies, executives have found the richness and vitality of their channel alliances to be at the very center of their ability to compete. Supply channel management used to be defined as those collective tasks centered around optimizing physical assets such as plant and inventory, ensuring effective logistics functions such as warehousing and transportation, and achieving objectives associated with the marketing utilities of time, place, and transfer of possession. Today, the supply channel system forms the fundamental building block of SCM, providing the capability for the creation of radically new forms of competitive advantage that can only be attained through the "virtual integration" of coalitions of businesses that converge to form unbeatable reservoirs of innovation and unique competencies to meet the challenges of today's global marketplace.

This chapter focuses on detailing the supply channel functions and strategies necessary for the effective implementation of SCM. The discussion begins with a definition of supply channel management, its historical development, critical components and business principles, mission, design structure, and operations. Following this analysis, attention is then turned to an investigation of supply channel functions. Two areas are explored: basic channel functions, such as inventory movement, functional performance, and efficiency, and channel marketing functions, such as title flow, transaction flow, and information flow. Continuing on this theme, the elements necessary for the creation of effective supply channel strategies are then discussed. Among the topics reviewed are developing basic operational strategies, engineering strategic partnerships, and designing market-wining channel strategies. The chapter concludes with an overview of the issues confronting today's enterprise and the development of global channel strategies.

Defining the Supply Channel

Fundamental to the development of a theory of supply channel management is a concise definition of what is meant by the term. It must be pointed out at the outset that the term "supply channel management" is not identical with "supply chain management." As defined in Chapter 1, SCM is described as a *philosophy* of channel management that seeks to unify the total productive resources of a group of businesses that have joined in the search for collective methods of developing innovative solutions and synchronizing the flow of products, services, and information to create unique, individualized customer value. In contrast, the term "supply channel management" refers not to a concept but rather to the actual physical business functions, institutions, and operations strategies that characterize the way a particular channel system moves goods and services to market through the supply pipeline. This view of the supply channel is meant to be understood in the widest sense as including all enterprises and their activities that contribute directly or indirectly to the marketing of goods and services. SCM provides today's market leaders with innovative competitive strategies that enable them to leverage the resources and competencies of coalitions of companies to reinvent industries and create new competitive space; supply channels, on the other hand, provide the actual structures, cooperative relationships, and day-to-day management of sales, inventories, and deliveries on which a particular application of SCM is based.

Historical View of Supply Channels

Supply channel management is an outgrowth of one of the earliest activities of civilization—the movement and transaction of goods and services spanning geographical space and time. The supply channel is based on the existence of *surplus value*, either in the form of actual goods or in services. One can only imagine the first supply channels occurring in the most primitive times. One social group creates a surplus of grain, raw materials, or manufacture which it trades with geographically dispersed groups of other societies that have shortages of some needed commodity. In time, these exchanges are institutionalized and emerge as formal supply channels. By the era of mercantilism, over 300 years ago, the term "channels of distribution" already defined the flow of goods between one nation and another. This meaning of the term appears to have persisted until the end of the 19th century, when the outpourings of the Industrial Revolution and advancements in transportation and modern management techniques increased attention on the types of institutions influencing the functions of the supply channel.

It was, however, not until the early decades of this century that a formal analysis of the supply channel concept occurred. In 1914, Ralph Starr Butler defined the supply channel in one of the first "modern" books on the subject as

> the trade routes along which goods pass on their way from the manufacturer to the final customer. If the route is direct . . . , we have the manufacturer-direct-to-consumer selling method. If there are several routes leading from the production center . . . , we have the selling method that utilizes one class of middlemen—possibly the manufacturer-retailer-consumer method, or the manufacturer-agent-consumer method. And if these branch routes, each in turn, come to smaller distributing centers . . . , we have still another complication in our trade channels—the manufacturer-jobber-retailer-consumer selling method. [1]

Building on the work of Butler's and other economists, the appearance in 1922 of F. E. Clark's *Principles of Marketing* marked a watershed in the study of supply channels. He perceived the mechanics of supply channels as functioning according to three central marketing principles: the concepts of concentration, equalization, and dispersion. Materials and manufactured goods began the market-ing process by being concentrated in a central locus. The natural rhythms of supply and demand provided for the equalization of products along the dimensions of time, quantity, and quality. Finally, dispersion is basically the movement of the goods that had been concentrated out to the consumer, regulated by the natural interactions of demand and supply. In his definition of the supply channel, Clark stated:

> By the term "channel of distribution" is meant primarily the course in the transfer of title. The transfer of the goods may or may not be through the same channels. In some cases title to goods is passed several times while they are en route or in storage. But ordinarily the goods themselves pass along the same channel as does the title. [2]

Clark's definition was important because it redefined who was to be included or excluded from the supply channel by shifting attention away from the physical flow of goods to the transfer of ownership. During the ensuing decades, other economists expanded on the definition by adding in any channel organization that also provided supporting functions such, as physical possession, payment, ordering, ownership, negotiating, financing, promotion, and risking.

By the mid-1960s, the modern definition of the supply channel had been fully formed. According to Bucklin, the supply channel is characterized by the following three principles;

1. A channel of distribution is defined by the specification of a set of institutions which forms a link between production and consumption.

2. The link serves to connect production and consumption with respect to some specific product or products.

3. Invariably included in the set are the institutions which create the product and consume it. Consumption, in addition to its usual denotation, may also be taken to include further changes in product form.

The chief departure of these principles from past definitions is found in the character of the "link." Past authors focused on the issue of title; Bucklin, in contrast, reasserts the importance of the physical flow of goods through channel intermediaries alongside title flow. His definition provides the basic concept used to describe the supply channel in this chapter:

> A channel of distribution shall be considered to comprise a set of institutions which performs all of the activities (functions) utilized to move a product and its title from production to consumption. [3]

Supply Channel Components

The modern supply channel is composed of a series of closely networked internal organizations and independent companies that extends from primary and secondary suppliers at the beginning of the channel to the customers and their customers that mark the furthest extension of channel output. Historically, supply channels have been described as neatly arranged alignments of independent businesses. As Figure 5.1 illustrates, the supply channel has been traditionally organized into three primary constituents—manufacturers, wholesalers, and retailers—that reflect the basic movement of goods, services, and information through the channel. Despite the appeal of this model, however, the notion that each channel constituent is basically independent of other channel members and that channel functions proceed serially from one to the other has become, in today's environment, dramatically obsolete. In addition, the model also excludes suppliers and customers from the supply channel. In reality, new management and technology techniques, such as JIT, electronic commerce, interorganizational product design and process engineering networking, and computerized joint inventory, production, and transportation planning and scheduling, have broken down the old channel paradigm and substituted an expanded view of the supply channel as a closely integrated network comprised not just of core channel partners but also of extended links of suppliers and customers at the furthest reaches of the supply pipeline.

Figure 5.2 provides an illustration of today's modern supply channel. Again, the keynote of this new model is the close integration and cooperation of channel business components. This movement toward channel collaboration, evident in the explosion of partnerships, joint ventures, and coalitions of every imaginable kind marking today's business environment, has a wide spectrum of tactical and strategic objectives. Some companies seek channel partners to create economies of scale without investing in internal resources by leveraging similar capabilities out in the channel. Some aim at joint development of new "back-office" capabilities. Others boldly engage in collaborations with industry rivals in the development of generic technologies, even as they maneuver to compete with one another for market dominance. The most adventurous perceive the entire channel as a single organization where economies of scope can be achieved by synthesizing the physically distributed competencies of each channel member into a unified, "virtual" competitive business [4].

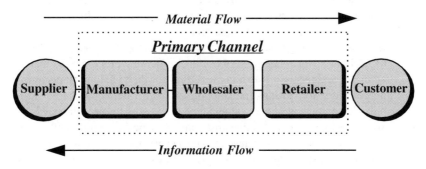

Figure 5.1. Supply channel constituents.

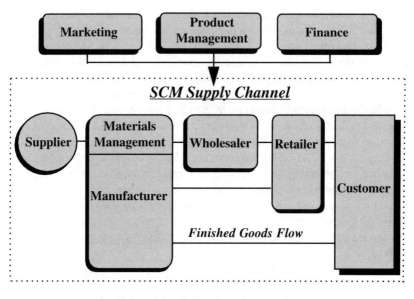

Figure 5.2. SCM channel structure.

The five supply channel participants, as portrayed in Figure 3.2, and their importance in channel management can be described as follows:

- *Suppliers.* The primary role of channel suppliers is to act as raw materials and component integrators and single point-of-contact vendors. In the past, the buyer/supplier relationship was adversarial and marked by distrust and competition. As each side tried to negotiate for the best deal, critical issues relating to quality, cooperation, and mutual success were dramatically reduced in importance to a singular concern over cost. In contrast, today's supplier is perceived as a critical component of a network *value chain*. Instead of being outside of the channel, leading-edge suppliers provide a wide spectrum of critical services formerly performed by entities within the channel. Among the value-added services being offered by suppliers and demanded by customers can be found the following: acting as a "one-stop shopping" source that can also deliver to customer point of use/line; providing procurement, manufacturing, and receiving/stocking functions for their customers; performing consolidation, kiting, and shipping points for items purchased from multiple sources; acting as value-added processors to meet the needs of customers looking to outsource upstream noncritical activities; and acting as quality inspectors and testers who certify the production or sale-readiness of the products they supply and distribute. Tools such as EDI and computer networking have also provided suppliers with the power to achieve lightening-fast cycle times, participate in customer product design, and collaborate on inventory planning and scheduling.

- *Manufacturers.* This channel participant is focused primarily on the development and production of products to be used by industry or by the end consumer. Occupying a gateway position in the supply channel process, it is manufacturing's responsibility to make available those products demanded by the marketplace at the right time and in the right quantities. The ability of manufacturing to produce goods in a timely fashion has brought about revolutionary new ways of controlling processes. The use of JIT, build-to-order, focused factories, pull production modes, and computerized networking techniques that have enabled the inventory requirements of the entire supply network to be rapidly transmitted up and down the channel have provided manufacturing with the ability to synchronize productive capacities with the needs of wholesalers, retailers, and customers downstream in the pipeline. Although it is true that some product commodities, such as wood, coal, and grains, often bypass manufacturing and move directly into the supply channel, manufacturing's role as a product originator places it at a critical juncture in the supply channel process. Some manufacturers, such as Ford Motor Company, Sony, and Apple Computer, assume the responsibility for assembling and managing a supply channel, whereas others depend on upstream channel constituents to perform that role.

- *Wholesalers.* Somewhere between manufacturers and retailers stands the *wholesaler*. Because wholesalers are less visible than other channel partners and can assume a number of different forms, wholesaling is more difficult to define. The traditional function of wholesalers in the supply channel has been to serve as middlemen, providing retailers with products originating from the manufacturer. Wholesalers exist because of their ability to act as a consolidator, assembling and selling merchandise assortments in varying quantities from a number of manufacturers. In addition, wholesalers also design and operate channel arrangements between customers and those manufacturers who do not have distribution functions. As the use of JIT and quick response techniques have grown in the 1990s, manufacturers and retailers have been expanding their logistics functions to include many of the tasks performed in the past by wholesalers. Although it can be expected that supply channel members will continue the search for ways of eliminating pipeline costs and improving customer service, wholesaling, as testified by the strength of such companies as Burgen Brunswig, SYSCO, McKesson, and ACE Hardware, is still an extremely important part of the supply channel

- *Retailers.* If suppliers can be said to be positioned at the gateway to the supply channel process, then *retailers* can be considered to be at the terminal points in the channel. Whether performed by a manufacturer, a wholesaler, or a retailer, the essential function of retailing is to sell goods and services directly to the customer for their personal, nonbusiness use. In contrast to manufacturers and wholesalers, retailers have a completely different set of business decisions and objectives. Among these concerns are identifying the target market so that retailing operations can be optimally leveraged, meeting customer expectations with the appropriate mix of product assortments, services, and store convenience and ambiance, and determining pricing and promotions that will provide competitive differentiation. It would be a gross error to regard retailers purely as passive channel members. Although many retailers are dependent on upstream channel suppliers, others like Sears, Best-Buy, Circuit-City, and Wal-Mart take a very active role in the development and management of the supply channel.

- *Customers.* In the past, customers were viewed as market segments into which were sold product lines consisting of standardized, mass-produced products. Today, customers are demanding to be treated as unique individuals and expect their suppliers to provide configured, variable combinations of products, services, and information matched to their special needs. The requirements for absolute quality, low price, short delivery cycle times, products characterized by superior design and robust functionality, and premium service have swept away the fiction of the *internal* and the *external* customer. As products and services make their way through the channel, each "customer" at each channel level needs to be closely integrated with the capabilities of each business component from beginning to end. World-class companies like Wal-Mart, Hewlett-Packard, and Milliken attempt to forge strategic links several channel echelons back to their suppliers and their suppliers' suppliers and forward along their chains of customers. By employing agreed upon reengineering and total quality management objectives, the goal is to realign the entire channel into a single competitive entity driven by a collaborative strategic purpose.

Principles of Supply Channel Goals and Structure

The five business components that constitute the supply channel each have a wide variance in individual operating objectives, productive processes, and activities within the supply channel. Outwardly, a form of loose cohesion between these diverse groups is achieved by each business entity performing its role in moving inventory down and market information up the supply pipeline. However, in the era of SCM, the most effective channels are not only those that can execute basic channel functions flawlessly but also those that seek to unify the operations and strategic goals of each channel member into a single, borderless competitive entity focused on providing total customer satisfaction. Achieving convergence among channel participants requires a conscious effort to develop a single channel mission, methods of designing and redesigning the channel to respond to new marketplace opportunities, and a clear understanding of how channel operations can be leveraged to closely integrate channel partners and provide for sources of competitive measurement.

The Channel Business Mission

The objective of the *channel business mission* is to unify all direct and indirect channel participants into a single competitive business entity. Although the nature of particular supply channels will vary by industry, they can all be said to have a basic business mission: to continuously create superlative, customer-winning product and service value at the lowest possible operating and investment cost. The elements of this definition of channel mission are revealing. The first key concept, *continuously create*, implies that the fundamental productive processes within the channel must be constantly focused and refocused on a ceaseless search for innovation both in the products and services the channel offers and in the way they are delivered to the marketplace. This element describes the market regions in which the channel is going to compete. *Customer value* refers to the key dimensions of quality, price, delivery, product robustness and configurability, and level of service desired by the marketplace. Continuously searching for ways to provide value to the customer forms a common performance measurement that unites the functions of all supply channel levels and directs them toward a common objective. This element describes who is the target of the channel. Finally, the *operating and investment cost* of achieving channel objectives provides the performance measurements and channel organizational mechanisms guiding operations functions up and down the channel. This element describes how the channel is going to compete.

A useful framework for developing and maintaining an effective business mission is to divide the channel customer service strategy into four distinct dimensions: the customer, the service strategy, people resources, and information and communications systems strategy. These areas are portrayed in Figure 5.3. The integration and effective functioning of each of these components is essential to the business mission of the supply channel. At the center of the supply channel mission stands the customer. It is customer wants, needs, and expectations that drive channel objectives and define channel structure and marketplace values. Pursuing superlative customer service has a significant impact on the following channel functions found within each channel participant:

- *Marketing.* Focuses on uncovering the products and services that tomorrow's customer will want. Market foresight teams will also be responsible for locating tomorrow's newest and most profitable industries and identifying the skills and competencies required for effective participation.

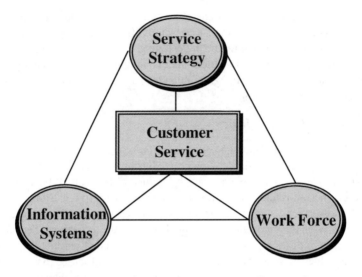

Figure 5.3. Customer service management framework.

- *Supplier Management.* Integrates with each upstream supply node to ensure the timely delivery of quality materials and components at the lowest possible cost.

- *Product Development.* In concert with the supply channel, develops products and services that possess unparalleled customer value in the shortest cycle time possible and at the lowest cost.

- *Production Operations.* Creates the highest quality products customers want through the use of flexible processes and quick response that enable custom manufacture while decreasing lot sizes and cost.

- *Demand Management.* Provides short order cycle times and responsive service and delivery through the use of computerized information systems and electronic commerce.

- *Channel Management.* Integrates products and information to ensure their rapid flow through the supply pipeline and out to the customer at the shortest possible time and lowest cost.

Clearly identifying the channel's customer service mission forms the foundation for the formulation and implementation of the channel's service strategy. The service strategy consists of two parts: an *internal* commitment to service centered around the formal channel mission statement and an *external* commitment to executing those service attributes promised to the customer through the sales cycle.

Positioned in the third area of the channel mission are the collective management and staffs comprising the channel. The fact that this area is connected to the service strategy means that there must be a shared vision if service values are to permeate not only each enterprise but the whole supply channel. In addition, the link between people resources and the customer represents the everyday communication that occurs between the channel's work force and the customer.

The final area of the customer service framework is the channel's information systems. As Figure 5.3 illustrates, the channel's information systems link the other elements of the service framework. People resources utilize the channel's systems to execute functions, record information, and communicate with each other and with customers. These *systems* are multidimensional and are composed of management systems that determine channel goals, internal and externally devised rules and regulations systems, the mechanical technical systems, and the channel's collective social systems that guide problem solving, teamwork, and service values. The customer must use the systems employed by their vendors to make their wishes, needs, and expectations known. Finally, information systems can be heavily influenced by the service strategy. Strategic objectives shape the channel's matrix of systems as they seek to engineer a customer-service-centered culture.

Channel Design

Once the channel business mission has been constructed, channel planners must turn their attention to *channel design*. When constructing a supply channel, two critical elements must be kept in mind. The first revolves around the concept of change. Perhaps no business segment is more subject to change than supply channels. At no time in history have the factors driving channel design been so dynamic and so volatile. The continuous shrinking of product life cycles, increasing customer demands for service, configurability, and quality, the explosion in information and communications technologies enabling cross-functional and interchannel boundary networking, the globalization of the workplace, and other factors have made it even more difficult in today's environment to crystallize the ebb and flow of a given set of channel relationships into anything greater than a temporary arrangement between channel constituents. Supply channels are inherently dynamic; today's supply channels are constantly in the process of reinventing themselves, searching for partners that will provide new sources of resources and core competencies, taking risks and sharing infrastructure costs, moving quickly to seize changing product and service opportunities, and engineering uncopyable skills and processes to provide unique customer solutions.

The second element of channel design revolves around the degree of the acknowledged dependence of channel participants. The more channel members are dependent on one another, the greater the need for cooperation and sharing of critical functions and processes. The less the dependence, the more individual companies must depend on their own competencies and vertically integrated resources to sustain competitive advantage. For example, what would Wal-Mart be like without its tightly integrated supply channel relationships, or McDonald's, or K-Mart? According to Bowersox and Cooper [5], channel structures are grouped around the following classifications, ranging from least to most open acknowledgment of dependence:

- *Single-Transaction Channels.* A great many marketing transactions occur in supply channels that are considered onetime and nonrepeatable. Examples of single-transaction arrangements can be found in real estate, construction, international trading, and the purchase of capital equipment. Normally, supply channels in such environments are transitory and founded for the sole purpose of facilitating a unique transaction. Once the requirements originally agreed to by the channel participants have been completed, the basis for the channel ceases to exist. Although single-transaction channel arrangements are usually unique and nonrepeatable, the success of a particular transaction may provide the basis for future channel agreements among former participants.

- *Conventional Channels.* This category of channel arrangement is also known as a *free-flow channel.* The main characteristic of this type of channel is the comparatively high degree of independence maintained by channel participants. The reason why they participate in the channel at all is to leverage the different forms of specialization offered by the various business nodes found along the channel network. The motive is purely opportunistic: They seek to attain channel efficiencies without becoming fully committed members of a networked channel system. Although conventional channels provide members with a some of the benefits of a networked channel system without having to engage in contractual or other agreements, there are certain drawbacks. To begin with, free-flow channels will, over time, be more volatile than dependent channel systems, as members move freely between multiple channels in the pursuit of short-term efficiencies and opportunities. Second, relations between members are typically adversarial, as members continuously maneuver for cost and price benefits. Finally, channel arrangements can be terminated rapidly by either party if and when once perceived advantages disappear.

- *Networked Supply Channel Systems.* This type of channel arrangement differs from single-transaction and conventional channels by the fact that channel members both acknowledge and desire interdependence. As such, these organizations feel that short-term ability to maximize operational efficiencies and marketplace exposure and their long-term capability to maintain competitive advantage depend on their participation and close integration with partner companies in a networked channel system. The basic advantage of this form of channel arrangement is the ability of individual companies to achieve a level of performance and capability for innovation far above what they could achieve acting on their own. Although there is the potential for conflict among channel members, the value of the synergy engendered by the networked channel provides strong inducements for conflict resolution and convergence of objectives.

Several networked supply channel systems, some functioning concurrently, are possible. The most common can be found in a *corporate channel* where the existence of a single networked channel entity, such as a Ford or a Wal-Mart, occurs by virtue of ownership. Another type of networked channel can occur in which member dependence is defined by a *formal contract*. Franchises, exclusive dealerships, and joint ventures belong to this type. Other networked channels can be described as *alliances*. This type of channel is typically voluntary in nature and usually not formalized by contractual agreement. The advantages of channel alliances can be found in members' sense of loyalty to one another and to a set of common strategies, ability to achieve order-of-magnitude efficiencies and draw upon extended resources, and desire to facilitate innovation and value-added processes for the good of the entire channel. The final networked channel system, *administered*, is neither contractual nor based on alliances but can be characterized as a recognition of a dominant company's channel leadership by the rest of the channel's members. Dependence on the channel is recognized as necessary if members wish to continue receiving the benefits of the channel. Prime examples of this type of channel network can be found in the channel organizations established by large retail chains such as Sears, Best-Buy, Wal-Mart, and JCPenney and their suppliers who often have long-standing relationships devoid of formal contracts or agreements.

Although the relative size and volume of business transacted by networked channel systems dominate channel structure choices, it would be erroneous to assume that single-transaction or conventional forms of channel management are in decline. These forms of channel constitute major segments of marketing transaction activity and must be given serious consideration when designing a supply channel. Still, the demands of today's global customer are forcing companies who historically resisted participation in dependent channel arrangements to revise their traditional policies. In addition, technologies and new forms of management methods have increased the ability of networked channels to provide order-of-magnitude advantages in the form of increased efficiencies and competitive advantage. In any case, when constructing a supply channel, managers must determine the desired level of control over the cost and quality of the functions performed by the channel as a whole, the size of the operating economies to be achieved, the degree of channel member loyalty and network stability to be established, and, finally, the ability of the channel to project to the marketplace an image of collective strategic direction that can be easily identified and measured.

The design of a channel structure is a fundamental activity that establishes not only the physical flows of goods and information up and down the supply pipeline but also the strategic opportunities that can be found by leveraging the collective resources and competencies of channel members. The true capabilities of the channel can only be unleashed when the operations functions and strategic goals of all members converge, even if the entity in the channel that originated the change or added the value is not the direct beneficiary or even incurs increased costs because of the innovation. For example, the apparel supply channel received a radical jolt when textile giant Milliken sponsored the "Crafted With Pride Council" in the early 1990's to introduce the concept of channel *quick response.* The eventual adoption of this strategy resulted in increases in channel costs due to smaller production runs, more timely service, higher transportation costs, and new services like EDI ASNs with fabric flaw identification. Despite the costs, these reengineering efforts aimed at channel integration efforts helped save the domestic apparel manufacturing industry, and the benefits in reduced retail inventory and markdown far exceeded the costs [6].

Channel Operations

The primary purpose of the integrated supply channel network is to facilitate the fundamental operations activities of manufacturing, distribution, and customer delivery. The objective of these three channel processes is to create unique and unassailable value to the customer that cannot be copied by the competition. In accomplishing this mission, the supply channel must accomplish the following critical customer-satisfying drivers: superior customer service, decreased channel cycle times, high levels of channel performance, and reduced costs. Today's best supply channels seek to leverage operational excellence, strive for quality and continuous improvement, and foster the creation of new, innovative techniques possessed by their members to achieve superlative, customer-winning processes unmatched by the competition.

The supply channel's contribution to customer service can be found in the matrix of concepts and practical deliverables it offers to the marketplace. The most basic of these customer needs centers around the order to delivery cycle. Among the most important of these can be found in relative order of importance the following:

- Completeness and reliability of deliveries
- High product/service quality
- Best value for the price
- Competent inside sales representatives

- Speed/frequency of deliveries
- Least total delivered cost
- Breadth and depth of available products/services
- Lowest price and availability of credit
- Use of electronic commerce technology such as EDI ordering capabilities

To these "basic" elements must be added several other customer service attributes that, although not deliverables, are immensely important to achieving the operational service goals of the supply channel. Among these can be found the appearance the whole channel would like to project to the customer through such tangibles as new facilities, state-of-the-art technology, commitment to quality, and highly qualified personnel. In addition, other elements such as reliability, responsiveness, and competence project to the customer a sense of dealing with acknowledged and measurable standards of service, a supply partner who can respond quickly and concisely to their needs, and confidence that their product and service issues will be satisfied by a supplier who possesses the necessary skills and knowledge. Finally, other channel service attributes can be found in the courtesy by which customers are treated, the feeling of credibility and honesty when dealing with the supplier, a sense of security and peace of mind when the transaction is completed, and the degree of communication and ability of the service provider to unearth and respond purposefully to the needs, desires, and expectations of the customer.

Perhaps the key to achieving superlative customer service is controlling the total time it takes goods and services to move from the point of manufacture to the point of customer delivery. The effective management of cycle times results in high customer service, greater operational flexibility, lower investments in working capital tied up in inventories, and lower operating costs. Managing the cycle times of the three fundamental supply channel operations processes can briefly be described as follows:

- *Manufacturing Process Cycle Times.* Production processes stand at the gateway to channel cycle time management. The ability of manufacturing to quickly convert raw materials and components into the finished goods the market wants requires the establishment of agile and flexible processes. The goal of supply channel manufacturing cycle time management is to improve planning and scheduling, continuously shrink setup and processing times, produce in lot sizes identical to actual customer demand, achieve the highest possible product quality, and move goods as rapidly as possible to the distribution functions in the supply channel.
- *Distribution Process Cycle Times.* The time it takes products to move through the distribution portion of the supply pipeline is directly related to the geographical size and number of transfer nodes found within the

supply network. Three critical elements dominate cycle time management in distribution: the speed of the transportation services used, the capacity of supply point nodes to move goods through their processing operations, and the speed and accuracy of information transfer. Concepts such as single-source suppliers, cross-docking, and electronic commerce have had a significant impact on cycle time reduction in this area of the channel.

- *Customer Delivery Process Cycle Times.* Cycle time management in this area revolves around the effective management of the order process. Order process management requires the channel to ensure the timely and accurate movement of both *internal* supply channel orders as well as orders that are delivered to the end customer. These two regions of order management are portrayed in Figure 5.4. Besides the timely delivery of product, delivery cycle time management is also concerned with triggering accurate resupply information back through the channel to the source of manufacture. The availability of replenishment information enables planners throughout the channel system to develop more accurate schedules, control channel movement, minimize wastes in transportation, and reduce channel supply node inventory imbalances.

Effective cycle time management enables the five components constituting the supply channel and their internal processes to be more closely integrated together. As inventories and information pass quickly through the network system, channel

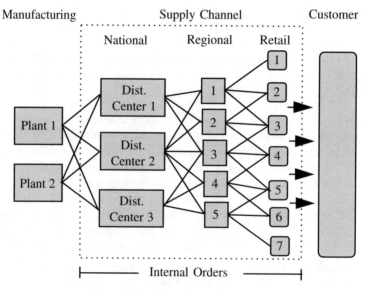

Figure 5.4. Internal supply channel.

members are better able to shrink costs and increase the entire channel's ability to achieve greater levels of competitive superiority.

Another critical area of supply channel operations is the development and monitoring of performance levels. The role of performance measurements is to assist managers in determining whether the supply channel system is achieving its objectives as well as providing metrics that will serve as the basis for performance improvement. Although each supply channel may focus on a specific performance objective, such as quicker customer delivery or lower operating costs, there are essentially six major performance areas to consider [7]:

- *Customer Satisfaction.* In this area can be found metrics associated with customer delivery, such as product availability, ease of order processing, order information, delivery timeliness, delivery completeness, number of customer complaints, and customer rating.

- *Quality.* This area of performance has several attributes. Some refer to product issues such as reliability, conformance to standards, durability, and serviceability. Other metrics focus on service issues such as errors in order contents, incomplete orders, late shipments, and poor inventory replenishment planning.

- *Asset Utilization.* This metric attempts to assess the competitive advantage gained for each physical asset possessed by the channel and considers such measurements as inventory turnover, return on assets employed, and working capital employed. Although this measurement focuses primarily on each channel participant's asset utilization, the collective productivity of total channel assets will have a direct impact on such issues as price and operating cost that invariably make their way down the channel pipeline.

- *Operating Costs.* This area of performance measurement covers the expense incurred by each member in operating the channel. Among the metrics to be monitored are cost of labor, transportation, maintenance, taxes and insurance, information services, and rentals.

- *Cycle Time.* As detailed above, the measurement of the time between operations cycles, such as production processing times, order processing, picking and shipping, and delivery, is critical to the competitive position of the entire channel.

- *Productivity.* This final metric attempts to measure performance against some critical channel benchmark. These metrics are fairly detailed and consist of such goals as orders processed per unit of time, shipments per facility, operating costs per asset and process unit, and network costs per sales unit.

Although it may be difficult to gather the data occurring throughout the supply channel necessary to compile the above metrics, accurate and complete performance measurements can significantly assist channel planners in making the right decisions concerning network design, operation, and investment.

The final area comprising channel operations is understanding total supply network costs. The objectives of the supply channel are straightforward: total customer satisfaction, high profitability, and low operating cost. Achieving these goals requires channel members to make correct trade-off decisions between costs and expected benefits. Supply channel operating and investment costs are complex and can consist of a matrix of possible cost and investment drivers. The critical components of supply channel costs are comprised of several internal fixed and variable cost elements, such as labor, facilities operation, inventory investment, and transportation, and short-term and long-term investment costs arising from facilities expansion and improvement, information and communications equipment, and other capital outlays. The key is to make decisions that provide not only individual enterprise but also channel wide competitive advantage.

Understanding Supply Channel Functions

The nature of a supply channel can be understood as consisting of a specific channel business mission, a particular channel design structure, and a defined set of operations processes and performance measurements that indicate how the supply channel plans to compete. What supply channels will actually do to achieve their goals can vary widely and is dependent on a multitude of factors. Some are distinctly *functional* in character and can be distinguished by the manner in which physical movement and transport of inventories through the supply pipeline occurs, categories of goods and services offered, flexibility of the manufacturing and distribution processes, speed and cost of the delivery systems, types of sales promotions and deals used to stimulate business, extent to which information and communications technologies are employed, and complexity of the channels of distribution. Other factors, however, are concerned with *strategic* objectives, such as the level of channel integration, types of channel partnerships, ability to assemble coalitions of supporting firms, willingness to share information and cooperate in joint ventures, and capacity to create a level of marketplace service unattainable by any one channel member acting individually.

Every supply channel consists of a field of unique combinations of operational structures and strategic objectives that together establish channel coherence and identity. Traditionally, academics and practitioners have concentrated on delineating the "functional" aspects of network management and have barely scratched the surface of the strategic possibilities to be found in the channel system. Most texts on channel management are concerned with the calculus of channel design, customer service output segmentation, and transportation and warehouse analysis. In order to gain a full understanding of the supply channel's place in SCM, it is perhaps beneficial to separate the supply channel into its functional elements and its strategic elements. In this section, the functional elements of the supply chain will be explored.

Supply Channel Functions—An Overview

In the ideal marketplace, there would be no need for supply channels. Customers would be able to purchase the products they desired in the right configuration directly from the manufacturer and have them in their stockrooms, display cases, or their own homes on demand. In reality, the vast majority of products must move through a distribution channel, beginning with suppliers of raw materials, component parts, and finished goods, and then often passing through an intricate web of company-owned distribution facilities, independent wholesalers, dealers, and brokers, until they reach the retail level or point-of-sale node. The marketing channel that emerges out of these relationships is forged by the transfer of ownership and the flow of goods as they move from producer to customer. What functions are actually performed and how they are apportioned among channel members differentiates one channel from another and defines the degree of channel competency uniqueness.

There can be little doubt that much of the particular set of functions performed by a channel are determined by its *institutional* approach. As stated previously, channels can range institutionally from single-transaction arrangements, to free-form channels where each member retains a significant degree of independence, to networked supply channel systems where all channel member functions and operating goals are closely integrated together. How supply channel functions are determined, however, can also be defined by the movement of product and information as it flows through the supply pipeline. As illustrated in Figure 5.5, some channels are single level: The manufacturer or distributor performs all the functions relating to marketing, products and services, transportation, and warehousing. Other channels are composed of multiple levels of suppliers, manufacturers, wholesalers, and retailers, each performing some or all of the marketing and logistics functions and incurring the cost of operations. The level of channel integration and cooperation depends largely on the nature of the product and market objectives. For example, a fresh vegetable supply channel requires a tightly knit linkage of channel partners who depend on speedy delivery, communications, title transfer, and financing flows to bring product to market.

In practice, it is almost impossible to separate the institutional from the functional characteristics performed by the members of a supply channel. Still, it is far more useful to describe channel agents by the activities they perform than by attaching institutional nomenclature, such as "producer," "wholesaler," or "retailer." Such an approach assists firms in focusing on the actual mechanics of the marketing channel rather than a strictly organizational view. This factor also recognizes that the institutional structure of a supply channel can remain

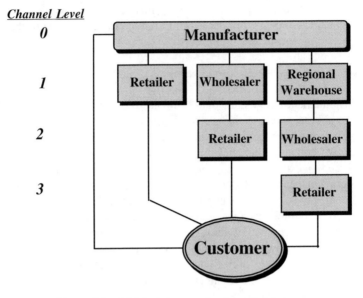

Figure 5.5. Multiechelon supply channel structure.

constant even though responsibility for performing channel functions may shift between channel members.

Basic Supply Channels Functions

Supply channels are formed to solve several critical functional problems described as follows:

- *The Physical Movement of Inventories.* The physical flow of inventory from supplier to consumer is the most visible function of the supply pipeline. The movement of inventory through the channel can be said to occur in six possible stages (Fig. 5.6). The supply channel flow begins with the acquisition of materials and components from suppliers. In

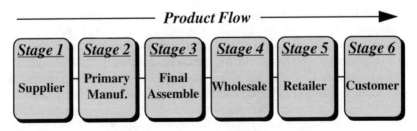

Figure 5.6. Supply channel material flow.

stages 2 and 3, raw and component inventories are transformed into finished or semifinished products through the manufacturing process. Stage 4 marks the beginning of the distribution process flow. There are three possible alternatives at this point. Product can be shipped directly to the customer; product can be distributed through a channel of manufacturer owned field warehouses; or, finally, product can be sold to a wholesaler, who, in turn, can sell goods to a retailer or directly to the customer. In stages 5 and 6, products are delivered to the retailer portion of the supply channel for eventual delivery to the customer.

- *The Process of Functional Performance.* Supply channels exist to increase the efficiency of marketing time, place, and delivery utilities. According to Bowersox and Cooper [8], the supply channel facilitates the flow of inventory through the pipeline by enabling three key marketplace functions. To being with, supply channels facilitate product and information exchange through the processes of selling and buying. Second, they facilitate the logistics process through the efficient management of transportation and storage. Finally, they facilitate financial and distribution processes by providing capital and credit, sorting of products into the quantities desired by the marketplace, shouldering financial risk, and providing for the collection, interpretation, and communication of market and product information.

- *Increasing Supply Channel Efficiency.* When product availability and delivery is immediate, the functions of exchange and delivery can be performed directly by the producer. As the number of producers and the size of the customer base grows, however, so does the need for internal and external intermediaries who can facilitate the flow of products and services through the marketing and distribution process. In fact, by streamlining information, marketing, and product flows in the supply channel, an intermediary can substantially reduce the number of transactions between producers and customers. For example, say that three producers trade directly with five customers. To calculate the number of trading links, the number of producers would be multiplied by the number of customers. As illustrated in Figure 5.7, this would mean that there would be a maximum of 15 exchange transactions.

By positioning an intermediary consolidation point between the producer and the customer, the number of transactions in the channel would be dramatically cut. As Figure 5.7 illustrates, if an intermediary is positioned between the producers and the customers, the number of possible transactions would decline from 15 to 8. The role of an intermediary grows in importance in facilitating channel efficiencies as the number of producers and customers expands. If there are just 50 producers and 2000 customers, the number of transactions without an intermediary

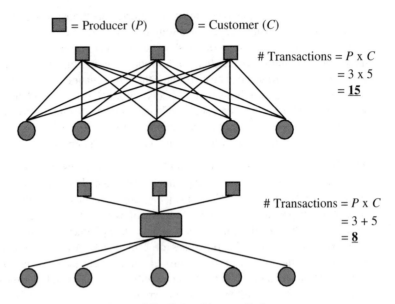

$$\blacksquare = \text{Producer } (P) \qquad \bullet = \text{Customer } (C)$$

\# Transactions $= P \times C$
$= 3 \times 5$
$= \underline{\mathbf{15}}$

\# Transactions $= P \times C$
$= 3 + 5$
$= \underline{\mathbf{8}}$

Figure 5.7. Role of intermediaries.

would be calculated at 100,000. With the presence of an intermediary, the number of transactions drops to 2050, or a 98% reduction in total channel transactions. Channel efficiencies can be further achieved by the addition of a second tier of intermediaries. Depending on the product and the marketing approach, intermediaries are a key part of the business strategy of many producers. It would be impossible to think of Coca-Cola selling its products directly to the consumer from the bottling plant! Whether a company operated and serviced its own distribution center or a wholesaler performs the channel management tasks, intermediaries reduce the number of transactions and consolidate flows of information and products through the supply channel.

- *Reducing Channel Complexity.* Besides structuring how products are to flow through the distribution pipeline, supply channels also decrease complexity in the performance of ownership transfer. Channel arrangements facilitate transaction management through the functions of *routinization* and *sorting*. Routinization refers to the policies and procedures that provide channel members and new entrants with common goals, channel arrangements, marketplace and profit expectations, and the framework for transactional efficiencies. Complexity is reduced by the normalization of past alliances among these channel partners and their preexisting arrangements. Sorting, on the other hand, refers to the process of removing complexity associated with product location and timing.

Through sorting, product configurations and quantities can be matched to meet the specific assortments and bulk sizes demanded by the marketplace. It is the function of supply channels to perform activities relating to assembling marketable assortments and packaged quantities and positioning them throughout the channel to satisfy the moment of customer demand.

- *Solving the Problem of Specialization.* As the supply network grows more complex, costs and inefficiencies tend to grow in the channel. To overcome this deficiency, many supply channels contain members that will specialize in one or more of the elements of supply channel management, such as exchange, transportation, or warehousing. The net effect of specialization is to increase the velocity of goods and value-added services as they progress through the supply pipeline by reducing costs associated with selling, transportation, carrying inventory, warehousing, order processing, and credit. Sometimes a dominant channel member, such as the manufacturer, may seek to eliminate a specialist partner by absorbing the function into the firm's operations. Vertical integration can be beneficial if it seeks to facilitate the flow of product, decrease cycle times, decrease costs, and eliminate redundancies.

- *"Informationalizing" the Channel.* Today's enterprise considers the flow of supply channel information to be such a significant corporate asset for competitive leadership that it must be viewed as a fundamental function. In the past, critical information concerning products and markets was often trapped within layers of organizational and operational redundancies that inhibited the timely flow of data up and down the supply channel. As the number of information-based products and services spiral, companies are beginning to leverage the supply channel as an information conduit that can be used to continuously create radically new methods of integrating suppliers and customers across global space and time. The exploding domestic and international digital "superhighway" of networked computers, the availability of on-line databases, the merging of PC, CD-ROM, audio and communications systems, and the rapidly emerging home and business fiber-optic cable TV system have revolutionized the role of the supply channel and presented companies with as yet unrealized opportunities for marketplace advantage and customer satisfaction.

Marketing Functions of the Supply Channel

The basic channel functions described above provide the physical framework and operations objectives that determine how a channel network approaches the marketplace and how it will deal with today's rapidly opening and closing windows of opportunity for product and service leadership. It is the role of the channel's marketing functions, on the other hand, to manage the mass of customer transactions that must be planned and executed on a daily basis. According to scholars in the fields of marketing and logistics [9], the distribution network performs the following marketing flows:

1. *Information Flow.* Marketing research about new and current customers, advertising, pricing, competitors, and other forces in the marketplace is fundamental to channel success. The timely and accurate transference of marketing information enables channel members to keep abreast of sales opportunities, product and service requirements, promotions, and other activities necessary to achieve the sales goals of the whole channel system. In addition, effective sales and procurement training ensures that the channel network is focusing on the most up-to-date products and services.

2. *Title Flow.* As goods move from one node in the supply channel to another, actual transfer of ownership takes place.

3. *Promotion Flow.* Promotion can take the form of both internal and external information concerning prices, marketing sales materials, advertising, delivery, and other elements.

4. *Negotiation Flow.* The transfer of ownership of goods from one independent business unit in the channel to another usually involves attaining agreement on price and other sales terms. Negotiation should always be supportive of the overall competitiveness of the supply system.

5. *Ordering Flow.* The actual placement of customer and inventory replenishment orders, as well as intelligence concerning marketing trends, provides critical information for the supply network. This information cascades backward from the end consumer through retailers, wholesalers, and manufacturers, ending up eventually with the components and raw materials supplier.

6. *Financing Flow.* As goods move through the channel network, capital must be acquired and allocated to finance the transfer of ownership. When a channel member assumes physical possession of inventory from the manufacturer, they are, in effect, financing the manufacturer by exchanging capital for inventory.

7. *Risk Flow.* Ownership of goods incurs risk on the part of the possessor. Inventories can become obsolete, damaged, or unsalable, leaving the possessor to incur the loss.

8. *Payment Flow.* The flow of cash payment proceeds backward through the supply channel. Often banks and other financial institutions are involved in payment for goods and services.

The marketing channel should be designed to facilitate and make as efficient as possible each one of these functional flows. Regardless of the physical structure of the channel network, each of these functions *must* be performed someplace by one or multiple business units within the network. Channel members can be eliminated or substituted; the nine marketing functions, however, cannot be eliminated and must be assumed by remaining business units either upstream or downstream in the distribution pipeline. A manufacturer, for example, may elect to sell direct to the end customer, bypassing the distribution middleman. Such a channel strategy would require the manufacturer to perform a number of forward and backward marketing functional flows previously provided by former channel partners. In the final analysis, the real value of the structure of a particular supply channel system is that it provides a form of synergy, permitting individual channel partners to achieve objectives they would otherwise be unable to attain acting individually.

Although the effective and efficient performance of channel functions can be said to significantly expand overall competitive advantage, its benefits can also be seen in the everyday performance of the various businesses positioned along the supply channel pipeline. For example, according to various independent researchers and trade associations, grocery supply participants can save an estimated $30 to $50 billion annually in reduced inventory costs across the entire distribution pipeline, and drug store health and beauty care segments can save an estimated $7 billion through effective supply channel management [10]. Besides costs, there are other benefits to be realized. In well-run channels, enterprises are able to shrink handling costs and transit times, speeding up the time to market from the manufacturer to the retail outlet. Rapid product turnover in turn decreases storage space, whereas increased communications and flexible manufacturing techniques enable customers to have products and packaging configured to meet their specific needs. Finally, an effective supply channel permits companies to increase profits and marketshare by reducing lost sales, eliminating unnecessary processing, and freeing capital tied up in channel assets.

Understanding Supply Channel Strategies

In the previous section, it was stated that the structure of "world-class" supply channels consists of bundles of functional performance characteristics and strategic goals that together shape channel direction, provide for channel cohesion, and enable channel business nodes to keep enterprise objectives integrated with overall channel objectives. To be competitive in today's global marketplace, companies must constantly seek to create and reinvent supply channels that provide superlative customer service, high performance, and low costs. They must, however, also tirelessly search for innovative strategic breakthroughs in the use of new information and communications technologies, the ability to create market-wining coalitions of supporting firms, and the capacity to leverage channelwide repositories of resources and skills that preempt the competition and germinate new competitive space. Companies that fail to devise effective supply channel strategies and continue only to focus on the same old paradigms of channel functional management are doomed to a debilitating treadmill where business objectives centering around "incrementalism" continually attempt to pare down the business in the search for a way out of the intense competition and exploding asset investment needed just to maintain par with industry leaders. The development of effective supply channel strategies can, therefore, be said to be critical to channel survival. In this section, the various strategic elements of supply channel management will be explored.

Supply Channel Strategy Overview

For the most part, managing the *functional* elements of the supply channel has traditionally been the central focus of channel planners. The prime objective has been centered around improving those operational elements associated with product and service marketing, pricing, promotions, channel design, and logistics activities such as warehousing and transportation. In creating functional channel arrangements, companies concern themselves primarily with the implementation of channel strategies composed of the following six steps [11]:

1. *Enterprise Positioning.* In this step, planners attempt to ascertain the enterprise's relative position within a prospective channel by determining the legal-social setting, level of channel complexity, degree of specialization and routinization, dependency requirement, and amount of risk.

2. *Design of Marketing and Logistics Performance Structures.* Charting out the types of performance expected from the marketing effort and logistics operations forms the second step in channel strategic design. The goal is to draft specific profit and cost benefits to be gathered from the channel arrangement.

3. *Planning and Analysis.* During this step, the possible channel approaches are reviewed and one or a matrix of intersecting channel partnerships are identified.

4. *Negotiation.* In this crucial step, the actual channel partnerships are formed. Key negotiating points revolve around issues regarding the roles to be assumed by channel members, level of commitment based on the relative risks involved, and decisions as to which channel members are expected to take a leadership role.

5. *Channel Management.* Managing the smooth functioning of the channel is the focus of this ongoing step. The critical factors are ensuring the

continuous creation of service value and the maintenance of contractual or implicitly agreed-upon channel mechanisms that promote cooperation and neutralize conflict.

6. *Accurate Measurement of Channel Performance.* The final step in the traditional supply channel planning process is the drafting of detailed profit and cost measurements designed to track channel success and identification of areas for improvement or adjustment.

Whereas the above steps necessary to the development of effective functional channel strategies appear to be comprehensive, it has become painfully evident in today's global economy that the arrangements that they produce constitute the bare minimum for channel success. The result of this tactical, pragmatic approach to channel management has been the creation of channels that can be described, at best, as chaotic jumbles of alliances whose prime motivation is to provide independent channel members with the ability to achieve short-term economies of scale and cost reductions. Instead of an overall integrated plan focused on assembling the complementary competencies of channel partners into a single competitive entity, many channels are composed of loose coalitions of individual businesses characterized by disconnection and misalignment, each channel node serving independent and often unrelated purposes.

Today's market leaders have begun to recognize that effectively leveraging the SCM philosophy requires viewing the supply channel not only from a functional but also, and more importantly, from a strategic perspective. For decades, academics and practicing business professionals have acknowledged that the business units constituting a given supply channel network can add significant incremental value up and above their ability to execute operations activities. To begin with, it has long been understood that supply channels not only satisfy demand by providing goods and services at the right place, quantity, quality, and price but they also stimulate demand through marketing and sales activities. In this sense, the supply channel is more than just a pipeline for the flow of goods and services from supplier to consumer: Its real utility lies in the ability of channel constituents to promote and actualize a level of customer service unattainable by any one channel member acting individually.

Furthermore, supply channels create strategic advantage by providing for other value-added activities beyond the fundamental functions of physical movement and marketing transaction. Customers today expect flawless execution of the four P's of marketing: product, price, place, and promotion. They, however, also want continuously improving quality, shortened cycle times, product and delivery flexibility, and solutions that answer their unique requirements. In addition, they are seeking partnerships, communications and data integration, participation in shaping product development, manufacturing and distribution processes, and dependability. The result of this emphasis on the *value* channels can add to the customer beyond operations activities has been dramatic. For example, J. Robert

Hall, former CEO of Nabisco Specialty Products Company, sees the supply channel as a defensive and an offensive competitive weapon. The superlative performance of operations functions, such as satisfied customers, on-time deliveries, accurate invoices, and others, he states, provide companies with effective defenses to fight off competitors and retain market share. However, a market-wining offense requires companies to be constantly reinvented around a new view of channel management as "action oriented, entrepreneurial, able to position risk as a way to learn, able to really communicate, and absolute intolerance of bureaucracy." The goal is the creation of focused, highly flexible channels that not only understand the customer's business but are actually coparticipators [12].

Development of Strategic Partnerships

The central focus of strategic supply channel management can be found in the existence and continued development of channel partnerships. Although it is true that supply channels, by their very definition, consist of arrangements of businesses linked formally or informally in the pursuit of a common objective, the requirements of today's channel structures are distinctly different from those of the past. In the era of mass-production, the goal of channel management was one dimensional: dominate the marketplace by offering standardized products with the lowest price delivered at the lowest operating cost. In contrast, in the era of *supply chain management*, supply channels are multidimensional. Although traditional marketing and operations issues are still important, today's channels have become more concerned with the development of cooperative relationships of all sorts, even with historical competitors, that will allow individual companies within the channel to leverage the skills and competencies of their channel partners to overcome resource shortages inhibiting the performance or creation of new sources of channel value. Japanese game company Sega, for example, has engineered cooperative deals with AT&T, Time Warner, TCI, Pioneer, Yamaha, Hitachi, and Matsushita. Sega's partners will give the company access to the technologies that will be needed to download computer games over a cable television network, provide video games with lifelike graphics, create "virtual" amusement parks, and much more [13].

This dramatic change in the nature of supply channel management is the result of distinct changes occurring over a decade ago. Beginning first as enterprise-level movements designed to integrate and focus production first and then support processes around a total customer service and quality paradigm, today's concern with forging linkages several echelons backward and forward through chains of suppliers and customers represents the extension of these strategic objectives to the entire supply channel. In essence, there are three levels of supply channel partnerships that can be explored. They are detailed as follows [14]:

1. *Multifunctional, Within-Company.* For many companies, just being able to unite their *internal* supply channels can provide significant sources

of competitive advantage. On this level, traditionally separate functions within the company's supply channel, such as manufacturing and distribution, are integrated to form multifunctional product and service strategy teams. The goal is to assemble operations teams composed of members from sales, marketing, finance, manufacturing, product development, and logistics and collectively charge them with the task of long-term decision making regarding specific products and markets. A unified team approach ensures that all the key business elements—demand management, manufacturing capacity management, purchasing, inventory and storage, transportation, and finance—can maximize the enterprise's resources and service to the customer.

2. *Next-Echelon Partnerships.* At this level of channel partnership, functional teams from one company coordinate directly with the corresponding supplier or customer teams found within the supply channel to facilitate information transfer or to devise innovative ways of approaching the marketplace. For the most part, these linkages are limited to addressing mutually critical issues such as production/supply problems, quality and warranty issues, product design, information transfer, and others. The development of Boeing's new 777 commercial airliner is an example of a next-echelon partnership. In the past, the company had developed aircraft with virtually no input from its customers and suppliers. On this project, however, Boeing established some 235 all-embracing teams that included customers, suppliers, and support teams.

3. *Multi-Echelon Partnerships.* This level is the most sophisticated of the three possible supply channel partnerships and consists of intersecting channels of suppliers and customers cooperating together to realize optimal response to market requirements and to invent new areas of competitive space. Coalitions at this level are formed to respond to four critical strategic requirements: (1) the need to acquire quickly resources and competencies beyond the capacities of individual firms; (2) the ability to join with potential competitors, thereby reducing the threat of future rivalry or denying the expertise of a partner to a competitor; (3) the capacity to gain access to new markets, particularly foreign markets; (4) the capability to have partners share in the risks associated with the development of new technologies or expansion into unfamiliar markets.

The development of supply channel coalitions provides significant benefits that could not be gained by a single company or even a group of companies acting alone. To begin with, close channel partner cooperation enables companies along the channel to shrink the cycle time from new product/service idea to marketplace entry without imposing severe trauma on the organization. Partnerships reduce development costs, risks, and the consequences of failure. Second, partnerships remove the adversarial and control relationships characteristic of past channel management strategies. Intrachannel and interchannel cooperation requires members to recognize their mutual dependence on one another for joint success or failure. Third, close channel relationships accelerate technology transfer and broaden the resource pool available for competitive innovation. Cooperation enables cross-functional enterprise teams and fosters information sharing to mutual advantage with customers, suppliers, partners, and, when beneficial, even with competitors. Finally, supply channel partnerships open dramatically new competitive possibilities by providing radically new solutions to customers, leveraging focused functional teams across company boundaries, creating new mechanisms for the open exchange of information necessary to intercompany project collaboration, and fostering a spirit of communication and commonality of purpose that provides the opportunity for channel members to pursue closely integrated market-wining strategies.

Developing Effective Supply Channel Strategies

The application of the SCM concept to supply channel strategies requires a redrafting of the traditional paradigms formerly used for channel design. Past channel strategies focused around the enterprise's ability to produce and distribute mass-produced products to the marketplace regardless of actual need. Such a channel strategy focused on the practices of manufacturing and was driven by the following points:

- Rigid mass-production systems designed to produce a few standardized products
- Long supply cycle times
- Large batch sizes
- Capacity based on annual volumes
- Volume-driven technology
- Multiplicity of competing suppliers for the same parts.

Supply channel designs supporting such a system were inventory push systems and were marked by low overall delivery reliability, the uncoupling of the schedules driving manufacturing plants and suppliers, functional silos within the supply chain, and the absence of formal partnerships [15].

Today's supply channel strategies, on the other hand, require the reversal of the classical doctrines of the traditional channel. Inventory must be pulled through the supply pipeline. Reliable delivery systems, synchronization of network supply and demand, and flow replenishment systems are valued as optimal operational paradigms. Above all, closely integrated supply chain partnerships are viewed as essential for competitive survival not only to provide for the basic functions of product movement and transaction but also for the establishment of whole economic communities composed of interacting channel organizations capable of providing unique goods and services to customers who themselves are members of other supply channel systems. As such, the real value of today's supply channel is to be found in its capacity to enable members to move beyond parochial objectives and toward the realization of interwoven webs of collectively shared marketplace visions activated through mutually supportive roles. When designing a channel strategy, therefore, planners must create structures that can not only respond to tactical product, delivery, and information requirements but also be responsive to new marketplace opportunities unearthed by market foresight teams that are capable of driving the entire channel to new competitive levels.

In designing effective channel strategies, planners must focus on two areas. The first is concerned with the functional side of the channel and utilizes the core elements of the traditional supply channel development methodology. Channel *service outputs*, such as product variety available, the time customers are willing to wait for products, lot-sized inventories, and spatial convenience, must be analyzed and agreed upon [16]. The existing channel system must be delineated, constraints and opportunities identified, the ideal channel defined, and gap analysis applied to measure the difference between the ideal and the real system. Finally, planners must reconcile the optimal channel system with the current one, set objectives, and begin the process of implementation [17].

Once the core objectives of each channel member have been defined, the process of integrating each individual objective into a cohesive channel strategy can then begin. The strategic planning process can consist of the following stages:

1. *Opportunity Analysis.* In this stage, the search for new channel coalitions and sources of innovation takes place. This is a "visioning" stage that requires channel partners to talk with one another and with their customers several levels both above and below the immediate enrichment-adding chain. A critical part of this stage is to determine total channel capabilities and opportunities to integrate channel member competencies and uncover radically new possibilities and competitive space. The search for new avenues of channel coevolution provides new meaning to the *value chain.* A well-known concept, the value chain refers to the set of activities necessary to drive a product or service from concept through to market delivery. What channel planners must do at this stage is begin the process of developing value chains composed not just of product and service offerings but also of combinations of whole channel networks possessed of the capacity to invent new market capabilities, technologies, and opportunities for satisfying the customer. The goal is to create sources of dramatic value, compared to what is already avail-

able, that captivate the interest of the customer by leveraging the financial and knowledge power of other channels who wish to network in the new market.

2. *Current Status.* After completing the opportunity analysis, channel planners must take a hard look at the existing channel network. One of the goals of this stage is to identify obsolete functions performed in the existing channel. Sometimes, this might mean changing channel functional roles, or off-loading functions to new partners who are more capable, or eliminating the activities altogether. On the positive side, this step presents the opportunity to uncover untapped capabilities and sources of innovation that can enrich the channel with minimum changes in channel membership. The objective of this step is to locate the most desirable companies and business partners possible: the best customers, the strongest suppliers, and the key channel networks. The new coalitions that arise must be capable of meeting any challenge from competitor channel systems.

3. *Strategy Implementation.* Creating the supply channels that will realize the strategic goals identified in stage 1 is a painstaking process. Activating the new channel means determining how each channel member will contribute to providing the best solution along the entire chain of customers. Establishing new coalitions is also always fraught with dangers both from within as well as from outside the channel. Success or lack of success can generate new dynamics of power as members or even whole value chains jockey for alternative positions of power or push the channel in new directions. Channels are also vulnerable to rival channels who can split even the most established network by sponsoring new and innovative ways to approach the marketplace gained through access to superior competencies or technologies.

4. *Strategic Renewal.* As the marketplace changes, so must channel arrangements and assumed functions. Similarly, the rise of new, more efficient, more innovative network systems can render a once powerful supply channel obsolete. Maintaining the viability of any channel means uncovering new sources of innovation that continuously provide dramatic improvements to channel operations as well as to strategic positioning. According to Moore [18], this can be accomplished through a four-step process:

    1. Continuously create a system and sequence of closely integrated channel relationships that result in the establishment of something of real *value* relative to what other channels may offer.

    2. Establish *critical mass* as the channel system expands across available customers, markets, allies, and suppliers.

3. Lead *innovation* and *coevolution* within the channel systems that already exist.

4. Ensure that each channel member and the channel as a whole sustain *continuous performance improvement* rather than become obsolete.

Determining both the form and content of supply channel strategic plans often requires channel partners to rethink fundamental assumptions about how the channel has been functioning, what its critical values are, and how it approaches competitive and innovative opportunities. In this sense, the strategies followed by today's best channels are unique and spring forth from the particular sets of aspirations, cultural values, and operational structures characterizing each company as well as the collective channel community. Devising effective strategies is at once a creative outpouring of the entire network focused on achieving continuous and unassailable competitive advantage while serving as a guideline for purposeful action.

Global Supply Channels

When supply channel historians look back at the period of the last 20 years of the 20th century, perhaps one of the most salient developments will be the emergence of the global economy. For several decades after the Second World War, most manufacturing and distribution enterprises remained within their own national boundaries. Although some of the world's largest companies, such as Coca-Cola, Ford, and Procter & Gamble, had historically engaged in a significant international trade, governments were fearful of exporting technology and wealth that might drain the nation in the face of the Cold War. In many cases, whole markets, such as Eastern Europe and China, were closed to US and European companies. Today, the end of the Cold War and the concept of "the global community" has accelerated the growth of the international marketplace and the integration of the world's economic activities. The former Soviet Union and the emerging nations of Eastern Europe are now openly soliciting economic assistance, trading status, and investment from the West. The European Community is moving closer to full elimination of trade barriers. Perhaps the most fertile area of economic growth, the Pacific Rim, is seeking favorable trading status and the import of foreign goods.

Critical Global Issues

There are several strategic, tactical, and operational reasons for this radical change in the world's business structures. Some are part of the economic realities of the time, such as the maturing of the world's highly industrialized nations as the era of mass production concludes, the growth of global competition as companies compete in an international arena for labor, capital, materials, and finance, the explosion of strategic alliances and joint ventures spurred on as much by economic changes as by the growth of regional trading blocks, and the blurring of the traditional distinction between domestic and international logistics as information and communications technologies, advances in transportation modes, and removal of transit restriction and tariffs make trading across the globe as easy as shipping products across one's own neighborhood. In summary, supply channel management has become the key to tomorrow's competitive leadership. Bender [19] has succinctly described the conditions propelling globalization as composed of three interconnected areas. The first consists of *strategic* reasons such as the following:

- Attempting to leverage shrinking product and process life cycles and recover development costs by selling products on a global basis.
- Denying marketplace sanctuaries to competitors. Companies can sell at a high profit margin to captive markets, making it affordable to sell at a lower margin to more markets.
- Avoiding government-directed protectionism, as found in many developing countries.
- Balancing production and investment with the differing economic growth patterns and economic cycles occurring across the globe.
- Profiting from global financial systems, communications and media, and market demand homogenization.
- Establishing early presence in emerging markets.
- Maximizing opportunities arising from symbiotic relations between suppliers and customers based on long-term commitments and close relationships.

The second area fostering business globalization is associated with *tactical* issues such as the following:

- Capitalizing on foreign trade to increase profits. Companies participating in international trade are likely to grow faster and be more profitable than companies that focus solely on national or regional markets.
- Participating in countertrade agreements. About one-third of all international trade involves countertrade (bartering), rather than cash transactions.
- Achieving stabilization by matching product and investment with global business cycles.
- Obtaining economies of scope by maximizing marketing, production, and logistics advantages through international trade.
- Reducing costs by transferring products across national boundaries that reduces taxes.

The final area focuses on *operational* issues and is concerned with the following:

- Reallocating manufacturing and distribution capacities to match global market demand.
- Reassigning production, purchasing, processing, sales, and financing to take advantage of different rates of international exchange and inflation.
- Accelerating the learning effect. As a company learns more about the global marketplace, costs associated with manufacturing and distribution processes decrease.
- Exploiting automation's declining break-even point. As the volume of product and processes increase, technology costs are recovered much quicker.

Developing Global Channel Strategies

Embarking on an international supply channel program, regardless of the benefits, can be a complex affair. Even a mature supply channel must expend significant effort developing products and services that will appeal to a global market, engineering new core competencies that will provide the mechanism for the performance of global channel functions, and creating channel distribution networks that will handle the flow of products and services through the new global supply pipeline. Achieving these objectives requires that channels first gain access to critical (because of market size or opportunities for growth) national markets and distribution channels. Next, it must have the competencies, products, and perception of value add that will be able to preempt possible rival channels. The best global channels, such as Coca-Cola, Procter & Gamble, Unilever, and Hewlett-Packard, provide brands that are recognized the world over and generate a sense of customer security and well-being when they are purchased. Finally, a successful channel strategy requires superlative, highly integrated channel logistics, marketing, and performance measurement functions that ensure effective communication and feedback. Unless thoughtful preparation is undertaken to gain a presence in global markets or preempt competitors, the strategic initiative will fall prey to more resilient channels.

In addition to these strategic considerations, several practical issues also need to be considered. International transactions and shipments normally take longer to complete, often involve the use of intermodal transport, and are subject to foreign government regulations involving tariffs, environmental issues, and requirements that a portion of the material composition of the product and the labor force originate from the home country. Once inside countries, shipments are also subject to a different set of transport regulations. To compensate for longer lead times, international supply channels normally have to carry more inventory than is found in the domestic channel. Foreign nations also may have quite different distribution channels than the home country, along with different service and quality-level expectations and purchasing behavior. Finally, the amount of paperwork is enormous. This is mostly due to foreign customs agencies, laws, languages, and currencies.

With these realities of international trade in mind, the development of a comprehensive international supply channel strategy is critical for enterprises seeking to enter global markets. An effective strategy must first identify the nature and scope of the international trade effort, define the appropriate marketing and logistics strategies and operational objectives and structures, and, finally, develop meaningful performance metrics to measure success and point out regions for improvement. The goal of the exercise is to develop "world-class" international channel functions that provide the firm with the mechanics to optimize and align the supply channel system with each international target market. As illustrated in Figure 5.8, an effective international channel strategy is composed of the following five elements:

1. *Environmental Analysis.* The first step in the international supply channel strategy development process is defining the strategic dimensions of the existing channel. There are essentially three areas to consider.

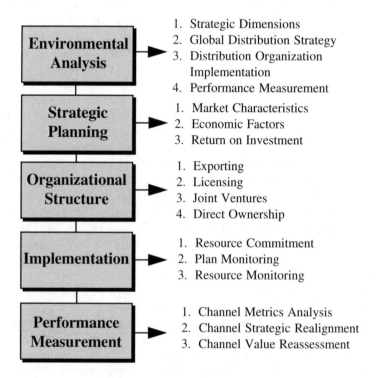

Figure 5.8. Global channel strategy.

The first is concerned with detailing the external environment of the firm. This analysis is organized around the macrodimensions of economic, sociocultural, political, and technical factors, and the microfactors of markets, costs, competitors, and government regulation. An effective analysis of the external environment should cover the world, ensuring that no relevant market, competitor, or trend is overlooked. The second strategic dimension to be analyzed is the channel organization. The goal is to pinpoint organizational strengths and weaknesses. A firm understanding of this strategic area is essential in dealing with identified opportunities, threats, and global trends. The final strategic dimension is coming to terms with and matching stakeholder values with perceived channel objectives. The goals of individual companies, and the stockholders, managers, employees, and customers behind them, often have conflicting values and interests regarding channel size and growth, profitability and return on investment (ROI), sense of social responsibility, and ethics. Before a global channel strategy can be constructed, planners must be sure that objectives are in alignment with the realities of the external environment, the capabilities of the existing channel, and the desires and assumptions of stakeholders.

2. *Strategic Planning.* Once the strategic dimensions of the target environment have been identified, planners can proceed to detail the nature and scope of the channel's overall strategy and then to define the unique characteristics of each national market in detail. Candidate countries should be rated on three major criteria: market attractiveness, competitive advantage, and risk. *Attractiveness* features can be described as such factors as language, laws, geographical proximity, stability, cultural similarity, and other microfactors. *Competitive advantage* focuses on the firm's overall marketplace strategies, presence of competitors, product life cycles, and the such. *Risk* is divided into two types: asset protection/investment recovery risk, which is concerned with the possibility of foreign government nationalization or limitations on the transfer of invested resources, and operational profitability/cash flow risk, which is concerned with the possibility of local economic depression, currency devaluation, strikes, and other factors. Once these criteria have been reviewed, choices must be made based on expected return on investment. This step includes estimating market potential and possible risk, forecasting sales potential, estimating costs and profits, and, finally, determining the rate of return on investment [20].

3. *Organizational Structure.* In the past, international supply channels have suffered from poor organization, lack of training, and the absence of interorganzational power and influence. Physical distribution was often seen in a supporting role and not regarded as a key element in the

enterprise's international marketing strategy. Without effective channel organizations, however, supply networks cannot hope to optimize on global opportunities and deter possible competitive threats. The actual structure organizations can assume will vary. In defining the strategic approach to the international marketplace, firms have four possible alternatives available:

- Exporting. The least involved form of international channel is the export of products into foreign markets carried out by some form of contracted international trading house or intermediary that is outside the boundaries of the formal supply channel. Exporting can be executed *indirectly* through a variety of specialized brokers or *directly* through foreign merchants or wholesalers.

- Licensing. Licensing can be defined as a contractual arrangement by which a supply channel (the licenser) in one country agrees to permit a company or supply channel in another country (the licensee) the right to use a manufacturing process, trademark, patent, technical assistance, trade secret, merchandising knowledge, or other skills. In exchange for these rights, the licenser will receive a fee or royalty. The objectives of a licensing agreement are straightforward: The licenser is able to gain access to a foreign market with a minimum of risk and capital expense; in turn, the licensee receives the right to distribute brand name products or access to proprietary processes either to found a business or to add to existing product lines. The best known examples of licensing are Coca-Cola and McDonald's.

- Joint Ventures. Unlike the first two strategies, the decision to execute a joint venture with a foreign company or channel directly involves a supply channel in the joint ownership and management of a foreign enterprise. Joint ventures may occur when a channel invests in the manufacturing and distribution operations of an existing foreign company or existing channel, or the two parties may join to found an entirely new supply channel.

- Direct Ownership. The direct ownership of a supply system in a foreign country represents the highest level of control and involvement a channel can have in the pursuit of foreign trade. Instead of working through an intermediary or a venture partner, the channel assumes all the responsibilities for facilities, personnel, marketing, and product distribution in the foreign country.

4. *Implementation.* Once the international supply channel strategy has been completely structured, it must be implemented. This step entails obtaining and committing current resources to executing market, market cluster, product, and logistics operations plans. The key element to

successful implementation is the presence of effective organizational structures for ongoing channel resource management.

5. *Performance Measurement.* A critical element of global supply channel operations is the existence of a comprehensive program of performance measurements. In this step, actual results are constantly compared with expected output. Performance measurements provide information that enables continuous channel strategic alignment with the external environment and organizational and value assessment assumptions established at the beginning of the strategy formulation process.

International Supply Channel Strategy Summary

The mega-opportunities currently occurring in many industries, including information and communications technologies, chemicals, pharmaceuticals, machine tools, autos, and others, have become inherently global in nature and require the formation of effective channel systems that span global space and time. As Hamel and Prahalad have so succinctly pointed out:

> No single nation or region is likely to control all the technologies and skills required to turn these opportunities into reality. Markets will emerge at different speeds around the world, and any firm hoping to establish a leadership role will have to collaborate with and learn from leading-edge customers, technology providers, and suppliers, wherever they are located. Global distribution reach will be necessary to capture the rewards of leadership and fully amortize associated investments [21].

Today's best supply channels are constantly searching for new alliances, assembling new competencies, and exploring possibilities in markets both at home and in the four corners of the earth. The goal is to use intersecting channel network systems effectively that permit global leaders to create and direct the development of tomorrow's markets by leveraging the resources for competitive innovation to be found both within companies and primary supply channels and outside across global partners.

Achieving this goal will require companies to find solutions to the following key issues:

1. Enlarging the boundaries of the supply channel to embrace global opportunities. This requires channels to possess the ability to balance total channel resources with the needs of the global marketplace.

2. Monitoring and managing the constant changes occurring in the global marketplace. Trends include shifting attitudes toward tariffs, administrative procedures, restrictions on intercountry transportation modes, and warehouse storage, as well as managing the pace of rapidly changing marketing and logistics strategies, new product introduction, and in-

creased information linkages. By responding effectively to change, multinational supply channels can increase the speed and reliability of global delivery, reduce overall transportation costs, and reduce the size (quantity) of shipments.

3. Extending and tailoring the supply channel to meet the distribution structure of each foreign nation or geographical region. Distribution channels in Europe, for example, are very mature and require local distribution centers, local management of transportation owing to the number of transit countries, and knowledge of local customs, export/ import, and tariff regulations. Trade with Japan requires short delivery cycles and local inventory to meet planned and random demand patterns. Distribution in the Pacific Rim, on the other hand, requires using local freight-forwarders and lead-time planning to counter delays due to the lack of a fully developed logistics superstructure.

4. Implementing information systems that provide for worldwide inventory, transportation, and customer service.

5. Aligning channel competitive values and global logistics capabilities. The logistics strategy used to operate the channel network must be carefully integrated with the overall competitive values pursued by all segments of the supply channel system.

6. Benchmarking distribution channel performance with that of international competitors.

The companies and their channel partners that will succeed in the highly competitive international marketplace of the latter half of the 1990s will be those who can construct agile and flexible channel network systems that will provide the foundation for the continuous search for new products and services, the deployment of new functionalities, the reinvention of current capabilities or the acquisition of new competencies, and the reconfiguration of customer interfaces in the search for global market leadership.

Summary and Transition

There is much confusion in today's literature over what is meant by *supply chain management* (SCM) and *supply channel management.* Often these two terms are used as if they were interchangeable. In reality they have very different meanings. SCM describes a *philosophy* of managing supply channels that seeks to leverage the collective competencies of networked business to develop innovative marketplace solutions while synchronizing the flow of products, services, and information occurring across the supply channel to create sources of unique, individualized customer value. In contrast, the term supply channel management refers not to a concept but rather to the actual physical business functions, institutions, and operating strategies that characterize the way a particular channel system moves goods and services to market through the supply pipeline. SCM provides today's market leaders with innovative, competitive strategies that enable them to leverage the resources and competencies of coalitions of companies to reinvent industries and create new competitive space; supply channels, on the other hand, provide the actual structures, cooperative relationships, and day-to-day management of sales, inventories, and deliveries on which a particular application of SCM is based.

Even though the actual objectives and structure of individual supply channels will vary, they all possess certain common characteristics. To begin with, in its most complex form, a typical channel is composed of five distinct participants: raw material and component parts suppliers, manufacturers, wholesalers, retailers, and customers. Not all of these channel elements are necessarily independent businesses. For example, some functions, such as manufacturing, wholesaling, and retailing, can actually be performed by the same corporation. Regardless of how the channel network is constructed, convergence of members can only be brought about by a clear focusing of channel functions and objectives. Achieving effective channels requires three management activities. The first can be found in the creation of a detailed *channel business mission* that unifies all direct and indirect channel constituents into a single competitive entity. The second activity involves structuring flexible and agile *channel designs* that permit the smooth flow of goods, services, and information as well as employing the collective resources and competencies of channel members to create new areas of competitive space. The third and final management activity requires the development of superlative *channel operations* that provide superior customer service, decrease channel cycle times, promote high levels of channel performance, and reduce costs.

How supply channels approach the marketplace depends on a particular combination of *functional* activities and *strategic* objectives that, together, establish channel coherence and identity. Channel network functions can be logically separated into two areas: basic and marketing. *Basic functions* are comprised of fundamental operations activities that can be found in all supply channels. Among basic functions can be found the following: physically managing the flow of inventory and services from the supplier to the customer; providing for functional performance by facilitating channel requirements for buying and selling, storage and delivery, and financial and information transfer; increasing channel transaction efficiency by streamlining channel structure; reducing channel complexity by routinizing and normalizing channel functions and arrangements; removing channel inefficiencies by leveraging specialized channel members who can facilitate channel management; and enabling the quick and accurate flow of information up and down the channel network. Whereas basic functions establish the physical framework and operations objectives of the channel, *market functions* manage the mass of customer transactions associated with such activities as title transfer, promotions, negotiation, ordering, finance, and payment.

Whereas the management of the physical functions of the supply channel is critical to overall performance, channel planners have come to realize that the development of channel strategies are equally important in achieving marketplace success. Supply channel strategy can be seen as consisting of two parts. In the first, the basic steps necessary to determine the channel's design and level of complexity and dependency requirements, types of performance expected from channel marketing and logistics efforts, use of other channels, and measurements needed to guide profit and cost decisions are executed. After this *functional* strategy has been drafted, planners must then shift their attention to exploiting the resources and competencies existing in present and potential channel partnerships. There are essentially three forms of channel partnership that can be implemented. In the *multifunctional, within-company* approach, the goal is for a corporation to unite their internal supply channels in an effort to maximize the enterprise's resources and services. In the *next-echelon partnership,* the goal is for functional teams across company boundaries to coordinate directly. In the final and most complex form of channel partnership, *multi-echelon partnerships,* channels of suppliers and customers cooperate to realize optimal response to market requirements and to invent new areas of competitive space.

Another key area of supply channel strategy is exploring the opportunities to be found in the creation of global channels. Global channels enable supply networks to enlarge the boundaries of their marketplace, exploit cost and sales opportunities in new and growing markets, balance production and investment with the differing economic growth patterns and economic cycles occurring across the globe, and profit from global financial systems, communications and media, and market demand homogenization. Creating international channels provides many distinct challenges in the form of government, differing national cultures and values, administrative efforts, benchmarking channelwide performance, and aligning channel competitive value and logistics capabilities.

Converging these often divergent aspects of global channel strategy requires a series of comprehensive activities that begin with the search for new channel coalitions and sources of innovation. The critical part of this stage is to determine total channel capabilities and opportunities to integrate channel member competencies to uncover radically new possibilities. At this point, planners must identify obsolete channel functions as well as untapped capabilities and sources of innovation resident within the existing supply channel. Once this analysis has taken place, the new channel systems that have been constructed can then be implemented. The final step in global channel strategy development is constantly renewing the mix of channel partners and systems to meet the realities engendered by changes to technologies, channel arrangements, and markets.

Notes

1. Ralph S. Butler, H. F. Debower, and J. G. Jones, *Marketing Methods and Salesmanship*, vol. III of *Modern Business*. New York: Alexander Hamilton Institute, 1914, pp. 8–9.

2. Fred E. Clark, *Principles of Marketing*. New York: Macmillian, 1922, p. 5.

3. Louis P. Bucklin, *A Theory of Distribution Channel Structure*. Berkley: University of California, 1966, pp. 3–6.

4. Steven L. Goldman, Roger N. Nagel, and Kenneth Preiss, *Agile Competitors and Virtual Organizations*. New York: Van Nostrand Reinhold, 1995, p. 29.

5. Donald J. Bowersox and M. Bixby Cooper, *Strategic Marketing Channel Management*. New York: McGraw-Hill, 1992, pp. 102–108.

6. Alan J. Dabbiere, "Working in Concert Toward World-Class Logistics," *APICS: The Performance Advantage* 6 (9) (September 1996), 44–49.

7. Christopher Gopal and Harold Cypress, *Integrated Distribution Management*. Homewood, IL: Business One Irwin, 1993, pp. 135–142.

8. For a full discussion on the following issues see Bowersox and Cooper, pp. 15–22.

9. See Louis W. Stern and Adel El-Ansary, *Marketing Channels*, 3rd ed. Englewood Cliffs, NJ; Prentice-Hall, 1988, pp. 11–14; Philip Kotler, *Marketing Management*, 6th ed. Englewood Cliffs, NJ: Prentice-Hall, 1988, pp. 530–532; Frederick Webster, *Industrial Marketing Strategy*. New York: John Wiley & Sons, 1984, pp. 189–190; David F. Ross, *Distribution: Planning and Control*. New York: Chapman & Hall, 1996, pp. 177–179.

10. Andre Martin, *Infopartnering*. Essex Junction, VT: omneo, 1994, p. xxi.

11. Bowersox and Cooper, pp. 431–432.

12. J. Robert Hall, "Top Dog's View of Supply Chain Management," *Inbound Logistics* 16 (6) (June 1996), pp. 23–30.

13. This and other examples of channel cooperation can be found in Hamel and Prahalad, pp. 205–213.

14. Richard J. Schonberger, *World Class Manufacturing: The Next Decade*. New York: The Free Press, 1996, pp. 140–153.

15. See the excellent summary in Rhonda R. Lummus and Karen L. Alber, *Supply Chain Management: Balancing the Supply Chain with Customer Demand*. Falls Church, VA: APICS Educational & Research Foundation, Inc., 1997, pp. 53–55.

16. See the classic definition in Louis P. Bucklin, *Competition and Evolution in the Distribution Trades*. Englewood Cliffs, NJ: Prentice Hall, 1972, pp. 18–31.

17. For an excellent discussion of the traditional channel design process see Louis B. Stern and Adel I. El-Ansary, "Marketing Channels," in *The Logistics Handbook*, pp. 118–137.

18. James F. Moore, *The Death of Competition*. New York: HarperBuisness, 1996, p. 83.

19. Paul S. Bender, "International Logistics," in *The Distribution Management Handbook,* (James A. Tompkins and Dale A. Harmelink, eds.) New York: McGraw-Hill, 1994, pp. 8.2–8.4.

20. Philip Kotler, pp. 388–389.

21. Hamel and Prahalad, p. 30.

6

Supply Chain Inventory Management

The effective management of supply channel inventories is perhaps the most fundamental objective of supply chain management (SCM). Up to this point, the focus of SCM has concentrated on how channel strategies, partnerships, network designs, and operations management plans can provide today's enterprise with the ability to leverage channel network resources to activate business processes and core competencies that merge infrastructure, share risk and cost, reduce design time to market, and exploit technology tools to anticipate and create new vistas for competitive leadership. Although these strategic topics have dominated the discussion, it must be remembered that SCM has an equally important operations side at the core of which resides the management of supply channel inventories.

The basic goal of SCM is to resolve the fundamental dilemma that resides at the very core of inventory management. In the supply channel environment, inventory is necessary to satisfy the sales and revenue objectives of marketing and customer service; however, too much inventory or the wrong inventory is destructive of the well-being of the entire channel. Inventory ties up capital, incurs carrying costs, needs to be transported, and can deteriorate or become obsolete over time. When it is improperly controlled, inventory can become a significant liability reducing profitability and sapping away the vitality of strategic channel initiatives targeted at revitalizing new markets or inventing whole new industries. On the other hand, the value of a properly managed inventory far exceeds its cost. Product availability at the time, location, quality, quantity, and price desired by the customer not only provides benefits in the form of immediate profits but also secures long-term customer allegiance and market segment leadership. When it is effectively controlled, inventory management enables the realization of channel marketing, sales, and logistics strategies and provides the lubricant for the smooth flow of material and service value from supplier to customer.

This chapter explores the elements necessary for the effective management of supply chain inventories. The discussion begins with an overview of the flow and functions of supply pipeline inventories. The goal is to illustrate how buffer stocks staged strategically throughout the supply network can provide a fundamental source of competitive advantage and service value. Next, the chapter turns to an analysis of the principles of supply chain inventory management. The key points focus around maximizing service value, managing channel inventories as if they belonged to a single supply network, inventory and time-based competition, and the use of information as a substitute for inventories. At this point, the chapter turns to a review of channel inventory planning and ordering methods. The use of statistical models and *distribution requirements planning* (DRP) in creating supply channel replenishment linkages are explored. Attention is also paid to continuous replenishment concepts that seek to link the channel through computerized automatic replenishment techniques. The chapter concludes with a discussion of supplier management. Conceptual principles and practical techniques and benefits are detailed.

Elements of Modern Supply Chain Inventory Management

The management and strategic deployment of supply chain inventories is one of the essential pillars of channel competitive strategy. The prime purpose of channel inventory is to provide internal network partners and external customers with the required product in the quantities and at the time and place required. However, whereas there is wide consensus as to the benefits of stocked inventories, there is also universal agreement as to the destructive affects excess inventories can have on total channel profitability. Ensuring the proper balance between the value of holding inventory and the costs incurred requires channel inventory planners to have a complete understanding of the scope, functions, and expected value of inventories as they appear across the supply channel network.

What Is Supply Channel Inventory?

Inventory can be found throughout the supply channel in various forms and in response to various purposes. As illustrated in Figure 6.1, the physical flow of inventory through the distribution supply channel can be said to occur in six stages. The supply pipeline materials flow begins with the acquisition of materials and components. For the most part, this stage is the reserve of product manufacturers who purchase raw materials and component parts that will be later transformed into finished products that ultimately will be sold to the customer. The key inventory management issues during this stage revolve around vendor selection and qualification, contract partnerships, supplier scheduling, transportation and delivery, and vendor performance measurement.

In stages 2 and 3, purchased inventories are transformed into semifinished and finished products through the manufacturing process. Stage 2, termed *primary*

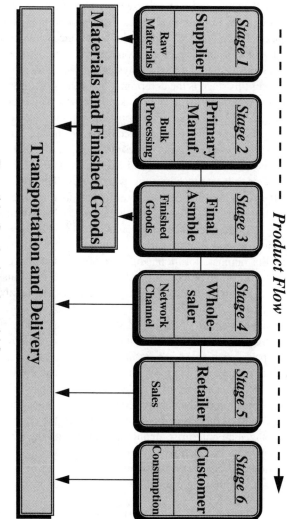

Figure 6.1. Supply channel material flow.

manufacturing, is focused on the conversion of basic raw materials into a wide variety of components and sometimes finished goods through the process of fabrication. An example would be a manufacturer who converts steel rods into nuts, bolts, and twist drills. In stage 3, purchased or fabricated components are *final assembled* into specialized or consumer products. Production processes at this level can follow one strategy or a mixture of three strategies: make-to-stock, assemble-to-order, and make-to-order. The choice of manufacturing strategy is determined by such factors as product characteristics, depth of marketplace demand, strength of the competition, transportation, and warehousing requirements. Critical inventory issues in these two stages focus around implementing JIT and quality management techniques to reduce the occurrence of stocked component and lot-sized inventories and inventory produced by scrap and rework.

Stage 4 marks the beginning of the physical distribution process flow. The most critical part of this stage is the determination of the structure of the channels of distribution by which products will make their way to the customer. As is illustrated in Figure 6.2, there are three possible supply channels that can be employed. In the first, product is sold directly by the manufacturer to the end customer. The second channel type represents a more complex, multiechelon channel comprised of field warehouses and customers. The warehouses could be company owned or they could be public warehouses. This type of channel is much more expensive and difficult to control than factory direct distribution. Finally, in the last type, the channel is characterized by extreme complexity and consists of multiple manufacturing plants, field warehouses, retailers, and customers. Critical inventory issues include channel marketing strategies, total logistics costs, the number, location, and size of distribution centers, channel inventory levels, and order processing and customer service. Finally, in stages 5 and 6, products are delivered to the retail portion of the supply channel for eventual delivery to the customer.

The number and ownership of the above stages of channel inventory flow can vary widely by channel network. In some supply pipelines, the flow is dominated by a single company, such as a Sears or an Abbott Laboratories, who performs most of the functions necessary to move products to the end customer. A wholesaler, such as a W. W. Granger or a SYSCO, on the other hand, occupies a relatively narrow position on the supply pipeline among other channel members who perform specialized functions. In the final analysis, the choice of how many inventory flow stages an enterprise maintains depends on the strategic decision as to *positioning* in its competitive environments. In considering the level of vertical integration and channel positioning alternatives to be pursued by each channel member, the following critical issues would need resolution:

1. What are the channel boundaries that an enterprise should establish over its activities? Large corporations tend to expand vertical integration in an effort to remove channel redundancies and leverage economies

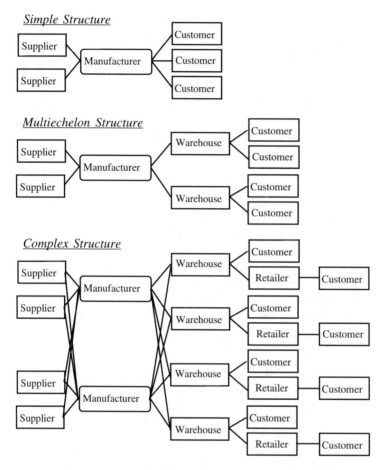

Figure 6.2. Channel structures.

of scale, whereas small companies and niche players require a network of strong players positioned at key points in the network. Key management decisions revolve around the relationship with other firms in the supply channel that are outside of its span of control.

2. Under what circumstances should a channel member change its boundaries or its relationships, and what will be the effect on its competitive positioning? Changes in markets, technologies, channel direction, government regulation, and the status of supply network alliances can alter channel equilibrium and cause inventory costs in the supply pipeline to grow disproportionately in relation to its customer service value. In addition, channel members should always be searching for new opportunities to eliminate inventory costs by either assuming channel

inventory management functions or off-loading to new partners who can perform functions more efficiently and at lower cost.

3. What is the desired balance between inventory and customer service necessary to meet channel marketing objectives? As a general principle, it can be stated that the higher the stocked inventory and the wider the variety of production selection, the higher the customer service level, and vice versa. Although JIT and quick response have provided inventory managers with new techniques designed to increase service while reducing inventories, determining the level of inventory and customer service to be achieved by the firm is a fundamental decision facing today's enterprise.

4. What is the desired balance between inventory investment and associated carrying, transportation, and replenishment costs? As inventory grows in response to better customer service, the carrying cost will grow proportionately. Although inventory costs may vary widely at different points throughout the supply network, the capabilities of the whole channel depends on keeping inventories and associated costs as low as possible. Inventory bottlenecks, wherever they may occur in the pipeline, simply diminish channel flow efficiency and risk market loss.

The Magnitude of Channel Inventories

One of the best ways to understand the impact of inventory not only on a given channel system but on the economy as a whole is to examine inventory statistics. Stocked inventories can represent anywhere from 40% to 70% of a typical enterprise's sales dollar. Effectively managing this huge investment is critical to the financial well-being of the entire supply network. The following figures provide a quick reference to the sheer size of the financial investment in inventory and related logistics costs. In 1985, US firms alone held $896 billion in manufacturing, wholesale-retail trade, mining, construction, and service industries inventories. In 1990, inventory investment had grown over a trillion dollars. By 1995, it stood at $1.2 trillion. This figure was equivalent to 17.2% of the entire US Gross Domestic Product. In addition, US firms spent $473 billion for transportation and administrative costs associated with moving this inventory through the supply channel pipeline. Inventory carrying costs alone accounted for over $311 billion [1].

This growth in inventories has placed a tremendous burden on channel costs. According to management consulting firm, The Colography Group, from 1975 to 1993 the relative cost of carrying inventories measured as a percentage of total physical distribution expense rose by 88.2% while the cost of maintaining warehouses to store those goods rose by 17.6%. If current trends continue, inventory and warehousing expenses could account for as much as 71% of a typical company's supply channel costs by the year 2010. Countering these rising costs is the function of advancing technology and changing transportation methods. A key factor is recognizing that the typical supply network must be diligent in creating different inventory and distribution strategies for high-value and low-value products. By calculating the optimum point where the cost of carrying inventory meets the cost of shipping it, inventory planners can begun to effectively apply different forms of transportation that will match sales with inventory value, including the use of expedited services. "Word-class" supply channels have already begun the search to leverage computerized tools and targeted transportation modes to lower stocked inventories while continuously increase customer serviceability [2].

Function of Channel Inventories

The fundamental function of channel inventory is to act as a *buffer* that decouples the supplier from the discontinuousness of customer demand, on the one hand, and limitations in vendor capabilities to ensure delivery timing and quantity availabilities on the other. Optimally, supply channel nodes would like to carry as little inventory as possible, preferring to move replenishment order receipts directly to the shipping dock to be matched just-in-time with customer orders. In reality, the supply pipeline needs a certain level of inventory to buffer it from the uncertainties of supply and demand.

Inventory control literature has traditionally identified six general functions of inventory to be found in the supply channel: Cycle stock (or lot-size) inventories, safety stock, anticipation inventories, transportation inventories, hedge (or speculative) inventories, and work-in-process inventories. A review of these functions is as follows [3]:

- *Cycle* (or Lot-Size) *Inventories.* This class of inventory is the result of ordering requirements that force channel members to purchase and manufacture inventory in batches that exceed the original demand quantity. The basic reason why cycle stock inventory exists is because of economies realized by trading-off the cost of ordering or producing and the cost of carrying the inventory. In addition, the frequency of item order cycles also may require the stocking of inventory in large lots. As the rate of the receipt of customer orders rises for a given item, so does the frequency of inventory replenishment order placement for that item. Finally, manufacturing functions often will produce lot-size inventory due to the cost of setting up a production line or a machine and gains in productivity attained by producing larger inventory quantities than required.

- *Safety Stocks.* This type of inventory is held in reserve to cover for unplanned fluctuations in customer demand and the uncertainty of supply. If demand remained constant, suppliers could rely on cycle stock to guarantee that there would always be sufficient stock on hand to meet demand. However, because products subject to *independent demand* can expect random periods of above-average demand to occur, cycle stocks function best when there is a safety stock buffer. The amount of safety stock for a given item depends on the degree of random variation in demand, the lead time required to replenish stock, and the service reliability policy established for the item at the stocking point. The larger the safety stock, relative to demand variation, the higher the percentage of customer serviceability. Figure 6.3 illustrates the function of safety stock.

- *Anticipation Inventories.* Often inventories will be built in advance of demand to enable suppliers to respond to seasonal sales, a marketing promotional campaign, or problems in vendor supply. A sporting goods distributor, for example, may begin to warehouse winter sports equipment during the summer months to take advantage of sales discounts from manufacturers and to avoid higher prices and potential stockouts as the winter season approaches.

- *Transportation Inventories.* Inventories in this category can be defined as products intransit (for example, in ships, railcars, truck transport) from one node in the distribution network to another. Transportation inventories exist because *time* is required to physically move stock through the supply channel. Suppliers must plan to have additional inventories on hand to cover demand while inventories are intransit.

- *Hedge Inventories.* Often suppliers will be provided with the opportunity to purchase large quantities of raw materials or components to take advantage of temporarily low replenishment prices or other opportunities. The key elements in purchasing speculation inventories are a knowledge of price trends, risk of spoilage or obsolescence, and handling commodity futures. The utility of hedge stocks is measured by the resulting percent of profit or return on investment.

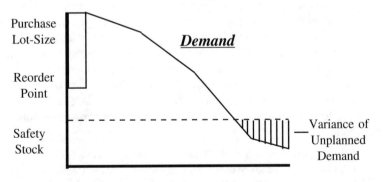

Figure 6.3. Safety stock function.

- *Work-In-Process Inventories.* As the manufacturing practice of mass customization grows in importance, producers have more and more come to see work-in-process inventories as performing a critical role. By building up goods to a certain level of completion and then holding final configuration to the latest point in the supply channel or until a firm customer order is taken, suppliers can significantly minimize the risk of carrying finished products throughout the supply pipeline. Obviously, the key to such postponement strategies is the ability of supply channel nodes to convert generic components and subassemblies and items stocked in bulk quantities quickly and inexpensively into finished products as close as possible to the time and place required by the customer.

The primary function of inventory buffers are to ensure product availability to meet customer demand as it occurs in the marketing channel while minimizing total inventory carrying costs. The existence of repositories of inventory located at strategic pipeline points provide essential buffers protecting network nodes from the occurrence of unplanned variance in demand and supply originating from other channel partners or to enable them to exploit quick-response and value-added processing opportunities. Where inventory creates customer value and provides cost savings to channel enterprises, it is an *asset.* On the other hand, channel inventories contain several serious drawbacks. To begin with, buffer inventories can potentially incur more cost in the form of obsolescence, material handling and storage, and low cycle time than the value they create. Second, they hinder the speed by which products move through the supply network on their way to the customer. Finally, buffer inventories mask the true nature of channel demand and supply, conceal channel inventory management inefficiencies, and gloss over costly channel inventory and capacity imbalances. Regardless of their utility, effective inventory management requires the whole channel to continuously search for new ways to diminish or eliminate buffer inventories altogether. Poorly managed buffers are a serious liability sapping the financial strength of the entire channel and limiting opportunities to expand channel system competitive leadership.

Understanding Channel Inventory Value

One of the central tenents of JIT is that all forms of inventory are undesirable and should be eliminated. In reality, as detailed above, this ideal is almost impossible to achieve. In today's business environment, companies can expect to continue to experience uncertainties in supply and demand caused by seasonality and rapid and sudden changes in the marketplace, breakthroughs in technologies, lower transportation costs for bulk quantities, and disconnection in channel planning, all of which seek to use inventory to cover for discontinuities in demand at multiple levels in the channel pipeline. However, simply because these elements seem to be a necessary part of the manufacturing and distribution environment does not mean that individual companies and the supply channel as a whole should not seek to continuously eliminate or at least diminish the conditions that cause inventory to occur in the first place or to ensure that, when necessary, only minimum levels of inventory are stocked.

Because inventories in some form or another are sure to exist at various points in the supply channel, the goal of effective channel inventory management is to ensure that stocked inventories are a continuous source of competitive advantage. Simply stated as a fundamental postulate, the challenge of effective channel inventory management is to identify new methods in which the ratio of value-added to cost-added elements of stocked inventory can be continuously improved. In its most basic form, channel inventory management can then be defined as the purposeful deployment and redistribution of raw materials, component parts, work in process, and finished products across an integrated supply network for the purpose of providing value to the customer.

The actual measure of the performance of channel inventories can, therefore, be determined by understanding the two separate yet correlative meanings associated with the concept of "inventory value." For the supplier, inventories are valuable to the extent to which the properties of cost associated with them diminish as their value increases. That value can be simply stated as the level of satisfaction attained by the customer as measured by the availability of products to fulfill the immediacy of customer demand, the ability to deliver products completely and accurately at the agreed-upon time, and the length of the cycle time it takes to process an order. Customers, on the other hand, perceive the value they receive from the goods they purchase as providing them unique solutions to their needs or desires or enabling them to pursue new opportunities, the benefits of which far exceed the original cost of the purchase. In this light, the question becomes not how much inventory is stocked throughout the supply pipeline, but what customer-satisfying processes are being used to increase its value as a source of competitive advantage while decreasing its cost.

The principle of inventory value can be seen in thousands of businesses that find the value of stocked inventory provides them with an unbeatable source of market leadership. Take, for example, the McMaster-Carr Supply Company, a billion-dollar global distributor of industrial products. The company consists of 4 mammoth sales and distribution centers with over 1000 experienced professionals. At any one given time, the company stocks over 175,000 separate items at a cost in the hundreds of millions of dollars. The value of this huge investment in inventory, and the services and physical assets necessary to support it, can be found in McMaster-Carr's unswerving commitment to ship a customer's order, regardless of the size or value, within 8 working hours of receipt. The fact that the company can boast that it has achieved a 97% success level in meeting this target bears testimony to its keen ability to manage enormous physical assets, complex channels of distribution, and unparalleled customer service to achieve significant value for its customers and for its continued business success.

On the other hand, for most manufacturers the ceaseless search to continuously wring inventory out of every production process is the result of an entirely different way of looking at inventory. McMaster-Carr sees the value of inventory as enabling them to respond almost instantaneously to the customer's requirement. Manufacturers, on the other hand, see inventory as providing value only at the moment it is needed in production. Manufacturers respond to the customer not by having inventory available, but by having flexible production processes capable of changing quickly, with little notice and the bare minimum of confusion and difficulty. For example, Baldor Electric, an Arkansas manufacturer of electric motors, has engineered production processes that permit it to profitably make thousands of motor models in small quantities. The plant is a marvel of kanban devices and quality techniques expressly designed to limit inventory queues and facilitate throughput at each stage of production. The entire manufacturing cycle time is reported at a stingy 5 days [4].

Obviously companies such as Ford, Deere and Company, TRW, Eaton, and PepsiCo have radically different perceptions of the value of inventory than do McMaster-Carr, W.W. Granger, and Wal-Mart. Manufacturers who methodically measure long-range competitiveness by the frequency of their inventory turns would collapse under the cost of maintaining a huge inventory of materials and components "just in case" they might be needed at some future time. In contrast, distributors would quickly go out of business if their customers had to wait 5 days for delivery of a much needed item. The real issue at hand is not whether companies should strive to totally eliminate their inventories, but rather what is the value that a specific level of inventory can contribute to a company's competitive positioning. Managing this inventory value is, therefore, the proper function of supply channel management and the central focus of the SCM concept.

Principles of Supply Chain Inventory Management

In the past, the overriding objective of the supply channel was the creation of mass-production and mass-distribution infrastructures that acted as a conduit for the flow of standardized goods and services from the manufacturer to the mass market. Based on the assumption that the marketplace consisted of a few archetypal customers, whose needs and desires had been determined by marketing analysis, the supply system's role was to push noncustomized products utilizing mass-production-era advertising, media, and distribution channels out to the consumer. Buyers then searched available suppliers to locate product and service offerings that came closest to matching their needs or desires. In such a supply system, consumers had little linkage with producers whose inflexible manufacturing systems focused on moving large, unsynchronized batches of rigidly defined products through manufacturing and distribution processes. Custom-made products, where there was participation by the customer in the design and manufacturing process, were more expensive and the reserve of specialty shops who utilized costly production methods.

In today's global marketplace, the demands on the supply channel to streamline inventory flows and cut inventory costs while improving customer service has radically changed traditional concepts of channel inventory management. It is not that the potential found in mass markets has disappeared. In fact, the growth of relatively unrestricted regional trading blocks, such as NAFTA (365 million people, $5,900 billion GNP), EUC (350 million people, $5,000 billion GNP), and China (1.2 billion people, $393 billion GNP), point to the fact that global trade is capable of generating unprecedented and massive demands. What has changed, however, is the way productive processes are being organized to leverage the dramatic decline in product life cycles, velocity of customer response, universal commitment to quality, and ability to customize products at the last stages in the distribution cycle to create unique solutions that provide superior value to the customer. Unlike the passive supplier-customer relationship of the past, today's dynamic supply channels demand a high level of interaction and close linkage between producer and consumer to ensure the production and timely delivery of premium, customized goods based on the configurability of a declining number of mass-produced components.

In responding to today's global business environment, supply channels must understand the following four critical channel inventory principles:

1. All channel inventories must be managed to maximize service value.

2. Channel inventories must be managed as if they were part of a single distribution network unit, not the management of inventories at each channel supply node.

3. The key to successful channel inventory management is the management of the time it takes products to move through the supply pipeline on their way to the customer.

4. The availability and manipulation of timely information about customer demand, inventory status, and replenishment supply is the best substitute for repositories of inventory stocked throughout the channel.

Inventory as a Source of Service Value

Supply channel systems stock inventory in order to optimize service opportunities. When channel inventories are properly positioned in the right quantities and at the right places in the supply network, they can provide an unbeatable source of service value. In yesterday's marketplace, supply channels counted on flagship products and brand names to provide them with competitive advantage. Although products and brands are still critical, today's customer also requires their suppliers to possess flexible and agile productive and distributive processes enabling them to expand the range of the customized and private products they offer their customers as well as increase value-added services such as reliability, low cost, and timely delivery. Increasing these service value offerings requires supply networks to tirelessly search for new avenues of channel efficiencies to increase the velocity of the flow of products and transactions as they make their way through the supply pipeline.

Channel inventories can assist in providing value to the customer through the following five service elements:

1. *Lowest Cost for Value Received.* Reducing inventory and other process costs enable companies to maintain market leadership by keeping prices low, ensuring depth of product assortment and quality, and expanding on capabilities to mass produce customized products. Some of the techniques used to achieve this service element are creating channel and cross channel partnerships that shrink inventory buffers and inventory flows, the use of alternate channel formats like warehouse/wholesale clubs, power buying, and manufacturer direct, and the pursuit of JIT contracts that guarantee fixed prices and service levels.

2. *Improved Channel Efficiency.* By removing excess supply network buffers, reengineering distribution processes, implementing JIT techniques, and streamlining inventory flows, supply channels can significantly diminish total pipeline costs while ensuring the right product is in the right place to capitalize on marketplace opportunities. Above all, service efficiencies increase product access. Access means the degree of ease by which customers can purchase products or contact sales and service functions. Access can also mean the availability of goods within a time limit generally accepted by the industry. Finally, access can mean the speed by which after-sales replacement parts and services can be delivered. Customer convenience and access to goods and services are fundamental to competitive advantage.

3. *Improved Quality.* Reducing the occurrence and resulting costs of inventory stock-outs, product defects, order fill inaccuracies, and other related inventory management errors can significantly decrease operating costs while increasing customer service. The key issue is service reliability. Market leaders must continually perform the promised service dependably and accurately each and every time. Reliability of service permits channel suppliers to "lock in" their customers who will gladly pay premium prices for delivered quality. The implementation of *total quality management* (TQM) techniques are essential in gaining and sustaining high quality levels.

4. *Supply Network Simplification.* Removing inventory flow bottlenecks and redundant channel functions simplifies and makes all supply channel activities transparent. Increased visibility increases the capability of channel suppliers to be more responsive to the demands of the marketplace. Simplification can take the form of intense process reengineering at the enterprise level, the discontinuance of channel functions whose costs exceed their value-enhancing capacities, and the use of contract services for inventory and transportation management and accompanying transaction processing.

5. *Improved Channel Inventory Information.* Accurate and accessible information concerning products, stocking levels, location, and order status is the primary tier of customer service value. Internally, information enables network suppliers to control inventory levels, ensure timely and accurate stock replenishment, and leverage price and delivery economies from upstream sources of supply. Externally, information reduces the occurrence of missed customer activities that have a rippling effect back through the entire supply channel.

The task of improving channel inventory service value has been significantly assisted by information technologies and new operations management methods. Andersen Windows, for example, had, until recently, considered itself a mass producer of standard windows manufactured in large batches. As demand for customized windows grew in the early 1990s, Andersen was faced with several serious customer service issues concerning design, pricing, production, and delivery. End products had mushroomed from 28,000 in 1985 to 188,000 a decade later, lead times and supporting inventories were excessive, and delivery error was on the increase. The solution was the implementation of an integrated information system that linked the showroom with the factory, permitted customers to easily design their own solutions, and provided instant pricing. Today, Andersen's next step in mass customization, termed "batch-of-one manufacturing," is targeted at expanding production flexibility and customer service while dramatically reducing the inventory of finished window parts. Currently, all custom windows are built using some standardized components. However, with "batch-of-one," everything is entirely made to order. In the company's prototype plant in St. Paul, MN, no inventory is kept on hand, except for raw Fibrex (a wood and vinyl composition) and some hardware. Only a month is required between receiving an order and installing a finished custom window. By reducing inventory costs and increasing service values, simplifying ordering, production, and inventory control, improving delivery quality, linking sales and manufacturing, and reducing missed opportunities by increasing the accuracy and availability of information, Andersen has utilized new information and management techniques to gain significant improvement in all five of the service elements detailed above [5].

One Inventory in the Supply Channel Network

Perhaps the most serious impediment that has limited the effective use of supply network inventories is the long-standing practice of considering each channel inventory node as occupying an independent position in the supply pipeline. Channel managers have traditionally focused solely on the costs and performance of each stocking location, meticulously measuring inventory costs, turnover ratios, and service levels. Whether the channel consists of a single corporation that passes product through a series of semiautonomous divisions or a chain of separate companies or a combination of both, the result has been the same: Inventory is planned, managed, and measured at each channel node without reference to the plans of other channel members. Because few companies have had the foresight to look beyond their own boarders and work together as a single integrated supply network, little effort is spent ensuring that inventory value is being increased for the good of the entire distribution pipeline and not just for the benefit of a single position holder in the channel. Such a disconnected and myopic view means that the real demands of the customer and downstream supply partners become clouded and lost at each channel supply node as critical information concerning actual inventory requirements becomes distorted by local operational objectives and performance priorities.

The collective impact on channel inventories caused by this uncoordinated, disjointed management system can be dramatic. As visibility to actual channel needs is lost, demand uncertainty tends to be covered by adding inventories at multiple levels in the pipeline. Today, most channel entities attempt to solve uncertainty by generating their own forecasts and determining inventory levels independent of the demand and supply requirements of other channel partners. What is worse, not only is each network nodal forecast uncoupled from other channel forecasts, but unpredicted shifts in past demand levels have taught each supply node to stock extra inventory in the form of safety stock as a hedge against uncertainty. Masked channel demands, excess supply node buffers, modulating overcapacity and undercapacity conditions, poor coordination—all these factors add cost to the inventory value ratio, inhibit the effective allocation of inventory to meet customer demand wherever and whenever it occurs, and undermine the process of adding value to inventory as it flows through the supply channel system to meet customer demand.

In order to respond to the increasing velocity of satisfying customer requirements in the 1990s, many of today's world-class supply channel networks are abandoning the fragmented view of pipeline inventory in favor of a new operations paradigm that considers the management of all channel inventories from supplier through to the customer as if they belonged to a single integrated supply network. Pioneering companies, such as General Motors, Kellogg, and Nabisco, are gaining significant strategic and competitive advantage by linking suppliers and retailers in tightly integrated networks extending from raw materials to the point of actual customer delivery. In other industries, from health and beauty care to consumer electronics, channel members have increasingly sought to implement quick response (QR) systems that dramatically reduce inventory carrying costs and increase product flowthrough. The concept has spread to mass merchandisers and price clubs whose life blood depends on smoothly operating distribution systems to shrink or eliminate costly inventory buffers and radically compress the time it takes for critical replenishment inventories to be delivered [6].

Viewing channel inventories as if they were a single integrated supply function is the foremost challenge of channel inventory management. Physically, inventory can be found in various forms and at various stages anywhere in the supply pipeline. However, the mere presence of inventory scattered among channel supply repositories provides little visibility as to how that inventory is to be properly configured and delivered to exactly the right distribution node in time to satisfy customer demand. If the goal of attaining and sustaining channel inventory service value is to be realized, then it is imperative that pipeline managers seek to realize the following goals:

- *Channel Strategic and Process Integration.* Not just point-of-sale nodes but all channel strategies and processes, beginning with suppliers and manufacturers and progressing through wholesalers and retailers, must be integrated with the needs of the marketplace. Achieving strategic and tactical integration is, by far, the most difficult of the challenges facing channel constituents.

- *Increased Flexibility.* The effective management of inventories requires flexible and agile processes that accelerate and add value to materials as they flow through the network. Flexibility permits channel members to reduce finished goods buffers by postponing product differentiation to the lowest level in the channel and to exact customer order specifications.

- *Lower Costs.* By considering all channel inventories as functioning as a single supply pipeline, unnecessary volumes of stocked inventories can be removed throughout the network. In addition, increased postponement capabilities will also shrink costs by enabling supply nodes to reduce finished goods overstocks and distribution point stocking imbalances while increasing product variety.

- *Faster Response Times.* The goal of the supply channel is to satisfy customer requirements through the timely delivery of products. Every day that inventories spend in the pipeline adds carrying costs. Every day of lead time required to get the right product in the right place means slower response to customer requirements. As the importance of delivery speed in today's global environment increases, the combination of high costs and lack of responsiveness risks competitive disaster.

- *Telescoping the Supply Pipeline.* World-class supply channels pay close attention to the length of the supply pipeline. As channel networks grow in length, so inevitably do transit times and buffer inventories. Correspondingly, as the size of inventories grow, so does total pipeline length. Buffer inventories have traditionally been used to protect upstream and downstream stocking nodes from fluctuations in demand and supply. In contrast, today's market leaders seek to continuously shrink pipeline size and shave time and inventory from the channel network through the use of JIT, supplier management, and information technology techniques.

- *Channel Performance Measurements.* Metrics that document independently the performance of each channel supply node yield little valuable information about the performance of the channel as a whole. As all network supply nodes are inextricably bound together in the management of the flow of value-added products to the customer, they should all share the same customer satisfaction goals.

Achieving actualization of the concept of a single-channel inventory requires the elimination of blockages and fissures in the network pipeline that lead to excessive inventory buffers and lengthening lead times. Several key examples come to mind. Manufacturing and value-added processing often have excessively long setup and changeover times that result in unnecessarily large lot sizes and accompanying lack of responsiveness. Another area is the presence of bottlenecks at key transit and consolidation points in the pipeline. Bottlenecks inhibit the velocity of the supply channel flow and are a fertile area for inventory buildups. Inadequate visibility as to the true nature of demand and supply occurring at the distribution ends of the pipeline provide another source of pipeline blockage. Poor visibility enables non-value-adding activities to increase, creating sources of hidden costs and decreasing response time so important to exploiting new marketplace opportunities. Finally, poor information can hide channel demand and supply requirements. Often the replenishment ordering systems used by supply channels are the culprit. As demand enters the channel, requirements are concealed by supply nodal safety stocks as they are passed serially down through the supply network. Actual inventory requirements are invisible to the next supply partner until the reorder point has been triggered.

The advantages of working toward the realization of the concept of a single-channel inventory can be demonstrative. In 1994, Amoco Petroleum Products' Product Supply and Logistics Department began a program designed for overhauling its supply channel has helped the company reduce inventory carrying costs by 10%, increase service factors, and improve on delivery reliability. Amoco sees supply chain management as more than just logistics. Each segment on the value chain is interlocked, and true supply chain integration is understanding how each channel partner links to other channel members. The company's computerized scheduling software, OPDSII (Operations Planning Data System) connects internal and external suppliers (refiners and pipelines), distribution centers, and marketing and inventory management departments. The system displays lifting and closing inventory across the nation and enables marketing to change the forecast in response to market conditions and inform the entire supply chain of the changes. The concept of a single-channel inventory system has enabled Amoco to pursue its objective of supplying the best value-added service to its customers in the industry [7].

The channel inventory experiences of Amoco and other industry leaders have radically altered the traditional view of pipeline inventories. Working toward creating a single-channel inventory vision can provide the following benefits:

- Instead of a combination of many different and often discordant supply node inventory plans, a single view of channel inventory permits network members to develop a single strategic inventory plan that will govern total channel inventory investment and channel stocking levels.

- A single inventory plan provides visibility to inventory requirements for channel members on all levels of the supply pipeline. Better and more responsive inventory planning will result in lower total working capital and lower total operating costs while meeting targeted service levels.

- Because all channel constituents can now follow common plans that permit them to integrate their inventory and operations capacities, the whole channel can better pursue highly competitive customer service values.

Time-Based Competition

Almost 40 years ago, in a pioneering article in the *Harvard Business Review*, Jay W. Forrester developed a model which demonstrated the impact of time on the flow of products and information through a supply channel system. In the article, Forrester tracked the effects of time delays and decision rates within a simple business system consisting of a factory, a factory warehouse, a distributor's inventory, and retailers' inventories. By adding the time it took demand information and actual supply inventory to cycle through the supply system, Forrester calculated that it would take about 19 weeks for replenishment goods to move from factory to the point of customer purchase. However, if an unexpected and temporary 10% rise in demand occurred, the ramping up of productive processes that would follow would take the supply system more than a year to return to its previous position of stability. What distorted the system so badly was *time*: the delay between the time an event occurred and the time when the supply system received and was able to act on the new information. The longer the delay, the more distorted the picture of marketplace needs becomes, producing confusion, waste, and inefficiency [8].

In today's highly competitive marketplace, the ability of a supply channel system to both deliver and react to change is measured not in weeks and months, but in days and sometimes hours. As product life cycles shorten, customers seek to reduce all forms of inventory, and markets become increasingly more volatile, supply channel systems have become more acutely aware than ever of the need to compete through time-based management. In the past, supply channels countered demand variances by identifying key supply nodes, increasing buffer inventories, and searching for ways to gain cost economies. Today, world-class supply networks have determined that the best approach is to reduce the amount of time the supply network consumes when moving product and critical information through the distribution pipeline. Strategies focused on time, such as flexible manufacturing and rapid-response systems, QR and efficient consumer response (ECR) in retailing, and increasing process innovation, have become critical operating concepts for competitive survival. Supply channels strive to be as close to their customers as possible, utilizing new tools such as the Internet and mass-media communications. New organizational models and management techniques today depend more on providing fast response rather than low cost to attract the most profitable customers.

At the core of time-based competition can be found the management of supply channel order lead times. Network order lead times can be viewed from two perspectives. The first, *customer order* lead time, is concerned with the time it takes to process a customer order from the moment demand is identified to the time it is delivered. Customer demand can be regarded as the prime mover that sets the whole logistics process into motion. As such, the quality, speed, and accuracy of the order processing function will have a fundamental impact on channel competitive positioning. Order processing functions that provide for the speedy and accurate transference of goods and marketing information will facilitate the customer service function and serve as the foundation for competitive advantage. Effective customer order lead-time management requires efforts to continuously remove bottlenecks, inefficient processes, and fluctuations in the volume of orders.

In addition to customer orders, it can be argued that the scope of lead-time management should be expanded to encompass the *internal* orders that drive inventory replenishment through the entire supply network. Internal orders can be defined as purchase orders, manufacturing orders, interbranch orders, and channel partner resupply orders. Internal order lead times are composed of a series of sourcing and procurement activities that begins with the identification of a replenishment requirement and concludes with receiving and order closeout. The objective of internal supply orders is to ensure the existence of targeted stocking levels of inventory throughout the supply channel. Procurement lead time is measured by calculating the number of days of inventory in the pipeline, whether it be raw materials, components, work in process, or finished goods, as well as time spent in-transit between channel nodes, in manufacturing, and in value-added processing.

Despite the organizational logic of viewing customer and internal supply orders

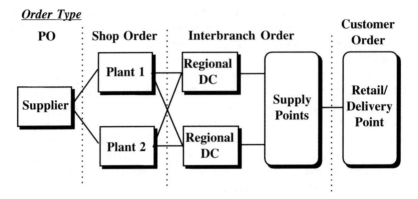

Figure 6.4. Order replenishment chain.

as separate from each other, it has become increasingly apparent that the lead times associated with each are so intertwined that they should all be managed as if they were extensions of a single process. As illustrated in Figure 6.4, the requirements for quick response to customer demands mandates that procurement and production orders be seen as the beginning stage of a continuous process of inventory lead-time management that ends with the customer order shipment. The need for close integration between these order types is driven by the following realities:

- "World-class" customer satisfaction means that not just the demand management function but the *whole* supply channel must be focused around superlative product delivery at the lowest cost.

- The timing of inventory procurement and interbranch product availability must be driven by the customer, not by inventory plans uncoupled from actual market demand. In this sense, it is actual customer demand that *pulls* products through the channel network just in time to meet customer requirements. Uncoordinated channel plans that do not connect suppliers and customers result in a *push* approach that adds excess inventory buffers and extends pipeline lead times.

- Well-coordinated and timely response to channel inventory demand has become as fundamental a competitive benchmark as product quality. Supply channels that can integrate the diversity of the supply network around a single standard of total customer service will always retain the competitive edge.

The demands of customer quick response require each supply channel member to investigate how it can contribute to reducing pipeline lead times and, by extension, total inventory cost. In way of summary, some of the most important methods of reducing pipeline inventory lead times are the following:

- Closely integrating supply channel replenishment and customer service strategic and tactical plans
- Reducing administrative delays by streamlining paperwork and operations procedures
- Reducing the size of the pipeline
- Eliminating channel network bottlenecks
- Pushing the customer order point as far down the channel as possible in order to prevent premature product differentiation.
- Shrinking batch-size and lot-size production practices
- Implementing continuous process improvement measures designed to eliminate demand and supply uncertainties.

The fundamental goal in achieving customer pipeline lead-time reduction is the creation of channel capabilities that enable quick recalibration of inventory and related logistics capacities to support the total channel service values demanded by the marketplace.

Substituting Information for Inventory

One of the fundamental postulates of supply channel management is that as uncertainty concerning the status of demand and supply grows, so does pipeline inventory. In fact, more than 50% of the inventories held by the average company exist as safety stock to cover for the lack of knowledge concerning what is really needed to respond to channel demand requirements. This lack of demand visibility cascades through each level of the supply network, adding buffer stocks at each channel node. Raw material and component suppliers keep excess inventories on hand because of the inability of manufacturers to provide detailed information about their future material requirements. In turn, manufacturers stock excess components because of poor material requirements planning and shop-floor scheduling, large lot sizes, lumpy demand, and unplanned scrap and rework. Finally, distributors and retailers are forced to stock added finished goods to prevent from being blindsided by radical shifts in customer demand. The solution to breaking this cycle of using inventory as a means to counteract uncertainty is to increase the timeliness and bandwidth of the information about what products are really needed and when across the length of the entire supply channel system. The practice of substituting inventory in the place of clear information about total channel network demand has become a sorry anachronism in the midst of what has become the Information Age.

In the past, supply channel nodes rarely communicated with each other concerning inventory needs except when a replenishment order was launched. Today, breakthroughs in information and communications technologies provide channel partners with the potential to share demand and supply information across the breadth of the entire pipeline in a "real-time" mode. The existence of planning, scheduling, and communications systems capable of processing massive amounts of data very quickly and cost-effectively reflects a radically new way of managing the supply channel. The principle is simple: The more information channel distribution nodes have about total supply network needs the better they are able to produce products to respond to the *pull* of demand requirements in the quantities and at the time they are needed. As the level of demand uncertainty decreases, so does the need to stock safety inventories. In this way, timely and accurate supply channel information concerning customer and interchannel demand becomes a substitute for inventory buffers.

The key to the whole system is the integrative capability of information technologies to provide channel nodes with critical data about what raw materials, components, and finished products are needed, and when they need to be received. According to Andre Martin [9], today's computer business systems enable whole supply channels to pursue high levels of network integration by providing answers to the *universal supply chain replenishment equation:*

- What is the channel going to sell?
- Where will it be sold?
- What inventories are available in the channel?
- What is the status of replenishment orders at each supply node?
- What open requirements can be found at each supply node?

Conceptually, Figure 6.5 illustrates the network decision linkage. At each level in the supply channel critical decisions concerning each of the questions constituting the channel replenishment equation must be made and translated into demand and supply requirements necessary not only to satisfy local channel needs but also those of successor and predecessor network partners. Ultimately, as the information linkage draws each supply node ever closer together, lead times and buffers shrink as only the inventory necessary to meet channel needs is made and transferred through the channel.

For such a supply network to work, two conditions are necessary. The first centers around the implementation of common computer system communication and application processing capabilities. The commercial availability of business planning and control information systems, such as enterprise resource planning (ERP), logistics information systems (LIS), and distribution requirements planning (DRP), provide "real-time" visibility to demand and supply conditions throughout the supply network. These systems enable channel nodes to utilize common scheduling and simulation applications capable of calculating current and projected demand requirements and demonstrating the impact customer, production, and inventory stocking decisions will have on the costs and profitability of the entire pipeline. In addition, data-collection systems featuring bar codes, radio frequency, and voice recognition will enable supply links to process inven-

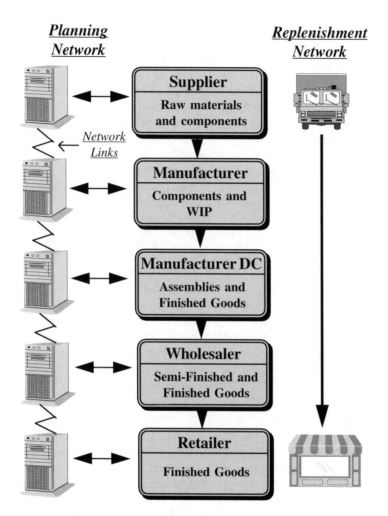

Figure 6.5.   Channel information and supply network.

tory movements at blinding speeds. Finally, new data communications technologies, such as wireless and fiber optics, digitized transmission, and EDI via Internet and Intranet, will reduce information transfer to seconds, further cutting lead time out of the pipeline and networking channel constituents ever closer together.

The second condition necessary for the creation of highly networked supply pipelines is the availability of repositories of critical information concerning channel customers, inventory levels, and costs. The following are key information elements:

- *Total Channel Customer Requirements.* This element consists of a broad range of information relating to the nature of what products are needed and how they are to be delivered to each customer. The content of this information must, in detail, show the exact requirements of both external and internal customers concerning which products and groups of products are needed and when. It must also provide enough data for the creation of an aggregate view of the total range of requirements placed on the supply channel as a whole. Critical sources of customer information can be gathered from the following: *product requirements information* which includes data concerning specific individual products, product families, and product configurations; *product cost information* which includes pricing, contract, and value-added cost data; *product ordering information* which includes product availability, volume and mix, order minimums, and order and communications data; *product delivery information* which includes delivery and cycle time, receiving, completeness requirements, return options, and routing data [10].

- *Replenishment Requirements at Each Channel Level.* Since inventories exist at virtually every level in the supply channel, visibility to inventory requirements as product is pulled by customer demand through the supply pipeline is one of the most critical sources of channel information. Inventory replenishment information is composed of the strategies guiding inventory cost/customer service trade-off objectives, visibility to the actual physical stocked inventory, inventory ordering methods such as statistical order point and DRP, and the accuracy of forecast elements that govern the collective management of all channel inventories. In addition, this information must be in a form that can be communicated rapidly and accurately to all network supply nodes.

- *Availability of Logistics Capacities.* Information concerning the available capacities possessed by each channel supply node is an often overlooked, yet critical element of channel information. Essentially, information in this area can be divided into data concerning the financial capacity of channel members to absorb levels of inventory cost, capacity of transportation modes to respond to intrachannel order replenishment movement, capacity of warehouse space in detail and in aggregate to hold stocked inventories, and capacity of labor and equipment for such activities as receiving and putaway, and picking and shipping. Although extremely difficult to compile, information concerning supply channel capacities is critical in avoiding bottlenecks that prevent the smooth and timely flow of inventory through the supply pipeline.

- *Total Network Performance.* Network performance information is composed of a matrix of data concerning customer service issues such as order cycle time and fill accuracy, inventory measurements such as accuracy, turnover, and cost, and delivery metrics such as timeliness, availability, and cost. The essential role of network performance information is to provide channel members with data to more effectively manage operations uncertainty arising from functions *internal* to the supply network. Channel operations uncertainties are composed of timing issues, such as lead times and ordering cycles, and accuracy issues, such as availability and ordering information.

Managing Supply Channel Inventories

Inventory constitutes perhaps the single largest financial investment to be found in the typical supply channel. The task of inventory management in controlling this enormous investment requires the close cooperation and communication of all levels of the supply pipeline. Effective inventory management permits each firm, and by extension the entire supply network, to pursue targeted customer service levels and profitability objectives and to leverage capabilities and core competencies to sustain marketplace leadership. The key to achieving channel inventory objectives is determining and making visible network demand so that planners positioned along the channel can replenish stocks in the optimum quantities to respond just in time to the upstream pull of customer demand. The goal of the whole process is to facilitate the actualization of customer service objectives while executing inventory planning and control functions designed to achieve least total cost and high productivity for inventory, transportation, warehousing, and the work force.

Basis of Channel Inventory Management

In the ideal world, there would be no need for buffers of inventory staged at strategic nodes in the supply channel. Customers would be able to receive the goods they wanted simply by triggering ordering mechanisms that would define desired finished goods configurations, transmit the exact product specifications to the factory where they would be made, and arrange for delivery as close as possible to the moment of order request. In reality, channel inventories are a critical element of the demand-satisfying process. Inventory can be found in the supply network for a variety of reasons. As discussed earlier, inventory buffers can exist because of batching or lot-size economies, timing issues such as geographical movement or the duration of the manufacturing process, planned overstocks due to seasonality or speculation, and availability uncertainties due to variances in inventory demand and supply.

Although the nature and quantity of inventory buffers will vary depending on the positions they occupy in the supply network, the content of stocked buffers can be said to satisfy two critical functions. The first function, termed *cycle* or *working stock,* can be described as inventory that provides the firm with the ability to respond to the average level of customer demand occurring during the period between replenishment order release and receipt. The second function of inventory, termed *safety stock*, is to provide additional inventory that is added to cycle stock in the event of variance in the normal distribution of demand. The critical factor is determining what is to be the size and location of cycle and safety inventories in the channel. This is a difficult task that requires the cooperation of all channel members who must now view their inventories as part of a collective inventory pipeline that extends from the raw materials supplier all the way out to the retail point of sale.

When developing strategies for channel inventory management, network partners must focus their attention on providing answers to two fundamental questions:

1. How can the causes that produce pipeline cycle and safety inventories be diminished or eliminated altogether?

2. When it is found that causes of pipeline inventory buffers cannot be eliminated, what is the optimum quantities of cycle and safety stock inventories necessary to sustain channel service levels?

In general, members of "world-class" channels continuously seek to examine their own inventories and cooperate with other pipeline partners for ways to reduce or eliminate buffer stocks, particularly at interorganizational and interchannel transaction interfaces. The following are methods that channel constituents can use in developing programs designed to ensure optimal levels of buffer inventories.

*   *Channel Demand Uncertainties.* Reducing demand uncertainties is perhaps the most fertile source for inventory reduction. Uncertainties in demand exist because of the inherent randomness involved in the marketplace. As demand variance at the point of sale occurs, the unevenness of requirements sent up the supply pipeline tends to result in predecessor network supply nodes increasing their cycle and safety inventories in an effort to prevent stock-out. As these inventory decisions cascade up the pipeline, demand lumpiness tends to escalate causing even higher safety inventories to be stocked. What is worse, intersecting supply channels, which service customers at any point along the supply network, are also infected with the demand uncertainty of their customers' customers, who, in turn, pass it on to their own upstream supply partners.

    Although it is impossible to eliminate demand uncertainty at the point of sale, today's technology has made it possible to significantly mitigate the impact of channel demand variance. The real culprit in the process has been the practice of each node in the channel developing their own forecast. The fact of the matter is that only the retail or point-of-final-sale channel node needs to forecast. The forecasts created by predecessor channel partners are not only unnecessary but actually harmful, as each channel level tries to outguess the requirements of each other. The key is identifying the *nature* of the demand in the system. Channel members at the end of the pipeline are subject to *independent* demand and need to apply forecasts in determining future requirements; all downstream suppliers are subject to *dependent* demand and should not forecast but merely calculate the needs as specified in their customers' total requirements. By creating closely integrated information partnerships, companies are finding that "real-time" data about requirements has reduced or eliminated upstream channel forecasting. According to the research of Martin [11], a number of top retailers and manufacturers have already developed partnerships that have eliminated 80% of their need to forecast for customers, and he projects by the year 2000 that a substantial number of wholesalers and manufacturers will have eliminated the need to forecast altogether.

- *Channel Supply Uncertainties.* Supply uncertainties arise due to replenishment response variances in the operational effectiveness of suppliers or to unforeseen delays in delivery of transportation functions. Reducing or eliminating supply variances can come from several sources. The development of information partnerships in which critical demand data can be passed "real-time" down the supply pipeline and notice of shipping data passed back up can significantly reduce supply variance. In addition, TQM and continuous improvement programs can assist in reducing errors that contribute to supply uncertainties. The same tools can also be applied to transportation. Following the lead of industry leaders such as FedEx and UPS, JIT techniques and the use of computerized tools such as EDI, shipment tracking, and the establishment of delivery appointment scheduling has significantly reduced delivery uncertainty.

- *Lot-Size Economies.* Lot-size requirements are a common phenomenon in manufacturing, materials handling, and transportation. Often minimum stocking requirements governed by box, container, pallet, truck load, and minimum production sizes dictate that product be transacted in batch rather than one at a time. The problem with lot sizes is that even when demand for a product is level, batch requirements will create demand that becomes increasingly lumpy as it works its way through the supply network. Solutions to the problem of lot sizing begin in manufacturing, where JIT and mix-mode production techniques provide the capacity to change over quickly from one product to another, thereby enabling manufacturers to run a mix that exactly matches customer requirements on a daily basis. JIT techniques can also assist in reducing transportation batches through the use of milk runs, consolidating several products from multiple sources for delivery to a single destination, and cross-docking.

- *Promotions.* Promotional campaigns, where products are sold in large numbers for a specified time for a discount, have traditionally caused havoc in the supply channel. Increased sales causes unplanned increases in inventory requirements that generate lumpy demand as they move through the channel pipeline. The solution is the creation of closely integrated information partnerships across the breadth of the entire supply network. As "real-time" information about demand speeds through the channel, suppliers have the ability to synchronize their production and logistics functions to manufacture and deploy the necessary products in the correct quantities, thereby avoiding the usual cycles of product shortage and overstock during and after the promotional period.

- *Seasonality and Speculation Stocks.* Dramatic increases and decreases in stocked inventories caused by seasonality and speculation (forward buying) are realities governed by the marketplace. Diminishing the impact of seasonality and speculation on channel inventories can be accomplished by tightly integrating demand and supply information and employing new production technologies that enable small-lot production, value-added processing, and shortened lead times.

The effective use of inventory can not only dramatically reduce costs but also provide for sources of unbeatable competitive advantage. For example, the LEAR Corporation is the world's largest independent supplier of automotive interior systems, with 1995 sales of $4.7 billion. In response to OEM customers such as Ford, GM, or Chrysler, who demand inbound JIT delivery of components, LEAR uses a 4 hour "live broadcast" to build loads in sequence for guaranteed delivery window times. At its Lordstown, OH plant, LEAR delivers fully assembled seats to GM assembly facilities at a rate of 60 seats per trailer, every 32 minutes, for 2 shifts during the course of a 20 hours work day. Also, because LEAR's customers do not stock inventories, the company must deliver seating in the proper sequence, every time, on time, or the line shuts down. As a JIT organization, LEAR, in turn, expects its suppliers to deliver stock to meet its own build schedules. LEAR plants turn inbound inventory between 120 and 214 times a year. Declining inventory and freight costs have allowed the company to also reduce the logistics and transportation costs per seat set. The savings are then passed on to their customers. Because of the tremendous savings gained by their low-cost inbound channel system, LEAR has the ability to build a dedicated seat assembly plant within 10 miles of a customer. These plants then provide local outbound JIT delivery within minutes to a JIT auto plant [12].

The Strategic Channel Inventory Planning Process

The effective management of channel inventories is based on the existence of strategic plans that synchronize customer demand with the manufacturing, distribution, and product flow process capacities of the entire channel. Such a plan seeks to identify the aggregate, long-term demand requirements occurring at each channel level and to measure them against critical supply factors such as product life cycles, overall market growth objectives, and competitive opportunities unleashed by technology enablers and the core competencies to be found in the "virtual corporation." The success of strategic channel inventory planning also requires the implementation of closely integrated and effective individual channel member tactical inventory plans. Individual tactical plans consist of the detailed objectives necessary to achieve targeted levels of customer satisfaction, superlative inventory accuracy, close control of inventory costs, and implementation of ordering techniques that provide for timely and cost-effective stock replenishment.

When developing the strategic channel inventory plan, network members must determine answers to such questions as the following:

- What is the aggregate level of inventory necessary to support expected customer demand?
- What service levels are being achieved by the competition?
- What is the total working capital needed to meet channel inventory deployment targets?
- What are the aggregate operating costs associated with channel service objectives?
- How large should channel supply node buffers be to achieve serviceability targets?
- What is the optimum ratio between channel inventory and transportation costs?
- What information and communications technologies should be implemented that will network channel members closer together and provide for "real-time" information?
- What are the supply channel's performance measurements?

The strategic channel inventory planning process, as illustrated in Figure 6.6, begins with a clear definition of the supply channel's aggregate inventory strategy. On this level, strategic inventory planning can be defined as planning that seeks to minimize total channel inventories so as to achieve optimum customer service. Such a strategy involves the identification and implementation of critical capacity changes in any or all components of the channel network (particularly in manufacturing and value-added processing), the positioning of distribution supply nodes, size and location of buffer stocks, product flow management processes, and rules and policies necessary to sustain marketplace leadership. Effective channel inventory strategies require the close integration of inventory goals, policies, and decisions and accompanying marketing sales, and logistics activities occurring at each channel level. In defining total channel inventory goals, channel partners

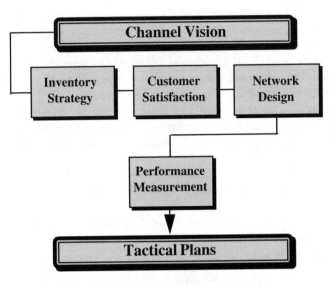

Figure 6.6. Strategic channel inventory plan.

must be able to respond to the potential dichotomies that exist between each channel level. By exploiting the natural linkages drawing channel businesses together, companies can overcome areas of possible conflict by leveraging information alliances that make transparent an awareness of interconnectedness, multiply efficiencies across the entire pipeline, and synchronize individual competitive goals and capacities.

The second step in the strategic channel inventory planning process consists of two activities: first, assessing the success of existing levels of channel customer satisfaction that have been achieved with current pipeline strategies, and second, what projected inventories will be necessary to capitalize on new marketplace opportunities. Determining customer satisfaction depends on a matrix of inventory management issues. Some cluster around indicators that measure the success of how the supply channel as a whole has presented and customers accepted product offerings, product pricing, the success of promotions and advertising, and the impact of customer service strategies that govern the sale from pretransactional to posttransactional activities. Other issues center around product delivery and quality. Timely product delivery is the result of having the right product at the right channel location or available through other media such as catalogs, brochures, or various forms of electronic communication. Often, supply nodes within the channel must decide on strategic trade-offs between service levels and inventory investment. To be effective, such trade-offs must be determined in light of total channel service objectives.

Ensuring the existence of effective network designs constitutes the third step in the channel inventory planning process. Network design has the potential of having a significant impact on channel inventory planning. Negatively, the number and location of pipeline partners or physical distribution points can be a severe drain on channel costs as excess buffers of inventories build up in the pipeline. In addition, poorly designed channels can actually inhibit the creation of customer-winning service value by supporting uncompetitive place and timing channel decisions. Well-designed network structures, on the other hand, enable each channel member to stock the proper inventory buffers necessary to support overall network service levels. Effective supply networks also make it easier for planners to determine changes to inventory levels due to the growth or contraction of markets. As the number of channel locations grows to meet rising customer demand, effectively organized supply channels permit planners to determine accurate inventory investment costs.

The fourth step in the strategic channel inventory planning process is determining the total working capital necessary to support targeted customer service objectives. Assessing the extent of inventory working capital calls for estimating how changes in pipeline manufacturing, distribution network design, and customer service will impact stocked inventory levels throughout the channel. A critical problem is developing a workable estimate of inventory cost given the nature of the typical supply channel. Although some channels, such as Wal-Mart's and K-Mart's, are dominated by a single corporation, others are composed of multiple numbers of diverse network partners, making it difficult to arrive at an accurate tally of inventory costs. In general, assessing inventory costs is concentrated around three principle areas as follows:

1. *Customer Service Objectives.* Supply channels that experience high levels of demand uncertainty, dysfunctional channel linkages, excessive lead time, and poor quality can expect to carry more buffer inventories than channels that have excellent point-of-sale forecasting, "real-time" linkages between demand and supply functions, and close working business partnerships.

2. *Production and Distribution Processes.* Normally, supply channels dominated by large-lot manufacturing processes and accompanying long distribution lead times can expect high inventory stocking costs. In contrast, increased use of value-added processing activities, mixed-mode manufacturing techniques, and shorter cycle times will result in lower network inventories.

3. *Network Design.* Supply channels that have unnecessary or poorly positioned customer supply nodes can expect to have larger inventory costs due to uncertain delivery and long in transit times. On the other hand, networks designed to have the right number of locations positioned to deliver optimum customer service can expect to have lower inventory investment due to shorter transit times, reliable delivery schedules, high

visibility to inventory status, and closely integrated information and communications systems.

The development of adequate inventory working capital requirements is critical in assessing the impact of total supply channel competitive objectives and targeted service levels. If the performance of supply channel inventories are not achieving targeted service goals, changes can be more effectively made if all channel partners can collaborate to increase or decrease inventories, depending on marketplace opportunities [13].

The final step in the channel inventory planning process is the development of adequate performance measurements that ensure that every customer, both external and internal to the pipeline, is consistently receiving the service value and quality necessary to meet demand requirements. Channel inventory performance can be measured from two interrelated perspectives: *customer service* and *inventory investment*. Performance metrics oriented around the customer assist in managing uncertainties due to timing and accuracy factors. Timing factors are focused around the time it takes for inventory replenishment activities to be completed; accuracy factors are concerned with order contents and quantity, order completeness, and inventory record accuracy. Performance related to inventory investment is concerned with tracking the impact of inventory costs across the channel network. "World-class" channel systems know that high inventory levels translate into increased probabilities of obsolete and damaged stock as well as loss of operating cash committed to pipeline inventories. Effective network performance measures require that channel members know both the costs in serving their own segment of the channel's market as well as the total inventory costs of the entire supply network.

Developing effective channel inventory strategies has become a fundamental source of competitive survival for today's top organizations. For example, the Case Corporation, a global manufacturer and distributor of heavy industrial and farm equipment, has developed a matrix of supply chain reengineering strategies, all targeted at maximizing customer service by having the right product at the right time at the right cost. In the past, Case operated channel inventories utilizing a push methodology that dumped products on the distribution pipeline. It was not unrealistic to have between 3- and 8-months supply on a dealer's lot. To be able to sustain the company's aggressive growth plans, it was quickly realized that Case had to control network inventories on a channelwide basis. One of the keystones of the plan was gaining worldwide visibility to inventory throughout the supply pipeline by implementing information systems that activated the necessary connectivity to draw together the company's network of manufacturing sites, warehouses, supply points, and some 4100 dealers and distributors in more than 150 countries. By developing strategic alliances with channel partners and third-party service providers, Case's efforts have significantly cut the company's and its dealers' finished goods inventories and spearheaded a 21% rise in operating earnings to a record $187 million in the second quarter of 1996 alone [14].

Channel Inventory Ordering Systems

Achieving the effective management of channel inventories requires the utilization of inventory replenishment methods that seek to ensure the availability of cycle and safety inventories across the supply pipeline while continuously searching for new avenues of cost reduction. Realizing channel replenishment objectives requires that each channel member be able to link their inventory requirements with their upstream and downstream partners to produce a single integrated system flexible enough to respond to any market opportunity before the competition. The key to leveraging pipeline stocks is determining when procurement orders are to be released and what the order quantities should be. Selecting the proper replenishment technique is the responsibility of each channel member. Creating the systems that link each supply node and providing for the "real-time" transfer of requirements from one end of the channel to the other is the responsibility of all channel members operating in collaboration.

Inventory Ordering Techniques

Before a planning method can be selected, inventory planners must determine the *nature* of item demand. According to Orlicky [15], the nature (or source) of item demand provides the real key to inventory ordering technique selection and applicability. The principle guiding inventory ordering can be found in the concept of *independent* versus *dependent demand.*

Demand for a given item is independent when such demand is unrelated to demand for other items. Sources of independent item demand come directly from customer orders and forecasts. For the most part, demand for products at the distribution end of the supply pipeline can be described as *independent* demand. Inventory is usually received as a finished product from the manufacturer or wholesaler, warehoused, and then sold directly to the customer. Conversely, an item is subject to *dependent demand* when it is directly related to, or derived from, demand for another part. Item dependencies can be described as *vertical*, such as when a component is required to build a subassembly or finished product, or *horizontal*, as in the case of an accessory that must accompany the product. Dependent demand is characteristic of most of the items in a typical manufacturing company. Manufacturers will purchase raw materials, components, or subassemblies that are never sold as received but are stocked and then issued in matched sets to build the finished products the firm does sell. This demand can be conceived as being created internally as a function of scheduling items to be converted into higher-level assemblies and finished products.

Determination of item demand status is the first step in selecting the appropriate inventory planning technique. In general, inventory planners can use one or a combination of the following two techniques: materials requirements planning (MRP), or statistical order point/order quantity. Items subject to dependent demand in a manufacturing environment are best planned using MRP. In this technique, a computerized system plans production components requirements in a priority-sequenced, time-phased manner driven by the demands placed on the finished assemblies. Simplistically, MRP provides a mechanism whereby demand in the form of forecasts and actual customer orders are first arranged in due date sequence. Then, by referencing each assembly's bill of materials (BOM), the demand is exploded through the BOM component structure. Finally, as each component demand is identified, time-phased by assembly due date, and back-scheduled by lead time, the system nets the demand quantity against available and on-order component inventory. By creating a time-phased schedule of demand, the inventory planner is provided with visibility to exactly *when* replenishment orders will have to be launched to avoid component stock-out.

In contrast, independent demand items are best planned using ordering methods that utilize forecasts to project demand into future periods. Such techniques plan products based on individual item demand *magnitude* without reference to other items. Inventory ordering decision models for items subject to independent demand can take many forms, but they are all related to one of the following:

1. *Visual Review System.* This method consists of a relatively simple inventory control technique in which inventory replenishment is determined by physically reviewing the quantity of inventory on hand. Procedurally, inventory is periodically visually reviewed, and items are ordered to restore balances to a preestablished stocking level normally determined by some workable heuristic.

2. *Two-Bin System.* Classically, this technique is a fixed-order system in which stocked inventory is carried in two bins (or some other form of container), one of which is located in the picking area and the other is held in reserve in a nonpicking location in the stockroom. Procedurally, when the picking bin is emptied, the reserve bin is brought forward to service customer demand. The empty bin serves as the trigger for replenishment.

3. *Periodic Review.* In this ordering system, a *fixed review cycle* is established for each item, and replenishment orders are generated at the conclusion of the review. The order quantity contains sufficient stock to bring the inventory position up to a predetermined quantity level. The review cycle can be established in days, weeks, months, or quarters, whichever satisfies the item's requirements. This method is also called a *fixed-cycle/variable-order quantity system.*

4. *Order Point.* In this ordering system, a fixed item stocking quantity is determined that is used as the order point. When the inventory position of a given item falls below this stocking point, reorder action must be taken to replenish the item back above its order point. The quantity to order can be manually determined, or some form of *economic order quantity* (EOQ) can be used. This method is also called a *fixed-order quantity/variable-cycle system.* Unlike the visual review, two-bin, and periodic review methods detailed above, the order point technique requires close perpetual inventory transaction control. As item receipts, adjustments, scrap, shipments, transfers, and so on occur, the firm's inventory control function must perform recordkeeping activities that provide planners with current inventory balances. An illustration of the reorder point system appears in Figure 6.7.

5. *Distribution Requirements Planning.* Whereas the replenishment review and reorder action mechanisms of the four methods discussed above are different, conceptually they are all closely related. Each attempts to establish the point *when* a replenishment order needs to be generated to prevent item stock-out in the face of normal demand and then to suggest an economic or target order quantity to be purchased. In contrast to these statistically or heuristically derived systems, DRP is a computerized management tool that plans inventory needs in a priority-sequenced, time-phased manner to meet customer and forecast demand as it occurs throughout the supply pipeline. The major advantage of the DRP method is that inventory replenishment action is triggered by matching supply with anticipated demand as it occurs over time. At the point where

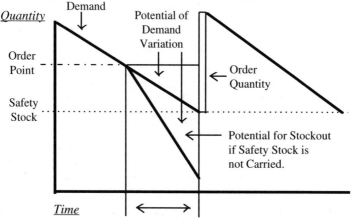

Figure 6.7. Order point system.

demand exceeds the supply, the system will alert the inventory planner to order the item according to a predetermined lot size and to have it available at the anticipated date on which stock-out is projected to occur. In addition, each time the DRP is generated, the system will resequence demand and supply relationships and suggest a new set of required order actions for the order planner.

6. *Just-In-Time.* During the past decade, the use of JIT techniques has been steadily growing among manufacturers and distributors. The basis of JIT inventory control resides around the *demand pull* of products through the factory or through the supply pipeline. As inventory is consumed, simple signal or triggers, such as colored lights, kanban, or golf balls, are tripped, alerting the system that replenishment must occur if stock-out is to be avoided. Successful use of JIT methods also requires the execution of strong buyer-supplier relationships to ensure total product quality and delivery.

Each of these inventory ordering techniques addresses the ordering needs of products that are subject to fairly continuous and *independent demand,* such as finished goods and service parts inventories. These techniques are *not* intended for raw material, component, and subassembly items involved in manufacturing. *Material requirements planning* (MRP) is the appropriate computerized planning technology to use when planning for raw materials and components consumed in production. In choosing one or a combination of these techniques, companies need to examine such elements as the level of planning and control desired, item cost, item physical characteristics, resupply lead time, continuousness of demand, dollar-value usage, storage and handling requirements, shipping characteristics, and the availability of data processing systems.

Creating Supply Channel Replenishment Linkages

The basis for the integration of channel inventory replenishment requirements rests on two fundamental principles. The first is centered around the simple fact that inventories and related information existing at each point in a supply channel system should, in reality, be considered as a collective resource with the single goal of achieving total channel customer satisfaction. In the past, most channel members operated as if they were isolated entities unconnected to the streams of products and information that flowed in and out of the channel systems to which they belonged. Today, "world-class" supply networks have begun to rediscover the natural interconnectiveness that links their members. It is not that these customer-satisfying processes were lost; they were always there, albeit invisible and unmanaged, producing fragmented output value. It is simply that each channel member was so focused on meeting internal company performance measures that they lost sight of the overall channel processes of which they were a vital part. Refocusing on the supply channel as a single process for the creation of service value has reawakened network members to the natural power that can be exercised when they treat pipeline inventories as if they were a single entity.

The second principle guiding channel inventory replenishment is the extension of the concept of *independent* and *dependent demand* to channel relationships. Because all channel members, from the raw material supplier all the way to the retailer, are really part of a single market process, it stands to reason that once demand is known or forecasted at the customer-facing end of the supply pipeline, it is the function of the inventory replenishment systems connecting each network level to respond as quickly and as accurately as possible by driving replenishment quantities back up the supply network. From this viewpoint, it can be said that only those channel nodes who sell directly to the customer are faced with *independent demand*. Once determined, however, the demand driven up the supply pipeline to all predecessor network members is, in actuality, *dependent demand*. As such, the inventory replenishment systems that restock the supply channel should be capable of satisfying the nature of the demand requirements occurring at each channel level while providing the necessary linkages so that timely requirements data are visible on a "real-time" basis anywhere in the supply pipeline.

Historically, once beyond manufacturing which uses MRP, supply channel organizations have heavily depended on *order point/order quantity* systems to determine and communicate replenishment requirements. Whether the order point is a statistical order point, minimum/maximum, periodic review, or some form of joint replenishment, they are all built around a common assumption. They all seek to arrive at the optimum moment to launch a replenishment order by referencing an on-hand inventory quantity calculated as the anticipated demand during the lead time plus a safety stock quantity. The replenishment order point (OP) is classically expressed by the following formula:

$$OP = \text{Anticipated demand during lead time} + \text{Safety stock}$$

For example, if the average historical usage (sales) for a given item is 100 units a week, the replenishment lead time from the vendor is 2 weeks, and the safety stock is 50, then

$$OP = 100 \text{ (usage)} \times 2 \text{ (weeks)} + 50 \text{ (safety stock)} = 250 \text{ units.}$$

In other words, the order point consists of sufficient inventory to satisfy projected customer demand (*usage*) while waiting for stock replenishment orders to arrive (*lead time*), plus a quantity of reserve inventory (*safety stock*) to account for variation in supply and demand.

Once the order point for a given item has been tripped, an appropriate stock replenishment order needs to be created. The calculation of the exact quantity to order has historically been a hotly debated topic. Whether it be a simple or complex ordering technique, each attempts to strike a balance between the cost of ordering and the cost of stocking inventory. The basic formula for what is known as the economic order quantity (EOQ) can be expresses as follows:

$$EOQ = \sqrt{\frac{2 \times \text{Cost of ordering (OC)} \times \text{Usage (R)}}{\text{Carrying cost } (k) \times \text{Unit cost (UC)}}}$$

The key to using EOQs is to balance ordering and stocking costs. The more an item is ordered, the less the stocking cost but the greater the chance of stockout. Conversely, the less an item is ordered, the greater the stocking cost but the less the chance that available inventory will be insufficient to respond to customer demand. Choosing a replenishment technique is, therefore, more than selecting the appropriate models; it is also a strategic decision to choose an inventory planning concept that requires a firm to structure its customer service and inventory control around the order point/order quantity method.

Although statistical replenishment techniques for channel inventory management have long held the field, they possess several difficulties when applied to managing supply network inventory linkages. These problems can be outlined as follows:

- *Replenishment Calculation.* By its very nature, order point methods are designed to be used by each business enterprise *independent* of channel linkages. The formulas employed require each supply node to create their own forecasts, safety stocks, and lead times. These calculations are then referenced each time a supply node calculates its inventory position. Once order points have been tripped, order action sets off a chain reaction triggering response on predecessor channel levels until the inventory requirement has been filled. Such a system provides network members with little visibility into satellite inventory requirements. In fact, statistical techniques trigger replenishment orders on channel supply points with no advanced warning, independent of supply point inventory availability or the needs of other satellite warehouses in the network. In addition, the random arrival of resupply orders makes it difficult for supplying warehouses to plan for cost-effective picking and shipping.

- *Lack of Integration.* Statistical inventory control techniques lack the mechanism to effectively couple business functions and channel partners together. The lack of *internal* and *external* integration draws channel energies away from purposeful planning to be focused on reflex reactions to the problems caused by the decoupling of strategic objectives, operations plans, and operations execution functions. Statistical inventory methods determine inventory action based on summary demand information, provide limited simulation capabilities for charting alternative courses of action, are insensitive to capacity issues such as cost, warehouse space, and transportation, calculate inventory replenishment action in isolation from business, marketing, and sales realities, and provide little information that can be utilized to determine the performance of the firm and the supply network as a whole.

- *Response to Change.* If a single word can be used to describe the supply channel in today's business climate, it is *change*. However, statistical approaches to business management are, by their very natures, *reactive* rather than *proactive* to change. By depending on the occurrences of the past to predict the course of the future, statistical methods cannot hope to provide distribution managers with the timely information necessary to integrate the resources of their businesses and the demands of their marketing channels. The absence of effective *external* integration renders the distribution enterprise powerless to respond to the constant changes occurring in marketplace demand and supply patterns.

- *Excess Inventory.* Statistical methods of inventory replenishment force companies to carry excess inventory. The objective of stock replenishment techniques is to restore inventory quantities to a predetermined level sufficient to respond to anticipated demand. The optimal level of stocked inventory is necessary to compensate for the inability of planners to determine the precise timing and quantity of customer demand in the immediate future. Line points and safety stocks provide order points with even higher inventory thresholds before order action is triggered. On the other hand, academics, consultants, and practitioner alike are firm in the belief that ideal supply channel inventory levels and lead times are none at all. In contrast, statistical replenishment techniques require stocked inventory because they suffer from false assumptions about the demand environment, tend to misinterpret observed demand behavior, and lack the ability to determine the specific timing of future requirements.

- *Performance Measurement.* Traditionally, in statistical replenishment, the inventory size, cost, turnover, and service levels of each stocking point are measured separately and in isolation from the performance of the channel as a whole. However, uncoupled inventory and customer service plans, no matter how successful at a given channel level, may actually foster a decline in channelwide effectiveness. Statistical methods cause higher but avoidable inventory costs, slow down the flow of products to the customer, and conceal operational efficiencies and unbalanced capacities among channel partners.

The ability to overcome the drawbacks of the order point system and integrate the replenishment requirements of the entire supply channel occurred with the introduction of distribution requirements planning (DRP) in the mid-1970s. A computerized approach to inventory management that utilizes MRP-type logic to time-phase supply and demand, the architecture of DRP lends itself well to providing "real-time" visibility of all replenishment requirements as they exist at any time and point in the supply network. As illustrated in Figure 6.8, demand in the system is collected at each supply node along with the specific due dates and quantities of forecasts and open customer and supply orders. The system then time-phases these requirements by due date and, by using the supply elements of on-hand and resupply stock on order, nets out inventory availability each time an order due date appears. This means that the system subtracts the supply from the demand or adds on-order quantities to on-hand quantities, as the due dates of each are referenced through time. If sufficient inventory remains after each calculation, then there is no resupply action to perform. If, on the other hand, the result is a negative, the system alerts the planner that a potential stock-out will occur at that point in time so that a counterbalancing planned order quantity can be placed in anticipation of a future shortage.

The impact of an integrated channel DRP system can be dramatic. As Figures 6.5 and 6.8 illustrate, DRP provides the key mechanism linking together all nodes in the supply network. In statistical replenishment systems, the data regarding inventory requirements are usually so badly timed and soon outdated that inventory begins to backup at various stocking points in the pipeline. The result is that the right inventory in the right quantities does not find its way through the channel, causing stock-outs and excess costs. DRP, on the other hand, offers channel members a way out of the problem by illuminating demand and supply requirements as they exist at any moment in the pipeline. DRP provides the mechanism necessary to link total channel inventories and customer demands. At each planning level in the process, DRP is used to summarize inventory requirements. By employing the concept of *independent/dependent demand*, sales forecasts, customer orders, and promotions occurring at the highest level in the pipeline are input into the DRP after which time-phased requirements are then calculated. These replenishment orders are then passed up the channel and input into the gross requirements of each preceding channel level. As each supplier node up the network processes their DRP system, they form, in essence, a channel master schedule linking the needs of the customer at the retail end with the resources of the whole channel.

With the use of electronic communication tools like electronic data interchange (EDI), the velocity of the transmission of this information through the supply network can be greatly accelerated, thereby enhancing the quick-response needs of today's competitive marketplace. By linking the output of networked DRP planned order schedules, whole supply channels can achieve the following inventory management breakthroughs:

ACME Retailing

SS = 210
LT = 1 week
OQ = 600

	PD	1	2	3	4	5	6	7	8
Gross Reqs		210	210	210	210	210	210	210	210
Sch Recpts				600					
POH	650	440	230	620	410	800	590	380	770
Pln Ord Recpts						600			600
Plan Ord Rel					600			600	

ACME Distribution Center *EDI*

SS = 100
LT = 1 week
OQ = 400

	PD	1	2	3	4	5	6	7	8
Gross Reqs		100	100	100	700	100	100	700	100
Sch Recpts									
POH	425	325	225	125	225	125	425	125	425
Pln Ord Recpts					800		400	400	400
Plan Ord Rel				800		400	400	400	

A & D Wholesaling *EDI*

SS = 0
LT = 2 weeks
OQ = 350

	PD	1	2	3	4	5	6	7	8
Gross Reqs				800		400	400	400	
Sch Recpts				350					
POH	125	125	125	25	25	325	275	225	225
Pln Ord Recpts				350		700	350	350	
Plan Ord Rel			350		700	350	350		

Tops Manufacturing (*MPS System*) *EDI*

SS = 0
LT = .25 weeks
OQ = 100

	PD	1	2	3	4	5	6	7	8
Gross Reqs		350		700	350	350			
Sch Recpts		400							
POH	10	90	90	90	40	90	90	90	90
Pln Ord Recpts				800	300	400			
Plan Ord Rel				800	300	400			

MRP System

Figure 6.8. DRP channel calculation.

- Decline or elimination of channel uncertainties brought about by total pipeline integration
- General shrinkage in the amount of overall inventory in the pipeline accompanied by improved customer service
- Simulation of the volume of future inventory requirements at every level in the supply pipeline
- Ability to resynchronize supply and demand for the entire supply channel as sales at the retail level occur above or below forecasted levels
- Increased visibility to lumpy demand as it makes its way through the supply channel
- Ability to plan for warehouse, transportation, and financial requirements associated with the purchase, storage, and movement of network inventories.

The experiences of Digital Equipment Corporation (DEC) provides an excellent example of the application of DRP to link channel supply sources. DEC's supply network consists of a large number of internal suppliers (its 25 factories) and literally thousands of external suppliers. DEC's decision to adopt a DRP approach to running its Westminster, Massachusetts distribution center was born out of the realization that its old informal system could not keep up with customer demand or the unplanned waves of unscheduled demand that flooded warehousing and transportation operations at the end of each quarter. After extensive analysis, DEC decided on the development and implementation of its own DRP system that would enable the company to include distribution demand in its production planning so that the company's internal and external suppliers would be better able to synchronize the needs of distribution with the capabilities of its factories. In just 2 years, DEC's Westminster operation was transformed from a reactive to a proactive driver of the supply channel. The plant realized a 55% increase in revenue per person, a 51% gain in orders per season, a 25% drop in the cost per order, an impressive 37% overall drop in inventory, and, most importantly, a 98% customer serviceability level [16].

Continuous Replenishment

Although computerized tools such as DRP enable today's supply channel members to be tightly integrated, the acceleration of customer demand requirements has prompted companies to search for tools that can increase the speed by which inventory needs can be communicated through the supply network. In the 1990s this operating paradigm has been termed *continuous replenishment* (CR). The mechanics of CR are focused on increasing the flow of information across the distribution pipeline, which, in turn, accelerates the flow of product from the manufacturer to the point of sale. To increase velocities, product distributors must focus on the entire pipeline, not just isolated segments, identify and eliminate all constraints, and utilize effective planning tools that provide channel members with early warnings of impending changes in demand and supply and the means to react quickly to synchronize customer requirements with supplier capabilities. As the velocity of product flow increases through the distribution pipeline, channel partners will experience higher customer satisfaction, lower inventory investment, and lower logistics costs.

The implementation of CR requires the whole channel not only to move beyond traditional inventory ordering methods, but it also requires the adoption of a philosophy of management that focuses on an unending search for quality and the institution of new methods of integrating distribution pipeline constituents. Figure 6.9 illustrates the techniques associated with the CR philosophy. Many of these techniques dovetail with elements of JIT. CR requires that every node in the channel be committed to continuous improvement, the elimination of wastes everywhere in the channel, total quality, visibility management, and empowerment of the work force. In addition, CR requires techniques that eliminate channel redundancies, provide for improvements in value-added processing, such as producing in small lots and quick changeovers, and aligning supply capabilities and demand requirements.

Ultimately, the CR management model that emerges can be understood as a series of planning steps designed to provide for the ongoing synchronization of anticipated customer demand and supplier logistics capacities. Effective CR systems can only occur when data and knowledge concerning demand and supply are connected together. As illustrated in Figure 6.10, customer requirements both forecasted and actual must be passed completely, accurately, and in a timely

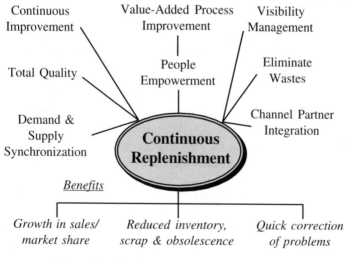

Figure 6.9. Continuous replenishment techniques.

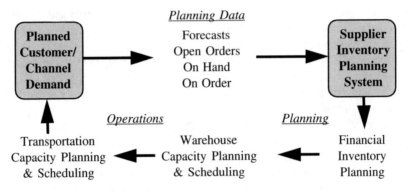

Figure 6.10. Continuous response model.

fashion to the supplier, who, in turn, can plan for the logistics capacities necessary for CR to occur. The goal is to remove response uncertainties from the equation by establishing a close-working partnership where customer and logistics imbalances are aligned before demand and supply information falls out of synchronization.

Perhaps the centerpiece of CR is the ability of suppliers to respond quickly to the *anticipated* inventory replenishment needs of channel partners with a minimum of delay in inventory planning and order release. In fact, some distributors want vendors and channel suppliers to respond *automatically* with the necessary inventory to satisfy customer requirements not on a weekly but on a *daily* basis. To be successful, such arrangements require the establishment of strong mutually beneficial partnerships, trust, absolute data integrity, superlative product and operational quality, and responsive computerized systems. With a few modifications, a standard DRP system can be created that is capable of responding with the detailed precision required by CR channel relationships. The following changes will permit microlevel inventory planning and automatic inventory resupply:

1. *Planning Buckets.* The system must have the ability to use *daily* time periods. This functionality permits planning and order action to occur daily if required.

2. *Forecasts.* The system must be able to calculate forecasts on a daily basis. In addition, the forecasting techniques employed must be able to use trending, smoothing, and seasonality calculations to arrive at the best forecast without seriously impairing computer hardware performance.

3. *Net-Change Regeneration.* To be effective, continuous replenishment DRP must be regenerated as often as possible. Running the DRP processor several times a week or even daily requires significant processing power. By regenerating only those products that have had a transaction

posted against them since the last regeneration will significantly decrease processing times and enable timely recalculation of requirements and order action.

4. *Automatic Replenishment Order Release.* Once planned orders have been created by the DRP generation, they must be automatically transformed into actual released resupply orders. This means that the system requires an additional subprocessor that will transform items with suggested planned order release quantities in the current bucket (action bucket) into actual vendor and interbranch orders *without* planner intervention.

5. *Automatic Purchasing Documentation.* For those items to be purchased from an outside supplier, the appropriate purchase order documents need to be printed or transmitted via computerized tools, such as EDI, directly to the vendor.

6. *Automatic Document Printing and Shipping.* For warehouses supplied internally through a Bill of Distribution, the DRP *autorelease subprocessor* must automatically allocate the required inventory, generate the necessary picklists and shipping documents, and ship the necessary planned orders to satellite warehouses to meet required due dates.

The strength of employing DRP for CR is the ability of planners to use the time-phased format to view immediate-term and short-term positioning of demand and supply requirements. This time-phasing mechanism makes DRP an ideal tool for implementing CR with both outside vendors and interbranch resupply. The system would function as follows. Buyers/planners would maintain the item forecasting module of the system, reviewing the applicability of trends, seasonality, and smoothing calculations and entering in variables, such as upcoming promotions, which will affect the forecast. After forecast review, the DRP processor would be run. When working with interbranch demand, order release would occur automatically without any further review. As demand filtered its way up to the last level in the supply pipeline, the resulting planned orders would be again analyzed, tailored if necessary, and then transmitted to the supplier who, in turn, runs the customer's time-phased requirements through their DRP or Master Production Schedule (MPS) system. By passing schedules of requirements up the distribution channel from point of sale back to the manufacturer, the system will streamline inventory, transportation, warehouse, and buying management by keeping all channel partners updated through DRP to DRP system electronic schedule synchronization.

Figure 6.11 provides an example of DRP continuous replenishment. The figure illustrates the DRP processing that would take place between a satellite warehouse and its supplying warehouse. Note that the planning periods consist of a time span of 1 day. Item forecasts in both warehouses have been generated that

ACME Retailing

		Day 1	Day 2	Day 3	Day 4	Day 5	Day 6	Day 7
SS = 5								
LT = 1 day								
OQ = Discrete								
Forecast		20	20	20	20	20	20	20
Resupply Ords								
Sch Recpts		20	20	20	20	20	20	20
POH	5	5	5	5	5	5	5	5
Plan Ord Rel		20	20	20	20	20	20	20

ACME Manufacturing

		Day 1	Day 2	Day 3	Day 4	Day 5	Day 6	Day 7
SS = 50								
LT = 3 days								
OQ = Discrete								
Forecast		50	50	50	50	50	50	50
Resupply Ords		20	20	20	20	20	20	20
Sch Recpts		50	70	50	50	50	50	50
POH	70	50	50	50	50	50	50	50
Plan Ord Rel		50	50	50	50	50	50	50

————▶ Purchase Order

········▶ Customer Order

Figure 6.11. DRP continuous replenishment.

estimate the expected daily sales, plus a small safety stock to provide for forecast error. Note that the *Planned Order Receipts* row has been removed. In an environment where replenishment receipts are planned every day, this row can be eliminated.

When the DRP generation takes place, the system performs the standard gross-to-net calculation for the item at each channel level. Net requirements are countered by the generation of planned orders in all but the current period. At this point, a subprocessor is executed that autoreleases the current period's net requirement into an actual replenishment purchase order. Released resupply orders are then created by the autorelease subprocessor that occurs at the end of the net-change generation. At the conclusion of each warehouse's calculation, the released orders are offset for lead time and passed to the supplying warehouse as *Actual Resupply Orders*. The gross-to-net requirements calculation then commences at this level and so on, until all levels in the supply network where the item is stored are reviewed. In the above example, the supplying warehouse's requirements are passed directly through EDI into the vendor's replenishment system.

Supplier Management

Successful supply channel inventory management in the 1990s mandates that customers and suppliers think of themselves as *business partners*. As lead times and product life cycles plummet and pipeline flow velocities accelerate, *partnering* in today's global business environment is no longer an option but has become a mandatory service strategy. Channel alliances are an increasingly important topic of discussion among academics, consultants, and practitioners. Scholarly research has been performed and studies conducted that analyze partnering success variables, partner selection factors, partnering success rates, and that even suggest models for partnering implementation [17]. Partnering can take many different forms, depending on the nature and requirements of supply channel members. Partner relationships can be found among competitors or noncompetitors and may exist for strategic or operational reasons. Whatever the formal arrangement, partnerships generally are cooperative alliances formed to improve operating procedures, efficiencies, and planning and product information exchange. In addition, partnering relationships may also be created to develop new products and technologies, moving beyond cooperation to include collaboration.

Defining Supplier Management

The concept of supplier management has grown to meet the changing needs of today's marketplace. As is illustrated in Table 6.1, traditional approaches to supplier sourcing, the procurement management process, and buyer/seller relationship values have undergone significant modification and accented the need for close-working business alliances. Like any partnership, the relationship between

Table 6.1. Traditional Purchasing Versus Supplier Partnering

Traditional Approach	Supplier Partnerships
Primary emphasis on price	Multiple selection criteria
Short-term contracts	Long-term contracts
Evaluation by bid	Evaluation by commitment to partnership
Many suppliers	Fewer selected suppliers
Improvement benefits shared based on relative power	Improvement benefits shared equally
Minimal involvement in design issues	Close involvement in design issues
Improvement at discrete time intervals	Continuous improvement
Problems are supplier's responsibility to correct	Problems jointly solved
Information proprietary	Information shared
Clear delineation of business responsibilities	"Virtual" organizations

buyer and seller must be open and honest; there must be commitment to using available resources to achieve common objectives; there must be an equal share in the risks and the rewards; and it must be a long-term proposition meant to weather the bad as well as the good times. Finally, partnership means redefining the usual ways purchasers and vendors think about product quality and reliability, delivery, price, responsiveness, lead time, location, technical capabilities, research and development investment plans, and financial and business stability. Channel partnerships have grown in response to the following marketplace realities:

- *Desirability of Supplier Relationships.* Partnering agreements have become a key source of competitive advantage. According to the study done by Ellram and Hendrick [18], buyers and sellers today are eager to pursue deeper channel partnerships that focus on continuous improvement, long-term ongoing relationships, and total cost.

- *Changing Nature of the Marketplace.* Today's focus on supply chain management issues, increased use of third-party and outsourcing alliances, supplier certification, and demands on the part of customers for participation in product design, production, and distribution functions have significantly facilitated the rise of closely integrated partnerships.

- *Increased Demand for Cost Control, Quality, and Innovation.* Customers are no longer willing to support suppliers who cannot meet critical standards for product and service quality, or who do not possess the capabilities to continuously unearth new sources of service value.

- *Increased Demand for Risk Sharing.* As the cost of innovation and operations flexibility explodes and the window for profits shrink in the face of competition, buyers and sellers have turned to business alliances as a means of sharing the risk of new product development and expanding operations costs.

- *Enabling Power of Information and Communications Technologies.* Continuous replenishment, electronic link technology, the Internet, open-systems architectures, and other technologies have provided suppliers and buyers with the ability to closely integrate replenishment techniques and to remove administrative redundancies and costly lead times out of the supply channel process. Technology has undercut the need for traditional purchasing devices such as requisitions, purchase orders, and negotiation.

- *Focus on Continuous Improvement.* At the core of today's successful partnering relationships can be found a strong commitment to partnering as a continuous improvement, dynamic process rather than a static business principle. Also, whereas effective partner relations might facilitate the achievement of mutual goals, effective partnerships recognize that partnering is an inherently dynamic process that must be continually renewed if vitality is to be maintained.

Based on these observations, supply partner management can be defined as an continuously evolving, value-enriching buyer/supplier relationship that requires a firm commitment to a common set of goals for an extended time period and a mutual sharing of information and the risks and rewards resulting from the relationship.

Developing effective supplier partnership involves a conscious effort on the part of both purchaser and supplier that begins first with the establishment of a consistent flow of internal integration and progresses to the use of techniques that promote closer external communications and new opportunities for competitive advantage. The process begins on the buyer's part by implementing a formal inventory planning system that aligns business, marketing, and sales needs with the firm's logistics capabilities. Following this is the establishment of a working interface between the scheduling system and the purchasing function. The goal is to ensure that replenishment needs are being met by the purchasing function. Once these steps have been completed, purchasers can begin to reduce the supplier base. A smaller vendor base will shrink communications and facilitate supplier performance measurement and alignment with enterprise goals. Upon completion of these internal tasks, purchasers will have the opportunity to institute vendor scheduling and capacity planning. These techniques will increase the velocity of requirements transmission between buyer and supplier, as well as ensure supplier capacity to respond to these needs. Finally, these activities provide the grounds for the establishment of a continuous search for new techniques, such as vendor certification, EDI, and joint participation in product design, that will enhance the partnership.

An excellent example of SCM supplier management can be found in Quaker Oats' supply chain practices. Quaker's supply strategy revolves around four key practices. The first is to be found in their continuous efforts to consolidate their supplier base. Over the past few years, the company has reduced the number of suppliers it uses by more than 50%. The second practice is the development of supplier partnerships. The focal point of partnership for Quaker includes developing meaningful performance targets and are characterized by a mutual commitment to continuous improvement in cost, quality, and throughput time. The third practice involves Quaker's efforts at improving supplier performance from the inside. A Quaker team moves inside the supplier's manufacturing process and, through opportunity assessments, develops a framework for identifying areas for value-added improvement and recommending the necessary actions. The fourth practice is a commitment to quality, which acts as the basis for the above practices [19].

Mechanics of Supplier Scheduling

One of the key tools enabling customer-vendor partnerships throughout the supply channel to achieve the benefits of supply partnering is *supplier scheduling*. Beginning as a replenishment tool between manufacturers and first tier suppliers, its use has spread to all channel network partners. Perhaps the best way to understand the mechanics of supplier scheduling is to contrast it with traditional purchasing. In the old method, replenishment order action is initiated at the unplanned occasion when the demand signal is tripped. Normally, a requisition is issued to the buyer who then generates a purchase order and sends it to the supplier. When problems with quantities or delivery arise, a series of communication flows going back and forth between customer and supplier ensues. Although the replenishment quantity and delivery date are eventually worked out, valuable time has been wasted: waste that adds cost to the transaction.

With the advent of computerized DRP systems, this traditional method of purchasing is no longer necessary. As discussed earlier, DRP provides inventory planners with a detailed window into future channel pipeline demand. Depending on the length of the planning horizon, the DRP processor will determine, for each product, the needed inventory replenishment quantities by required date. This *requirements schedule* can then be directly shared with the supplier. One of the possible changes that arises with the use of this concept is that the traditional roles of inventory planner and buyer found throughout the supply channel can be altered to leverage the *requirements schedule*. It is the responsibility of the inventory planners, now called *supplier schedulers*, to maintain the schedule and communicate directly with their scheduling counterparts at the supplier's facility. The goal is to ensure that the schedule of future requirements can be met by the supplier. The buyer's role then shifts from a concern with paperwork and expediting to value-added tasks such as negotiation, vendor selection, value analysis, quality improvement, and alternate sourcing. The schedule then becomes the crucible through which passes customer requirements and the realities of supplier delivery and quality problems. The scheduler must ensure that both are in alignment. With this arrangement, the buyers buy from the supplier, and the planners schedule the supplier. The mechanics of the whole operation are determined by the *supplier agreement* fleshed out in advance by the buyer and the supplier's sales force.

Benefits of Effective Supplier Management

The ability to closely integrate the entire supply network into a series of interconnected buyer/supplier relationships provides the whole channel with new opportunities to increase the value-added capacities of the supply pipeline from the raw-material and component supplier at the gateway to the supply channel to the final point of sale. Effective supply channel partnerships enable network supply nodes to mobilize channel competencies and resources to significantly reduce inventories and lower costs while greatly improving customer service. Other benefits are as follows:

1. *Material Delivery.* By using a detailed schedule of requirements, companies can plan receipt of the right material at the right time at the right

price. Furthermore, the schedule guarantees that customer demand and supply delivery will be synchronized to provide the highest serviceability at the lowest cost.

2. *Vendor Capacity Planning.* When customers provide vendors with a window into future demand, manufacturing capacities can be efficiently allocated to ensure timely delivery. Furthermore, supplier scheduling is essential for effective cost reduction and quality projects.

3. *Partnership.* Supplier scheduling provides the basis for the development of long-term "win-win" partnerships instead of short-term adversarial relationships where everyone loses.

4. *Reduced Inventories.* Supplier scheduling permits the purchase of products at the specific quantities required. In most cases, it eliminates the need for forward buying and assists in the reduction of obsolescence and spoilage. It also provides channel members a mechanism to control lumpy demand caused by seasonality and promotions.

5. *Shorter Lead Times.* Most of a supplier's lead time consists of customer order backlog. As the backlog increases, so does the lead time. Supplier scheduling eliminates the occurrence of backlog by detailing for the supplier the exact product requirements and due dates *before* they occur. Traditional methods of inventory planning dump orders unexpectedly in the supplier's production schedules, thereby scrambling priorities and overburdening capacities.

Summary and Transition

As organizations continue to search for new methods to increase customer responsiveness and supply chain agility, the effective management of channel inventories has come under the magnifying glass. In the past, the concept of the supply pipeline was dimly understood and the smooth flow of products and value hindered by decentralized methods of positioning network inventories. For the most part, the value of stocked inventories extended no further than the operational boundaries of the individual companies that, together, loosely constituted the supply channel. Each channel node established its own inventory buffers to dampen pipeline demand uncertainties and to cover for lumpy demand caused by promotions, seasonality, speculation, and lot-size production and distribution economies. In such an environment, supply channels were often clogged by unnecessary inventory stockpiles that slowed the flow of products from manufacturer to customer, ballooned inventory carrying costs across the breadth of the pipeline, ignored storage, handling, and transportation capacities, risked obsolescence and theft, and rendered impossible strategies targeted at increasing distribution flexibility.

In today's global marketplace, the demands on the supply channel to streamline inventory flows and cut costs while improving customer service has radically changed traditional concepts of channel inventory management. What has brought about this radical change is the way productive processes are being organized to leverage the dramatic decline in product life cycles, increased velocity of customer response, universal commitment to quality, and ability to customize products at the last stages in the distribution cycle to create unique solutions that provide superior value to the customer. Instead of the old paradigm that perceived the supply channel purely as a conduit for the flow of mass-produced and mass-distributed standardized products, today's "world-class" supply networks are guided by the following four critical channel inventory principles:

1. All channel inventories must be managed to maximize on service value.

2. Channel inventories must be managed as if they were part of a single distribution network unit, not the management of inventories at each channel supply node.

3. The key to successful channel inventory management is the management of the time it takes products to move through the supply pipeline on their way to the customer.

4. The availability and manipulation of timely information about customer demand, inventory status, and replenishment supply is the best substitute for repositories of inventory stocked throughout the channel.

Applying these principles requires a thorough understanding of the dynamics of supply chain inventory management. In the ideal world, there would be no need for channel inventories. Customers would be able to receive the goods they wanted simply by triggering ordering mechanisms that would define desired finished goods configurations, transmit the exact product specifications to the factory where they would be made, and arrange for delivery as close as possible to the moment of order request. In reality, channel inventories are a critical element of the demand-satisfying process. Inventory can be found in the supply network for a variety of reasons. Inventory buffers can exist because of batching or lot-size economies; timing issues such as geographical movement or the duration of the manufacturing process; planned overstocks due to seasonality or speculation; and availability uncertainties due to variances in inventory demand and supply. The ability to effectively utilize channel inventories requires cross-channel strategic inventory plans that synchronize customer demand with the manufacturing, distribution, and product flow process capacities of the entire supply pipeline. The strategic channel inventory planning process begins with the close integration of the inventory goals, policies, and marketing and production decisions occurring at each channel level, progresses through an assessment of current and expected customer service performance and identification of channel designs that optimize service values while shrinking inventory costs, and concludes with an analysis of the working capital and performance measurements necessary to monitor the effectiveness of supply channel inventory levels.

Achieving channel inventory service and cost objectives requires the utilization of replenishment planning and ordering systems that enable all levels in the supply networking to link their inventory requirements in such a manner so as to produce a single integrated system flexible enough to respond to any market opportunity before the competition. Although statistical replenishment techniques, such as order point and EOQ, have long held the field, the difficulty of applying them effectively to create the necessary network linkages has required channel planners to implement DRP techniques. A computerized approach to inventory management that utilizes MRP-type logic to time-phased supply and demand, the architecture of DRP lends itself well to providing the "real-time" visibility necessary to integrate demand at any time and distribution point in the supply network. By employing the concept of *independent/dependent demand*, sales forecasts, customer orders, and promotions occurring at the highest level in the pipeline are input into the DRP where time-phased requirements are then calculated. These replenishment orders are then passed up the channel and input into the gross requirements of each channel level. As each supplier node up the network processes their DRP system, they form, in essence, a channel master schedule linking the needs of the customer at the retail end with the resources of the whole channel. In addition to enabling today's supply channel members to be tightly integrated, DRP lends itself to implementation of electronic commerce tools such as EDI and the Internet that assist in the creation of continuous replenishment methods and increase the speed by which inventory needs can be communicated through the supply network.

The effective management of supply chain inventory requires today's customer and supplier to think of themselves as closely integrated business partners. In the past, the primary focus of the buyer/supplier relationship centered around an adversarial atmosphere dominated by issues of price, short-term contracts, lack of information and risk sharing, and poor product and delivery quality. In contrast, today's focus on SCM issues, increased use of third-party alliances, across the channel demands for cost control, quality, and innovation, and demands on the part of customers for participation in product design, production, and distribution functions have significantly facilitated the rise of closely integrated business partnerships. The benefits of supply partnerships are extensive. The ability to closely integrate the entire supply network into a series of interconnected buyer/supplier relationships provides the whole channel with new opportunities to increase the value-added capacities of the supply pipeline, beginning with the raw-material and component supplier at the gateway to the supply channel to the final point of sale. Effective supply channel partnerships enable network supply nodes to mobilize channel competencies and resources to significantly reduce inventories and lower costs while greatly improving the entire channel's ability to search for new competitive space that will provide lasting marketplace dominance.

Notes

1. Robert V. Delaney, *Seventh Annual State of Logistics Report*. St. Louis, MO: Cass Information Systems, 1996, Figure 15, Supporting Data I.

2. Helen L. Richardson, "Speed Replaces Inventory," *Transportation & Distribution*, 37, (4) (November 1996), 71.

3. This section has been adapted from David F. Ross, *Distribution: Planning and Control*. New York: Chapman & Hall, 1996, pp. 216–219.

4. This example is drawn from Richard J. Schonberger, *World Class Manufacturing: The Next Decade*. New York: The Free Press, 1996, pp. 167–168.

5. Justin Martin, "Are You as Good as You Think You Are?" *Fortune*, 134 (6), (1996), 142–152.

6. See the discussion in Andre Martin, *Infopartnering*. Essex Junction, VT: omeno, 1994, pp. xv–xxiv.

7. Sarah H. Bergin, "Recognizing Excellence in Logistics Strategies," *Transportation & Distribution*, 37 (10) (October 1996), pp. 52–53.

8. Jay W. Forrester, "Industrial Dynamics: A Major Breakthrough for Decision Makers," *Harvard Business Review*, (July–August 1958), 23–52.

9. Andre Martin, *DRP: Distribution Requirements Planning*, 2nd ed. Essex Junction, VT: Oliver Wight Limited Publications, Ltd., 1990, pp. 103–104.

10. See the discussion in Christopher Gopal and Harold Cypress, *Integrated Distribution Management*. Homewood, IL: Business One Irwin, 1993, pp. 109–113.

11. Martin, *Infopartnering*, p. 31.

12. Sarah H. Bergin, "Recognizing Excellence in Logistics Strategies," *Transportation & Distribution*, 37 (10) (October 1996), 56.

13. See the discussion in Gopal and Cypress, pp. 124–125.

14. Leslie Hansen Harps, "Case Corp. Constructs Logistics Model of the Future," *Inbound Logistics*, 16 (10) (October 1996), 25–32.

15. Joseph Orlicky, *Materials Requirements Planning*. New York: McGraw-Hill, 1975, pp. 22–25.

16. See the story in Martin, *Infopartnering*, pp. 56–65.

17. For a detailed analysis of the literature see Lisa M. Ellram and Thomas E. Hendrick, "Partnering Characteristics: A Dyadic Perspective," *Journal of Business Logistics*, 16 (1) (1995), 41–43.

18. *Ibid.*, pp. 41–64.

19. This first-hand overview of Quaker's supplier management practices can be found in Rhonda R. Lummus, *Supply Chain Management: Balancing the Supply Chain with Customer Demand*. Falls Church, VA: APICS Educational & Research Foundation, Inc., 1997, p. 21.

7

Supply Chain Quality and Performance Measurement

For over a decade, one of the most important subjects dominating not only manufacturing and distribution but all businesses from fast-food to health care has been quality improvement. Much of this concern has focused on product quality. Seminar courses and popular discussion abounds on the topic and interested practitioners, academics, and consultants can find literally dozens of books and hundreds of articles illuminating the topic from diverse directions. Despite its importance, however, product quality is only a single facet of the quality improvement philosophy. Increasingly, enterprises have also come to see that customer service quality is equally as critical to competitive advantage. However, although industry stories concerning such customer service leaders as Federal Express, L. L. Bean, Nordstrom, and Wal-Mart have traditionally received top billing, little has been written about the service quality requirements of the supply chain arrangements that support them. Although the quality programs of companies such as Milliken Industries, Hewlett-Packard, SYSCO, and Siemens have shared some press, for the most part the service quality efforts of companies up the supply channel who sell to the retailer, wholesaler, or manufacturer have been little explored.

The creation and maintenance of superlative channel service quality and value is at the very core of the concept of supply chain management (SCM). Satisfying the end customer can only take place when the entire supply channel from materials supplier to retailer are linked closely together in the pursuit of innovative ways to improve service value, reduce channel costs, and create whole new regions of competitive space. In today's global economy, it has simply become impossible to service the final customer successfully each and every time if the links in the supply channel that precede it are not also driven by the same focus on quality and value-added activities. Supply channels guided by the SCM concept have come to understand the final customer as merely the end point in a chain of customers. Whatever the level in the supply channel, each network node has a customer. Making the connection and ensuring that total service quality and value objectives are constantly pursued is perhaps the foremost goal of today's "world-class" supply channels.

This chapter seeks to explore the application of quality and improvement management concepts to supply chain management. The chapter begins with an overview of the origins of *supply chain quality management* (SCQM). The goal is to place SCQM solidly within the quality management movement of the last 15 years. Following this, a concise definition of SCQM is attempted as well as a detailed review of the eight dimensions of supply chain quality. At this point, the chapter shifts to an analysis of SCQM processes. Topics discussed in this section are generating continuous SCQM value, using quality analysis and statistical measurement techniques, and understanding the impact of SCQM on the channel's collective work force. Once the basics of SCQM have been detailed, the chapter turns to an exploration of how channel quality management can be utilized to create customer service value. Today's marketplace leaders have found that to be successful, the entire supply channel must be committed to quality and productivity techniques that enable quick response to ongoing changes in customer product and service needs. Such a quality philosophy means that the entire channel network must be diligent in measuring customer perceptions of service quality, identifying shortfalls, and responding decisively to service gaps. The chapter concludes by detailing the requirements necessary for the effective implementation of a channelwide program for continuous SCQM process improvement.

Defining Supply Chain Quality

To a larger degree than has perhaps been previously recognized, the concept of SCM has grown as a necessary response to today's continuous demand for increased channel service quality and value. In fact, *customer-centric* attitudes and concern for uncovering innovative breakthroughs in service quality shaping the marketplace today have provided the dominant objectives pursued at all levels in the supply channel. Historically, the strategies channel networks have used to improve customer service centered on expanding business capacities, gaining marketshare, penetrating new markets, and offering new products. Although critical, supply channels in the 1990s have found that these objectives constitute the bare minimum of competitiveness. In addition, customer strategies focusing on continuous improvements in quality and value-added services have become the keynote drivers of "world-class" channels. These attributes not only enable whole supply pipelines to meet and exceed the expectations of their customers, they also lead to increased profits through market expansion and gains in marketshare. In this sense, there is a direct relationship between perceived quality and return on sales or investment.

Origins of Supply Chain Quality

Supply chain quality management (SCQM) is a relatively new concept that has just begun to capture the attention of the business community. Prior to the mid-1980s, the typical enterprise considered their internal supply and distribution management functions, much less their *external* channels of supply, to be of little importance in the pursuit of service quality. Consequently, minimal human effort and management resources were invested in ensuring that critical service value processes possessed requisite quality standards. During the past decade, however, radical changes in the business environment have shifted attention increasingly toward supply chain quality. Several critical converging changes to be found in the rising level of customer expectations, product design and manufacturing management, government regulation, technology, functions and opportunities offered by channel partnerships, and exploration of new methods targeted at reducing total channel costs and accelerating the flow of value through the supply network have made the effective and efficient management of the supply channel today's hottest business topic.

The change in management thinking regarding the importance of supply chain quality can be said to have originated form three general trends. To begin with, the interest in SCQM stems directly from the quality movement begun in the manufacturing sector during the 1980s. As the quality of manufacturing processes increased, it soon became evident that the supply pipeline that fed materials and components into production and then acted as the conduit for the transfer of finished goods output to the customer also needed to possess the same level of commitment to process quality. Whereas total quality management (TQM) techniques could dramatically improve productive processes in individual companies, quality initiatives needed to be expanded to embrace upstream and downstream channel partners if their full potential were to be realized.

Second, the significant changes occurring as a result of deregulation in the transportation sector had germinated a growing recognition of the competitive advantages offered by extending the availability of value-added channel services. Companies that had previously considered themselves purely as producers, transportation specialists, or public warehouses began to expand and leverage their array of value-added services in an endeavor to uncover new sources of marketplace opportunity. Whether found in individual companies or in the explosion of third-party service providers who offer low-cost logistics solutions for firms who choose not to expand in-house services, the focus of the revolution in service value required a radically new understanding of supply channel systems and partnerships centered around the concept of service quality.

Third, the decision on the part of many organizations to expand logistics activities beyond the borders of the enterprise necessitated a reexamination of quality issues. The application of management methodologies, such as JIT and supplier management, and the availability of new technology tools from bar coding systems to integrated computer planning required more tightly networked channel members. Companies soon learned, however, that to be able to exploit opportunities to dramatically increase the speed of delivery to market and shrink inventories and costs they needed to closely define previously unformulated attitudes concerning the importance of supply chain quality. As it had in manufacturing, supply chain concerns over quality stemmed from many directions. Some companies experienced their quality efforts as slowly evolving from their experiences with the increasing pressures brought on by new competitive arrangements; others eagerly sought to elevate and promote their commitment to supply channel service quality as a marketplace differentiator. In any case, as today's enterprise has increasingly turned to their supply chain relationships as the newest area in the pursuit of marketplace leadership, quality issues have become a channel-level concern.

SCQM Definition

The concept and mechanics of quality management (QM) have been documented by such experts as W. Edwards Deming, Joseph Juran, Philip Cosby, and others [1]. It can be said that QM is comprised of a variety of topics, including the conformance and fitness for use of products, services, time, place (plants, offices, and operations), equipment and tools, processes, people, environment and safety, information, and performance measurements. QM is also concerned with quality assurance, or the certifiability that a product or service will fulfill stated specifications of quality, and quality control, or the operations techniques and activities employed to fulfill requirements for quality. Finally, value-based quality is the degree of product or service excellence in relation to price. The focus of QM is not so much on the products or services themselves, but on the *processes* that produce them; it is centered more on the rate of improvement of the process rather than on fixed standards of what constitutes product or service quality [2].

Based on these overall elements of business process quality management, SCQM can be defined as

> the participation of all members of a supply channel network in the continuous and synchronized improvement of all processes, products, services, and work cultures focused on generating sources of productivity and competitive differentiation through the active promotion of market-winning product and service solutions that provide total customer value and satisfaction.

When it comes to the management of the supply chain, SCQM can be said to be concerned with the following points:

- *Creating Customer Value.* The ability to respond to customer requirements is the most fundamental function of SCQM. Overall, customer value refers to three critical attributes. The first, *effectiveness*, can be described as the level of performance regarding such issues as in-stock availability, cycle time turnaround, convenience, service, and innovation. The second, *efficiency*, refers to the ability of the supply network to offer the desired product/service mix that is perceived by the customer as providing superior value for the price. The final element, *differentiation*, refers to the uniqueness of product and service combinations provided by the supply network. A critical corollary of customer value is that it must be equally extended to a supply channel's internal as well as its external customers. Ongoing customer value requires establishing quality goals that are to be extended to the entire channel as well as determining responsibility among channel constituents for creating and sustaining customer value [3].

- *Total Focus on Continuous Improvement.* SCQM provides today's supply chain with the ability to leverage the full range of available quality management techniques to continuously search for new ways to create as well as to improve customer service value. Far from being a one-time effort pursued by a single or limited group of channel members targeted at the creation of a new stabilized competitive state, continuous improvement requires that the entire supply channel be organized to shoulder the task of never-ending improvement. Such a philosophy re-quires individual supply network constituents to develop new forms of flexibility and self-knowledge concerning the value-producing elements (people, processes, and physical assets) of the business. Through effective experimentation, problem solving, and information transfer, individual and channelwide customer value can be continuously renewed to create unbeatable sources of competitive advantage. Over time, these patterns of behavior nurture the development of superior channel capabilities that serve to complement and enrich the channel supply partners of which they are composed.

- *Formalization of Quality Processes.* Today's SCQM initiatives require each channel partner to develop high-quality processes. This means that network members must synchronize their quality efforts and direct them toward the use of formalized quality management tools, management practices, and operating paradigms that guarantee quality processes as transactions and planning decisions cascade through the *chain of customers* [4]. Merely assuring conformance to specifications and defect preven-tion is insufficient in today's global environment to maintain competitive-ness: SCQM requires each channel supply node to follow a formalized process of product and service quality improvement that provides the best value to the customer.

- *Development of Quality Process Methodologies.* Formal SCQM can only occur with the consensus and collaboration of all supply network members in the development of a common channel quality process improvement methodology. The goal is to achieve commonality of means and objectives that draws the entire supply pipeline together as a single customer-satisfying force. Such a methodology will require interchannel management and operational commitment, definition of the service requirements for the entire chain of customers, charting current levels of performance, defining strategies to improve channel quality in order to exploit new marketplace opportunities, quality program implementation, and the creation of organizational and performance techniques that provide for continuous channel quality improvement.

- *Execution of Key Performance Functions.* Quality and productivity in SCQM means utilizing the combined resources of the entire channel network in the most efficient and value-enriching ways possible to provide market-wining, cost-effective customer service. Quality in this area can be stated as not only meeting but exceeding customer expectations concerning such critical operations activities as the ease of product ordering and order status inquiry, timely and reliable delivery and communications, error-free and complete orders, timely and responsive postsales service support, and accurate and timely generation and transmission of demand and supply information through the entire supply channel to support effective logistics, marketing, and inventory replenishment planning.

- *Developing Effective Performance Measurements.* The measurement of SCQM performance is a difficult process due to the large number of possible supply channel partners. Nonetheless, measuring the performance of the entire channel system through the use of statistical analysis and benchmarking is critical for the continued success of the entire supply network. Determining and tracking critical performance metrics can be significantly assisted through the development of a common database that will permit the convergence of total channel performance relating to such cost-intensive issues as inventory, logistics, marketing, order management, design engineering, production, and quality management. The more these and other metrics can be compiled and evaluated as they traverse the entire channel, the less will be the opportunity for individual channel member performance optimization at the expense of total channel performance.

As can be determined from these critical points, SCQM requires a tireless attention to creating customer value, a single-minded focus on conformance to process requirements not product/service "goodness," and an acknowledgment that the only performance standard for channel quality is zero defects. In the final analysis, SCQM considers the true measure of quality to be the cost associated with the inability of the supply channel to satisfy marketplace requirements. This cost directly reduces channel customer value and competitive advantage.

Dimensions of SCQM

Supply chain quality management can arguably be considered as the latest extension of the total quality management movement. In the beginning, QM had relatively little to say about the importance of channel quality. In the early 1980s, quality efforts were almost totally focused on products and the manufacturing process. The emphasis was on achieving "process certification," "conformance to requirements," "zero defects," "fail-safing and robust design," and "cost-based quality." Companies saw quality, for the most part, as an *internal* function directed toward guaranteeing the competitiveness of enterprise products and manufacturing processes. By the mid-1980s, however, the QM concept began to be applied beyond its original product/process orientation by including the quality capabilities of the company's supplier base. Renamed *total quality management* (TQM), this enlargement of quality management consisted of two stages. In the first, TQM was concerned with integrating other enterprise functions, such as marketing, sales, finance, and materials and physical distribution management with manufacturing in order to provide a single enterprisewide definition of quality that incorporated service value with product and process quality. In the second stage, TQM was further expanded to ensure the homogeneity of the level of product quality available from suppliers. Through various forms of quality certification, suppliers were now expected to participate in and provide levels of product quality that would facilitate JIT and product design methodologies and support the quality standards of their customers.

As TQM strove to radically change product-oriented processes and services, the requirements for QM in customer service were simultaneously growing. By the 1990s, it was evident that customers wanted more than just high-quality products: They also were demanding superior services that helped them ring costs from internal planning processes, logistics functions, and pipeline inventories while enabling them to develop new markets and innovative product/service solutions that would create radically new competitive space. As engineers had sought to employ quality techniques and philosophies to transform the manufacturing process, so marketing and sales managers turned to the same methodologies to create the service philosophies that would merge with the enterprise's production quality program to present a total approach to customer satisfaction. The goal was to create a single chain that linked superior product and service from the supplier at the beginning of the supply pipeline with the final customer at the other end.

This convergence of TQM and service quality management is the foundation of SCQM. In the past, companies strove to establish standards of product/service quality acceptable to the marketplace. Today, companies can no longer simply look internally to determine what constitutes quality but must construct their productive capabilities around the customer's definition of quality. Industry leaders know that they are able to meet the requirements and expectations of the marketplace only when they possess processes flexible enough to be tailored to meet and exceed the quality standards of each individual customer. Such a strategic objective can only be accomplished when not only the end customer but the entire chain of customers that constitutes each level in the supply channel receive and then pass on the same or higher level of quality. The quality standards required by SCQM, therefore, can never be characterized as a concern with a single dimension of quality. Successfully responding to the chain of customer value requires that the supply channel collective pursues the follow dimensions of quality [5]:

- *Conformance to Specifications, Durability, Serviceability, and Aesthetics.* These dimensions of quality are primarily attributes of product-oriented quality. They refer to how closely individual products function according specification, how long a product will last, ease of providing a service, and the beauty, elegance, or aesthetic appeal of a product.

- *Performance.* This quality dimension refers to the level or grade of value received from a product or service. Surveys reveal that customers feel that performance is the primary attribute of quality. For the most part, SCQM is concerned with the degree of service value found in such activities as order placement and ease of inquiry, timely and reliable delivery, accurate and error-free orders, responsive postsales support, and availability of accurate and timely information to support channel system planning and operations.

- *Quick Response.* Because of its crucial importance to the supply channel, the speed of response by which replenishment information and products are moved through the supply network has become a fundamental dimension of SCQM. Quick response enables the supply channel to shrink pipeline inventories and costs while providing more effective product and service response.

- *Value-Added Processing.* As quick-response systems seek to reduce channel costs, lead times, stocked inventories, and material handling, supply channel members have increasingly looked to product postponement methods to provide increased customer response and flexibility. The key to the process is to delay final-product differentiation to the latest possible moment before customer purchase. Adding final value at the appropriate place in the supply pipeline is accomplished through the use of such activities as sorting, bulk breaking, labeling, blending, kiting, packaging, and light final assembly.

- *Features.* This quality dimension in SCQM refers to added service elements that accompany the product or basic service. For example, most carriers such as Yellow Freight System and Ryder provide "real-time" shipping information, instant access to rating information, and a comprehensive array of information and communication technologies such as computer link, fax, and EDI that are added benefits to the basic service.

- *Reliability.* Of all the dimensions of SCQM, this quality attribute is viewed as the most important by the chain of customers. Although customers are becoming more demanding, it is not so much a question of speed as it is predictability. The reason is simple: Whether feeding a JIT assembly line or operating a major warehouse, consistent response and guaranteed service is the critical quality value. Con-Way NOW, an expedited service carrier serving the United States and Canada, guarantees delivery within 15 minutes of quoted delivery time. If a shipment is more than 2 hours late, the fee drops 50%, and if it is more than 4 hours late, its is free [6]. Such commitments to supply channel reliability can only be expected to increase.

- *Perceived Quality.* This quality dimensions refers to the impression of quality that the customer will receive above the intrinsic product or service. For example, when a product is purchased from Nordstrom's, the customer feels that the value of the product has been heightened by the knowledge that a "world-class" supply organization sourced, warehoused, and shipped the product to the store shelf.

- *Value.* This quality attribute constitutes the fundamental quality criterion that today's customer has foremost in mind. Although a product may possess superior quality elements, the real question is whether the price paid for the product or service value enables each level throughout the chain of customers to create additional sources of competitive advantage by further eliminating costs or providing the gateway to new innovations.

Although a few of the above dimensions refer directly to product quality, it is evident that the majority reflect the migration of quality management from a preoccupation with product to service quality, from TQM to SCQM. In the past, quality advantage was attached to superior products; today, only those supply channels capable of providing superior customer service value will be tomorrow's marketplace leaders. This shift in the QM paradigm indicates a recognition that today's enterprise must be able to respond to customer-based quality needs and that those needs can only be satisfied when the entire supply pipeline is committed to pursuing a level of quality that will satisfy the entire chain of customers, not solely the end customer. Additionally, SCQM must prepare channel members to be able to change production and delivery functions quickly as customer product and service needs change. This attribute is perhaps the most difficult to execute within a distribution channel, as it requires ongoing change in the organizational structure, planning values, and operating philosophies of each channel partner.

Furthermore, supply channels must respond to customer requests for products and services with 100% reliability. Whether it be product quality, pricing, shipping, billing, or after-purchase services, reliability and accuracy have become the benchmarks of the "world-class" supply network. Finally, not only products and services but the whole supply channel must be able to respond to the special needs of the customer. Rigidity in product design or features, service offerings, and availability causes customers to search elsewhere for solutions to their product and service needs.

Understanding SCQM Processes

The migration of channel quality management from quality control and quality assurance to TQM and, finally, to SCQM marks a decisive movement away from product-based and enterprise based quality to one that focuses on how the entire supply channel can increase service value for the chain of customers it serves. Although the application of SCQM can take many different forms based on channel circumstances, such as marketplace position, strength of the competition, management culture, and other operational characteristics, they all share a basic set of common attributes. To begin with, they are all centered around developing business processes that seek to continuously improve customer value. They all use similar analytical models and quality and productivity measurement techniques. In addition, they all seek to leverage the competencies and innovative capacities of their work forces to create new paths for competitive value. Finally, the resulting implementation of quality initiatives, regardless of industry, all bear common similarity. In this section, each of these characteristics of SCQM will be discussed.

Generating Continuous SCQM Value

According to Michael Hammer [7], when American companies began in the 1980s to adopt new methods of improving business processes, the two best known and most successful were TQM and business process reengineering (BRP). These management methods, one focusing on accomplishing small incremental improvements to processes, the other seeking fundamental restructuring of the organization in order to achieve order-of-magnitude breakthroughs in process quality, were instrumental in effecting the American industrial and economic renaissance of the mid-1990s. Each made it possible for all types of industries to remove excesses in manpower, materials, and information management that blocked process flows and inhibited the growth of competitive value. Each released untapped reservoirs of incredible opportunities for continuous improvement found as a result of closer integration between internal functions and external partners, significant acceleration in the speed and accuracy of competitive activities, and reformulation of the concepts of flexibility, quality, service, and cost.

In today's competitive business environment, companies have found that they can no longer simply depend on pursuing internal process improvement but must seek to extend TQM and BPR techniques to the supply channels of which they are a member. This realization signifies a dramatic broadening of the concept of process improvement. If a process can be defined as "a group of tasks that together create a result of value to the customer [8]," then improving any process to increase customer value requires that companies not only seek to improve the *internal* but also the *external* supply channel processes that serve the chain of customers. Just as narrow views of process improvement within a company result in departments working at cross-purposes and pursuing local optimization at the expense of the whole company, so the application of SCQM principles means that the community of channel members must seek to integrate process improvement across the entire supply network. When supply channels pursue a common process improvement program, they are better able to converge process functions to achieve superior customer value and avoid conflicting objectives and parochial agendas that splinter the search for sources of ongoing service value.

Establishing SCQM process designs and performance targets is a dynamic responsibility, not just for individual channel members but for the whole supply channel working in concert. Unfortunately, many quality improvement programs never achieve the expected benefits. All too often, process improvement is seen as a one-time effort directed toward the establishment of a new stable operating system. Organizing for continuous, never-ending improvement on both the company and the channel level, on the other hand, requires that organizational structures and relationships be capable of constant change and realignment. The challenge, then, for employees, managers, and channel constituents will be to lead process improvement on all levels of the channel. As will be discussed in the next section on customer service quality, this challenge requires each channel member to be constantly communicating with their customers and suppliers, benchmarking the competition, and working with channel partners to search for new avenues of customer value.

Once the need for process improvement is identified, the SCQM effort can take an incremental (TQM) or a radical (BPR) approach. Although there has been much discussion in QM circles concerning the compatibility of TQM and BRP coexisting as improvement paradigms, when applied to SCQM they easily converge into a single quality improvement program. TQM techniques focusing on problem-solving and root-cause process improvement are at the heart of SCQM. Utilizing statistical approaches that highlight areas to be targeted for improvement in what appear to be efficient processes, these TQM techniques seek to close performance gaps by enhancing the existing process. When the performance gap is so great, however, that the very design of the process is faulty, then BRP is needed. The BRP process, at this juncture, is a two-step process. To begin with, individual company and supply channel organizations must be realigned to accept the principle of process-centered SCQM. Second, once organizations understand the need for continuous improvement, existing faulty processes can then be replaced with superior ones. For the most part, the BRP element of SCQM represents a one-time shift in the operating philosophies of the supply channel from an uncoordinated, localized view of process value to a new perspective that embraces the concept of continuous total channel process improvement.

As will be discussed later in this chapter, welding TQM and BPR methods into a comprehensive channel improvement agenda requires the identification of critical channel network processes needing improvement, commitment of each channel organization to effect process change, choice of the range of techniques available in the TQM statistical improvement toolkit or BRP models, a closely monitored project plan to guide the channel project, and close performance review to monitor the success of the SCQM quality initiative. The goal of the whole process is the creation of an integrated channel network organization constantly reviewing performance in order to determine new quality levels. Today's "world-class" supply channels never perceive current quality goals as permanent, but rather continuously seek to work to assemble the collective competencies of the entire network to achieve ever-higher levels of customer value.

Using Quality Analysis Techniques

At the core of continuous improvement can be found several techniques that can be used to assist companies improve business processes. Beginning in the early 1980s as a set of tools to assist manufacturing quality, today their use has extended into purchasing, sales forecasting, and logistics management, and now into supply channel quality management. At the heart of quality analysis tools can be found quantitative techniques that are used to track statistical means, variances, process distributions, random sampling, and other methods of identifying process error. The common goal of the techniques is to provide statistical information that indicates when a process has deviated from the operating standard. Tracking process variances provide two critical sources of information: First, it indicates when a process has deviated from its quality limits; and, more importantly, it provides hard statistical data that pinpoints the source of the quality variation. When statistical data is coupled with knowledge of the process, companies have the critical information necessary to begin actual improvement activities.

Quality analysis tools can essentially be broken down into two families: process analysis and statistical analysis. Although quality control literature contains references to a host of different techniques and how they can be applied, there are approximately seven standard tools. *Process analysis* tools can be said to consist of the following three techniques:

- Cause and effect diagrams. Commonly called *a fishbone chart*, this technique seek to illustrate the relationship of possible causes of why an event occurred by examining its effects. The "worst case" is the spine of the fishbone chart, secondary causes (bones) are attached to the spine, and tertiary causes are connected to the secondary causes. The goal is to begin improvement on extremity "bones," eventually ending with an attack on the spine or central problem.

- Process flowchart. This technique attempts to provide a pictorial display of the process flows a particular product/service takes as it passes through all production steps and stages.

- Brainstorming. Designed as a technique for quickly uncovering problems, brainstorming involves using the opinions of a group of process experts who identify problems, pinpoint causes, develop potential solutions, and plan improvement actions.

Statistical analysis tools can be said to consist of the following four techniques:

- Pareto analysis. This technique plots variances (like defects, incomplete orders, late shipments) as they occur in the process flow. The worst variance (the longest bar on the Pareto chart) is the object of quality improvement.

- Histograms. A graph of contiguous vertical bars representing a frequency distribution (say, order placement to due date) in which groups of data elements (the number of orders) are marked on the x axis, and the number of occurrences in each group (lead time measured in days) appears as the y axis. The picture that emerges makes performance patterns visible.

- Run diagrams and control charts. A graphic technique that illustrates, through the use of ranges, averages, percentages, and other attributes, a chart of performance of a process over time. Results can be measured statistically for improvement action.

- Scatter diagrams and correlation. These graphic techniques attempts to plot the relationship between two variables. The x axis illustrates the degree (number) of error tracked (such as number of pick errors), and the y axis the base variable (such as number of years service).

The use of statistical analysis tools is at the very center of quality improvement. Similar to their use in individual companies, their value is no less important in examining supply channel processes for quality improvement. It has only been within the past decade or so that companies across the globe have come to realize that *quality is free!* Quality is everyone's business, extending from the raw materials supplier at the beginning of the supply channel to the retailer at the far end of the channel. Quality analysis techniques provide everyone in the supply network with a common set of statistical tools to reveal how processes really work, how they can be improved, and what the value-added impact improvements will have on rising levels of product/process quality and customer satisfaction.

Measurement Techniques

In the past, companies felt the only real performance measurements were those associated with financial numbers. Although there were many performance metrics coming from sales, inventory management, and logistics, the barometer that measured the success of a company was what could be found on the bottom line of the balance sheet. Today, it has become apparent that just running the company by the numbers, by top-down financial goals setting, not only indicates little about a company's prospects for marketplace success but actually can be destructive of its well-being. This is not to say that financial metrics, such as operations costs, sales revenue, profits, productivity, and others, are not critical. Financial numbers can assist in planning cash flow, managing assets, and pursuing long-term opportunities. However, when it comes to performance measures that are to be used for operations decision making in the short and medium range, they can provide little meaningful information.

Instead, global market leaders have found that the only meaningful measurements that can be used to assess a firm's actual strength and value-creating capacities stem from process management. This change in orientation means a shift in the organization from managing *results* to managing *processes* (how do things *really* work and what will happen). If the goal of individual organizations and the supply channel as a whole is to constantly create new sources of competitive value, then the fundamental object of performance measurement is not only monitoring and controlling costs but also improving the underlying productivity of product and service processes. By focusing on customer-oriented and root-cause-oriented measures of performance, by being able to track development costs to ensure that only market winners are promoted, and by opening a window on to the supply channel playing field so that accurate product and service performance can be tracked, today's best corporations can make informed competitive decisions for themselves and for the supply channels of which they are a part [9].

Up until recently, measuring supply channel performance had not been considered an important source of competitive information. Even within corporations, such as Sears and GM which historically have had large company-owned supply channels, cost and productivity measurements that viewed their distribution networks as a single system were not in existence. For the most part, the reason for this situation was the inability of management to combine the performance functions of differing business units into a total quality system. By its very nature, channel cost and productivity metrics cut across traditional organizational functions. The use of conventional accounting methods, which were useless in determining performance metrics that spanned multiple business units, normally sought to separate and absorb channel performance data into the overall performance of individual departments. The problem was compounded when trying to gather meaningful cost and productivity measurements for a supply channel that is composed of multiple independent companies. Although new accounting techniques, such as *activity-based costing* (ABC) and total cost analysis (which seeks to identify the total performance of the outputs of a channel system), have been created, gathering channel performance data is a complex undertaking. Charting the performance of each channel level is a murky affair where channel system metrics are obtained obliquely, and windows into the success of process improvement initiatives can be lost as data move from one end of the channel to the other.

The key to gathering channelwide cost and productivity improvement metrics is to develop an effective measurement system. In designing an effective performance system, the following four questions should be answered:

1. What is the goal of the measurement?
2. What data elements will provide a good measure of this goal?
3. What is the appropriate formula to calculate the measurement?
4. What are the sources of data?

The output of the measurement system should provide data that provides valid and complete data relating to costs and productivities, is compatible with existing performance data, covers all relevant factors, can be compared across time or different channel locations, and provides a useful guide to measure improvement [10].

Once a supply channel measurement system has been created, there are three major areas that should be used to track channel quality and productivity. The first area is *service quality*. This area seeks to measure the perceived value of services necessary to satisfy the requirements of the channel's chain of customers. The goal of service metrics is to track the effectiveness not only of performance issues, such as on-time delivery and order accuracy, but also of value-added services, such as financial and postsales services, that provide new regions of customer value beyond the core product or service. A deeper investigation of customer service value appears in the following section.

The second general channel measurement area is *productivity*. Effectively measuring channel processes involves assessing the cost and effort required to efficiently perform network functional processes. Simplistically, channel productivity measurements can be determined by calculating the ratio of the *input* resources consumed to produce a given *output* measured in the number of work units accomplished. As an example, what is the cost of a computerized channel inventory replenishment network system that speeds the pull of goods through the supply pipeline from manufacturer to retailer resulting in a targeted channelwide stocked inventory reduction? Accomplishing this productivity objective across the total channel requires the full cooperation of all members that might call for actions that focus on identifying and eliminating sources of process variance by using TQM techniques or by BPR methods that reengineer the basic process and improve resource utilization. The overall goal is to establish metrics that enable channel members to collectively increase process performance levels.

The final channel measurement area is *process effectiveness*. There are two main measures of process effectiveness: the service quality level of the chain of customers and key benchmarks of process effectiveness [11]. As was stated earlier in this chapter, the real objective of SCQM is the satisfaction, not just of the end customer but of the entire chain of customers positioned along the supply channel. These channel customers can consist of manufacturers, wholesalers, and retailers, as well as third-party logistics support partners. Ensuring quality channel process standards is critical: Superior end customer service value cannot be achieved without equally superior internal channel service quality. This principle becomes more apparent if the concept of stacked tolerances found in manufacturing processes is applied to product and service flows as they make their way through the channel. For example, if each stocking point in a channel consisting of five levels maintains 95% serviceability, the whole channel would be able to muster only a 77% serviceability level at the point of final customer delivery. Some internal channel quality processes include accuracy and timeliness in inventory replenishment requirements, length of cycle times for order management, total cost of transportation in the channel, and total inventory stocking levels and expected customer demand.

Besides measuring supply channel process efficiencies, today's intensely competitive global marketplace has prompted companies to compare their levels of performance with both their toughest competitors as well as companies renowned as the very best across industries. The goal is to analyze the product and service processes of industry competitors with an eye not just to match but to exceed the best performance of the process anywhere. The combination of these elements constitute what has become known as *competitive benchmarking*. The steps in benchmarking supply channel processes consist of the flowing activities:

1. Detail the structure of the channel process. Because the network of product and information flows in a supply channel can be very complex, this procedure seeks to identify the entire length of the pipeline process. This step will expose the place of each channel member in process performance.

2. Determine the effectiveness of the process at each customer/supplier interface. This step will require analyzing service levels beginning at the supply source and ending with the customer. Key critical activities to be monitored and the processes involved will have to be identified.

3. Isolate potential weak points in the channel network. The effectiveness of the supply channel is only as good as its weakest points. These critical points could be where the continuing reoccurrence of an event such as inventory stock-out or late delivery will jeopardize the performance of the entire channel process.

4. Identify generic channel issues. These topics relate to qualitative channel issues such as willingness of channel members to work together, channel commitment to principles of SCQM continuous improvement, acceptance of innovation and change, level of communication, and commonality of customer-oriented service value objectives.

5. Benchmark processes. In this step, the supply channel will have to identify and then benchmark pipeline weaknesses and strengths with both competing channels and with the very best across all industries.

6. Analyze results. All channel members will have to jointly agree upon what the key service values are that have emerged from the benchmarking process, and how well channel processes meet these value standards.

7. Establish action priorities. Finally, once the results of the benchmark have become evident, channel members must agree upon which processes are of strategic value and have a high relative impact on channel service quality and which have lesser importance. These findings are then used as the source for channel process improvement action.

An effective benchmarking program provides a unique perspective on the quality of channel processes. It enables channels to identify "best practices" and to creatively incorporate them into the SCQM improvement process. Furthermore, it provides channel members with a deeper understanding of their mutual dependence on each other by providing sources of quality information that exist in business relationships outside of their own supply channel network. Finally, it presents the occasion for the whole channel to gain insight into new methods of running and measuring their processes that previously had gone unrecognized. [12]

The Work Force and SCQM

The ability of SCQM to provide for continuous improvement in service quality and the development of new regions of competitive value is directly dependent on the collective efforts of the supply channel network's people resources. In the past, quality improvement initiatives were led by teams of managers and process experts who performed the analysis and then passed on their recommendations to line personnel whose task it was to implement the solution. The result of such a methodology rarely resulted in any lasting improvement. Even when the *quality circle* craze took improvement efforts by storm, their value was compromised by a lack of management commitment to invest in the resources, training and skills development, and QM processes necessary to ensure success. On their part, employees felt coerced into working on improvement teams and felt unsupported when the suggestions for improvements they did identify failed to be taken seriously by management. Whereas quality improvement was universally acknowledged to be the only path to competitive survival, traditional management styles and work cultures seemed to blunt even the best improvement project.

Today, the traditional "command and control" work culture has dramatically changed. As wasteful levels of management are eliminated, the work force has found themselves being trained, rewarded, grouped, titled, and managed differently. One of the central reasons behind this change stems from the realization that everyone, not just managers and specialists, is responsible for process improvement, both in individual companies and in the supply channels of which they are a part. This refocusing of the ownership of process improvement has also called for a distinct growth in teamsmanship, for the creation and participation of groups of employees or "associates" into multifunctional work teams organized around the flow of work, not simply around common functions. When it is done right, as it has in companies such as Milliken & Co. and Johnsonville Foods, workplace restructuring results in generating not a product-focused but a customer-focused organization led by groups of *process engineers* ceaselessly searching (and rewarded) to remove wastes and add value to the processes they work with daily [13].

Actual company and supply channel quality teams can take several forms. Quality management boards, typically composed of senior line and staff members, are normally charged with overseeing QM projects, assigning resources, creating quality task forces, and authorizing and guiding changes. Cross-functional quality improvement teams are composed of managers and associates from multiple functional areas who focus on improving processes that span departments. Natural work unit teams consist of managers and associates from a single functional area. Their goal is to improve processes within their own departments. Finally, quality improvement task forces are ad hoc teams that are assembled in order to work on specific aspects of a process improvement project. Task forces are composed of critical managers and associates who possess the right match of skills and experience to meet the needs of the project.

The use of the various quality improvement organizations for SCQM can be deliberate or indirect. It can be said that almost any improvement in process quality at the company level will basically improve total channel quality. However, to achieve the kind of breakthrough improvement in quality that would impact supply channel competitive advantage, it would be necessary to create quality improvement teams that span the entire channel and are composed of key associates of each channel member. The typical pan-channel quality organization would consist of a quality management executive board and special channel task forces recruited from each network member. It would be the quality board who would select those channel processes that would deliver the highest impact on total channel competitiveness, determine the criteria to be used to gauge process improvement, determine the resources to execute projects, assemble channel improvement task-force teams, and authorize and guide changes to channel processes. It would then be the responsibility of the channel quality task forces or functional teams to execute the process improvement project. An excellent example would be a channel quality team composed of information experts from each member company invested with the task of establishing or improving the use of EDI transmission throughout the channel system. Doubtless, creating such pan-channel quality teams is fraught with the usual political and organizational stresses. However, the significant customer value that they have the potential to create should more than offset the cost and trouble.

Supply chain quality management has been made possible by the dramatic shift in companies from being hierarchical and vertically based to organizations composed of networks of various process-centered organizations, some of them actually consisting of partner businesses, operating concurrently within a single enterprise. Such "virtual" organizations depend on pools of people organized around processes possessed of core competencies native to the process. A core competence can be defined as a "bundle of skills and technologies that enable a company to provide a particular benefit to customers" [14]. The potential of SCQM for quality improvement is to be found in the ability of networked companies to integrate the core competencies found within each organization to create channel-level sources of competency that will open whole new areas of competitive space. Linking the knowledge and skills of people across the entire channel system opens new vistas of innovative thinking to achieve levels of customer value previously thought impossible and to establish the framework for sustained quality improvement that wins customers and creates a source of unbeatable competitive advantage.

Creating Customer Service Value

A decade or so ago, the critical test of whether a supply channel was providing value to the customer could be determined by its success in delivering the "right product to the right place at the right time at the right price." As long as supply channel members had a firm understanding of the content of just what "right" meant, they were assured that when they executed operations functions that would enable them to optimize productivities and efficiencies, they would be meeting marketplace competitive standards. Today, although the old marketing utilities of time, place, and price have lost none of their importance, what constitutes the "right" way to respond to the customer has significantly changed. A single-minded preoccupation with economies of scale and scope in manufacturing and distribution have given way to operational flexibility. Standardized, mass-produced and mass-distributed products have been replaced by "mass customization" of both products and services. Instead of building supply channels to achieve volume and throughput, responsiveness to the customer, no matter what the cost, has become the fundamental principle of channel design. Performance measured around providing acceptable levels of product and service quality has been replaced by the continuous search for superior value that provides customers with unique opportunities for total service.

The level of service quality customers now expect from their suppliers has been, furthermore, significantly impacted by information technology and new management methods. Rising customer demands that their suppliers be sensitive to JIT, quick-response, supplier management, and continuous replenishment techniques have necessitated the creation of new dimensions of quality. The increasing use of EDI, computer network linkages, joint design team participation, value-added processing, and other value-added services have opened new vistas for partnerships and innovation that extend deep into the supply channel. On top of all of this, the growth of global sourcing, continuing decline in product life cycles, shrinkage in transportation lead times, and use of new technologies, such as interactive television and the Internet, have called for fresh perspectives on what constitutes customer value and how today's supply networks are to converge their competencies to enhance the service value they offer to their customers.

Defining Customer Service Value

Before the elements of customer service value can be explored, it is important to define its meaning and scope. Exactly what constitutes *customer service* value can vary from one company to the next, depending on the contents of the bundle of product and service benefits they offer to their customers. Although specific issues such as product, price, service, and quality are important, customers normally evaluate suppliers based on their overall satisfaction with the business relationship. In addition, the definition of what constitutes customer satisfaction can differ, depending on whether it is seen from the viewpoint of the supplier or of the customer. A very broad definition of customer service is the measurement of how well the entire supply channel meets the expectations of the marketplace in regard to product quality, service support, price, and availability. Fleshing out the details of effective service value means establishing quality criteria for tangible and intangible factors. Tangible factors include such measurements as product characteristics and quality, level and value of services, flexibility, availability of technology tools, ease of ordering, and timeliness, accuracy, and completeness of delivery. Intangible factors relate to how the customer feels about doing business with the supplier and includes ease of conducting transactions, friendliness and responsive attitude, and a feeling of dealing with a supplier who is looking out for their customer's well-being.

Customer service value is a complex concept with a variety of different possible definitions. In its simplest form, service value can be defined as the ability of a company to satisfy the needs, inquiries, and requests from its customers. In a landmark study by LaLonde and Zinszer [15], corporate executives were asked to define customer service in their organizations. The responses fell within three general areas. The first focused on *activity-related* factors. In this category were found customer service functions that could be directly managed, such as order processing, proof of delivery, invoicing accuracy, timeliness of delivery, and others. The second general area centered on *performance*-related criteria. In this group could be found metrics measured against a standard such as "percent of orders delivered to the customer within 10 days of order receipt" or "percent of orders received and processed in 48 hours with no back orders." The final general area defined customer service as an element within the overall *corporate philosophy* of the firm. Instead of detailed performance metrics, this view regarded customer service as an integral part of the overall business strategy, imbedded within the long-term *goals* of the enterprise.

Another definition describes customer service value as a process that

> results in a value added to the product or service exchanged. This value added in the exchange process might be short term as in a single transaction or longer term as in a contractual relationship. The value added is also shared, in that each of the parties to the transaction or contract are better off at the completion of the transaction than they were before the transaction took place. Thus, in a process view: Customer service is a process for providing significant value-added benefits to the supply chain in a cost effective way. [16]

In a similar vein, Band [17] defines customer service value as "the state in which customer needs, wants and expectations, through the transaction cycle, are met or exceeded, resulting in repurchase and continuing loyalty." In other words, if customer satisfaction could be expressed as a ratio, it would look as follows:

$$Customer \; satisfaction = \frac{Perceived \; quality}{Needs, \; wants, \; and \; expectations}$$

One of the most complex descriptions of customer service value can be found in Schonberger's "customer-focused principles" [18]. Instead of being centered around basic metrics, such as on-time delivery and customer satisfaction, Schonberger perceives customer service value to be composed of 16 critical customer-focused principles that span all aspects of customer management from customer/suppler partnership, product and operations design, and human resource management to quality and process improvement, information resources, and supplier capacity. The goal of the 16 principles is to assist companies in structuring practical organizations driven by a total commitment to customer value and eager to promote, market, and sell the firm's continuously increasing capabilities and competencies to create unbeatable marketplace value.

Elements of Service Value

Schonberger is but one of many researchers who have attempted to identify the fundamental elements commonly associated with customer service value. Perhaps the most concise list of service elements has been formulated by Zeithaml et al. [19]. They found that customers use basically the same criteria to rate performance regardless of the type of industry or the level in the supply pipeline they may occupy. The elements pertain both to individual companies as well as to the supply networks of which they are a part. These elements are listed below:

- *Tangibles.* This element refers to the *appearance* a firm's service value capabilities project to the customer. Often the image of quality a company wishes to communicate to the marketplace includes such tangibles as new facilities, state-of-the-art technology, highly qualified personnel, and the latest equipment. Tangibles are designed to give the customer a sense of confidence and assurance that the services they are receiving are truly "world class."

- *Reliability.* Once a company publishes their commitment to a specific level of customer service value, their ability to live up to that standard is the measurement of their reliability. Service leaders must continually perform the promised service dependably and accurately each and every time. Reliability of service permits suppliers to "lock in" their customers who will gladly pay premium prices for perceived quality.

- *Responsiveness.* The ability of a supplier to respond to the needs of the customer quickly and concisely lets customers know that their time and costs are important. Whether it be in rendering prompt presales service or a willingness to assist with product quality issues, a helpful attitude and timely service will always leave the customer with the sense of dealing with a winner.

- *Competence.* When customers purchase goods and services, they need to feel assured that the supplier possesses the required skills and knowledge to assist them when product or support issues arise. Firms that back up their products with cost-effective and competent services will always be leaders in their marketplace.

- *Courtesy.* Many a sale is won or lost based on the way the customer is treated in the presales and postsales cycles. Companies who do not respond to their customers with politeness, respect, consideration, and professionalism are destined to lose them to competitors who do.

- *Credibility.* Service leaders base their success on high standards of honesty, trustworthiness, and believability. Customers purchase products from firms that live up to claims of the best quality possible at the lowest price.

- *Security.* The delivery of products and services must be accompanied by a sense of security on the part of the purchaser. Security issues range from financial confidence to questions concerning the mechanical safety of products. A feeling of security frees customers from doubts and provides "peace of mind" for the products and accompanying services they purchase.

- *Access.* This element of customer service has several facets. Foremost, access means the degree of ease by which customers can purchase products or contact sales and service functions. Access can mean the availability of goods and services within a time limit generally accepted by the industry. Access can also mean the speed by which after-sales replacement parts and services can be delivered to the customer. Customer convenience and access to goods and services are fundamental to competitive advantage.

- *Communication.* The availability of sales and services staff to respond quickly and intelligently to customer questions concerning products, services, account status, and the status of open orders is the primary tier of a firm's customer communication function. Other communication forms such as printed literature, manuals, product and service newsletter updates, the Internet, and advertising form the second tier. Effective customer communication stands as perhaps the fundamental cornerstone for service leadership.

- *Understanding the Customer.* Unearthing and responding to the needs, desires, and expectations of the customer is the first element in effective sales and service. Firms that provide the products and services customers really want will always enjoy an edge over their competitors.

The above list of service dimensions is applicable to all types of businesses. Even though specific metric targets of service quality may vary from industry to industry, the 10 dimensions represent benchmarks by which firms can measure themselves.

Managing Customer Service Value

The task of identifying the basic elements of customer service value is but the first in a three-step process for developing a value-enhancing channel customer focus. Unfortunately, at this point, many companies turn inward, seeking to find new sources of customer value by reengineering processes targeted at achieving internal objectives such as cost reduction. Instead, increasing customer value in today's environment requires enterprises to turn outward, first determining the value requirements of the supply network's chain of customers, and second, developing a service strategy that encapsulates identified customer value requirements into a comprehensive statement of objectives around which the supply channel is to be redesigned. The resulting channel service value strategy should be detailed enough to guide practical action as well as possess a means for strategic renewal as the nature of customer value changes through time.

Abbott Laboratories, for example, has consciously sought to increase its capabilities and cut costs by assisting its customers to focus on opportunities for continuous supply channel efficiency. The giant health care products provider has created teams of specialists who work with the customer to analyze the total delivered cost of a variety of supply channel options for various product types. By consulting with the customer relative to the nature of the cost and service of possible alternatives, the customer can focus on answering questions such as: "What level of service do we need?" "What are the internal cost and process implications for each channel option?" "Is the cost of the service more than offset by its value?" Abbott's image of customer service extends to providing customers with more than just product value: By assisting its customers to identify their lowest total cost and supply channel service option, they are providing a key service value that provides them with another source of competitive advantage [20].

Identifying channel customer service value requirements is a complex task. Not only are the needs of every end customer different, but so are the needs of the chain of customers that constitute the supply pipeline. Developing service standards for these cross sections of the supply channel is a painstaking process that seeks to utilize both "hard" measurements as well as less precise judgments of what the *perception* of service value means to each customer. A general approach that seeks to address these issues consists of the following process:

1. *Segment the Marketplace.* This is a preliminary activity meant to separate the supply channel's customers into groups for the purpose of gaining relevant insights and market distinctions as a preliminary to customer value requirements identification. The process is iterative in nature, the results of which are designed to be reconsidered and revised as actual customer value demands are revealed.

2. *Identifying Customer Expectations.* Once meaningful market segments have been identified, key customer service values found within each segment can be isolated and merged to form a comprehensive list. Attaining this information is not always easy. Some companies, such as Ford, Boise Cascade, and Baxter's Hospital Supply Division are very explicit about their requirements and expectations to the point of detailing the requirements in policy manuals and brochures. In other cases where requirements are unclear, customer input must be solicited through personal interviews, telephone and mail surveys, and marketing-type focus groups. A negative technique is to chart current performance and "noise levels" to determine requirements.

3. *Benchmarking.* Besides actively seeking to identify service value from the customer, firms can also employ competitive benchmarking. Benchmarking is a quality management method in which a company compares its bundle of product and service offerings with those of its competition and with the "best of breed" in all industries. Whereas benchmarking

will shed little light on what actual customer expectations are, the technique does provide SCQM efforts with objective standards and enables companies to develop proactive approaches to identifying new sources of service value.

4. *Ranking the Importance of Customer Service Value Requirements.* Once a list of service value requirements has been compiled, it is important to rank each value in importance so that the critical value requirements are readily visible. Delphi groups, rating scale methods, trade-off techniques, and computer analysis can be used effectively to develop a usable scale of service component elements. The goal is to compile a list detailing which service values are market winners and which are baseline values.

The metrics that are gathered from these approaches are critical in highlighting for the entire supply channel, those service values that both the chain of customers and the end customer perceive as crucial. Applying management methods and technology tools to enhance these values and provide for new avenues of competitive advantage is the overlying goal of the process. The detailed measurements attained form the basis for the development of an effective channel service strategy.

The second element necessary for effective supply channel customer service is the development of an comprehensive service strategy that can energize the entire network around continuous service value improvement. Exploring ways to reconcile customer service value requirements and supply channel capacities is at the very core of a successful channel service strategy. As portrayed in Figure 7.1, developing a comprehensive strategy consists of six steps. The first step is to define and publish the SCQM principles that are to guide individual and channelwide service value decisions. The customer service leaders of the 1990s

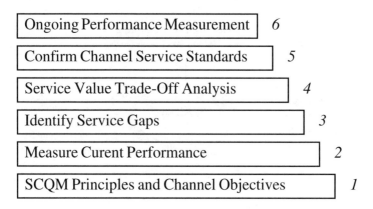

Ongoing Performance Measurement   *6*

Confirm Channel Service Standards   *5*

Service Value Trade-Off Analysis   *4*

Identify Service Gaps   *3*

Measure Curent Performance   *2*

SCQM Principles and Channel Objectives   *1*

Figure 7.1. The SCQM service strategy.

will be those supply networks that can weld channel members together in the pursuit of continuous and synchronized improvement of all processes, products, services, and work cultures. The ongoing ability to create value and deliver it to the customer can only be achieved by responsive "world-class" performers who are tireless in their examination of every aspect of the firm's operations in search of untapped sources of quality and customer satisfaction. Band [21] feels that value creation is strategic, systemic, and continuous. It is *strategic* because delivering quality to customers is at the very heart of the company's corporate strategy. It is *systemic* because the information, planning, and execution systems utilized by the organization must be continually refocused in the pursuit of customer value. Finally, it is *continuous* because the challenge of gaining and keeping customers in today's marketplace requires an unrelenting dedication to achieving continuous improvement in levels of performance. The elements of creating value for customers are portrayed in Figure 7.2.

The second step in the development of a channel service strategy is to measure current performance against the customer service value requirements identified in the process discussed above. Measuring the service effectiveness of the enterprise consists of two elements: determining customer needs and expectations, and quantifying the firm's current service practices and measuring the variance between existing levels and marketplace expectations. This discovery process should provide answers to the following questions:

1. Who are the supply channel's chain of customers?
2. What service attributes are pivotal in meeting their needs?
3. What service activities are currently being performed to meet these needs?

Once the channel's *external* services position has been detailed, an *internal* audit of actual service practices needs to be executed. This is a difficult task for

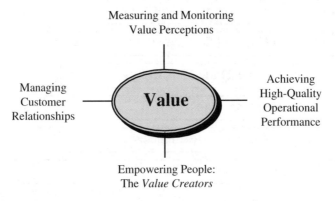

Figure 7.2. Elements of customer service.

the supply channel as a whole to perform. Performing an effective audit means that, as each channel member details its findings, the results need to be shared as much as possible with immediate channel suppliers and customers, if not with the channel as a whole. The audit should provide answers to such questions as the following:

- What is the prevailing corporate culture regarding customer service excellence?
- How do the firm's functional business units (Marketing and Sales, Finance, Logistics) perceive their role in providing customer service?
- How is customer service measured within each functional business unit?
- What are the performance standards and service objectives?
- What are the internal customer service performance measuring systems, and how are they integrated to provide the firm with a corporate viewpoint?
- What are the current performance metrics?
- How are these metrics used to increase service performance?

Answers to these question can arise by performing an analysis of existing service performance data and through departmental interviews. Performance metrics can be attained from internal data collected relating to the marketing, product, and service attributes detailed in Figure 7.3. Interviews conducted with managers, supervisors, and key staff members of each department can also assist in the internal audit. Elements such as organizational structure, performance measurement systems, problem-solving techniques, perception of customer needs, level of direct customer contact, and plans for service improvements will detail current service paradigms and interfaces with other functional departments.

The results of the external and internal services audits can be analyzed by positioning each attribute on a *services attribute matrix*. The model of the matrix appears in Figure 7.4. [22] The matrix permits the reviewer to rank each attribute by its relative importance to the customer and the firm's performance. The same method can be used for competitive benchmarking by ranking the position of the competition contrasted to the firm's performance. The matrix has five zones. The service rank for each attribute is measured by determining the relative importance of the attribute to the customer (or the strength of the competitor) to the actual performance of the firm. By intersecting the lines, service managers can see the strengths or weaknesses relative to customer expectations and the position of competitors. Once each service attribute has been applied to the matrix, management can begin the task of ranking each by level of importance. Attributes that illustrate a competitive vulnerability require high priority in the firm's service strategy. On the other hand, attributes that indicate irrelevant superiority are probably causing unnecessary service costs and should be elimi-

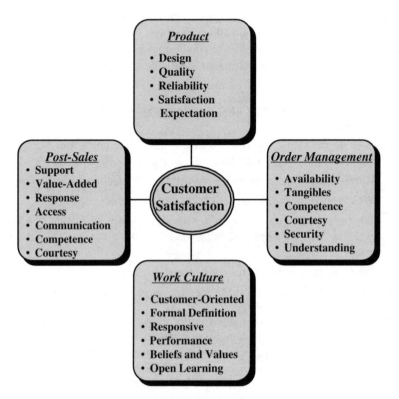

Figure 7.3. Areas of customer service management.

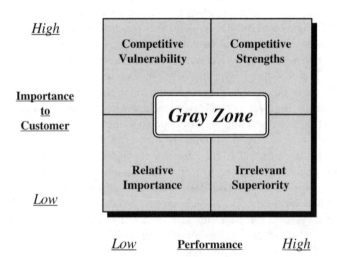

Figure 7.4. Customer service attribute matrix.

nated. Metrics that indicate that customer expectations (or competitive positioning) and the firm's corresponding performance intersect somewhere in the middle range means that the service attribute is located in the gray zone. This zone indicates that the level of importance and performance are not of significant strategic importance.

The third step in channel service value strategy formulation is to identify service gaps. Metrics illustrating performance *gaps* provide the firm with a clear understanding of the level of current performance as opposed to both customer perceptions and expectations and to advantages enjoyed by industry leaders. Once these gaps have been quantified, management can begin to redesign organizational and value structures, refocusing them on continuous improvements in processes and operational performance. Zeithaml et al. [23] have formulated a service quality model that highlights the gaps between customer expectations and actual service performance. The model, shown in Figure 7.5, identifies five gaps inhibiting "world-class" service delivery. The gaps are described as follows:

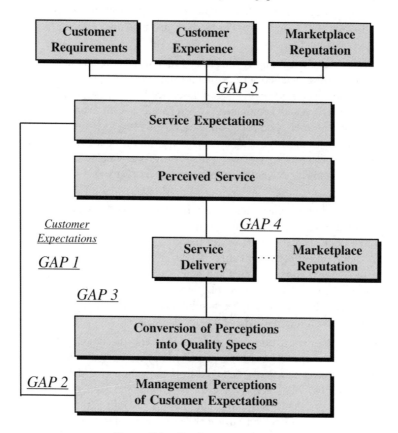

Figure 7.5. Service quality model.

- *Gap 1: Customer expectations and management perception.* Perhaps the most fundamental service gap can be found in the variance that exists between what customers expect from products and services and what the firm's management perceives as customer requirements. Often, there is wide congruence between expectations and perceptions. However, management might not always be completely aware of which service attributes constitute high priorities for customers. In addition, even when key attributes are identified, it might be difficult to quantify what exactly is the expected level of service.

- *Gap 2: Management perception and service-quality specification.* Although service managers might formulate correct perceptions of customer expectations, a gap can occur when a firm experiences difficulty translating those expectations into service-quality specifications. The most frustrating form of service-quality gap occurs when there are clear specifications but management is not committed to enforcing them. For example, a firm pledges timely field support for a product but does not hire sufficient staff to meet presales pledges, and then does nothing to remedy the service deficiency.

- *Gap 3: Service quality specifications and service delivery.* Even when firms develop detailed service-quality standards and programs for creating a *customer-centric* organization, service leadership is not a certainty. A gap may exist and continue to widen based on such elements as poorly trained customer-contact personnel, ineffective service-support systems, insufficient capacity, and contradictory performance measurements. When service delivery quality falls short of the standard, the result has a direct impact on what customer can expect (Gap 5).

- *Gap 4: Service delivery and external communications.* A fundamental element shaping customer expectations are service quality standards detailed in advertising media, promised by the sales force, and found in other communications. Take, for instance, a distributor that publishes a 24-hour order turnaround time but lacks the information systems for effective inventory accuracy and picklist generation to meet that objective. When the customer is told and expects next-day delivery of an order and does not get it, the gap begins to widen between what sales tells the customer and what logistics can deliver.

- *Gap 5: Perceived and expected service.* The discrepancy that arises between the expected and the perceived service(s) is detailed in Gap 5. The content of this gap is gathered from the cumulative shortfalls found in each of the four enterprise gaps. Determining the depth of this discrepancy is more than just totaling the metrics of Gaps 1 through 4; often the cumulative effect on customer perceptions of service delivery are greater than the sum of the parts. The source of customer service expectations can be found in word-of-mouth communications, past experiences, and customer requirements.

Metrics illuminating the gaps existing between the level of perceived and actual customer service performance are critical in orienting the supply channel's service value strategy. Once these gaps have been quantified, channel management can begin the process of focusing resources to satisfy targeted improvements in processes and operational performance.

The steps detailed above will assist supply channel members in determining the attributes of network customer service value and positioning it in relation to marketplace standards. However, before effective channel service strategies and measurement systems can be implemented, not only the current actual cost of selling and servicing the product but also the incremental cost of reaching the next level of service must be calculated. Obviously, attaining high levels of customer services cannot be achieved without corresponding growths in performance, not only in individual enterprises but also in the channel as a whole. In the past, it was assumed that as the service level increased so did the costs. As Figure 7.6 illustrates, companies often attempted to chart expected sales revenues and the cost of services and to calculate the optimum service level. Once the increased level of service costs had been justified, cost-service trade-off analysis could begin. Such cost trade-offs took place in one or several areas within the supply pipeline: warehousing and inventory deployment, inbound, interbranch, and outbound transportation, procurement and manufacturing processes, intra-organizational and extraorganizational communications facilities, information systems implementation, people empowerment, and customer service systems.

Although a critical element in the strategic plan, when performing service value-cost trade-off analyses channel members must resist falling into the fatal attitude of "we win, the customer loses." The object of the analysis is to pinpoint where in the channel service failures and excess costs occur and then to reengineer processes so that the level of value offered and the cost of the service make

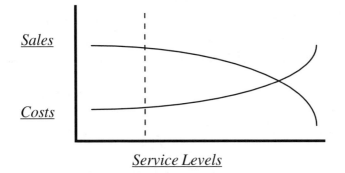

Figure 7.6.   Charting optimum service levels.

economic sense. Often this requires more forward thinking on the part of channel members. Take, for example, a strategy that calls for the elimination and consolidation of several channel warehouses into one. In the scenario, the consolidation is calculated to reduce inventory from $400 million to $100 million and improve cycle time and quality. The drawback is that customer service will suffer as a result of possible shortages on key items. Clearly, stocking additional inventory defeats the purpose of the consolidation. A possible solution would be to use expedited shipment services to ensure customers service would not suffer. Under the circumstances, spending an extra $500,000 for premium delivery service would be a bargain. The goal of service value-cost trade-offs analysis, therefore, is not to look for ways to optimize the customer satisfaction versus the anticipated costs but rather to rank and prioritize competitive service elements, and to refocus the resources of the firm on a continuous improvement campaign targeted at each element. Service leaders seek continuous improvement in cost, quality, response time, and flexibility simultaneously.

Step 5 in the channel service value strategy consists in setting service standards on which the whole channel can agree. Optimally, there is only one level of service to be pursued by the supply channel—100% conformance to expected levels of customer value. Hopefully, the process of uncovering and detailing the true nature of multiple levels of value required by the channel's chain of customers will be completed by this stage. To be successful, there must then be complete alignment of these identified service requirements and the capabilities of each level in the supply network. Setting such standards is a difficult task and requires the full cooperation of all channel members. Whether it is arranged by compromise or negotiation, the development of a common set of channelwide standards is fundamental in offering the customer consistent service value and warding off competitors. Possible key areas where standards are essential are listed as follows:

- *Order cycle time.* Measured by customer standards, this is the elapsed time from order receipt to order delivery.

- *Inventory available.* Percentage of a particular item available when ordered from stock.

- *Order reliability.* Percentage of orders delivered on time.

- *Ordering convenience.* Ease of ordering, both person-to-person and computer-to-computer.

- *Order flexibility.* Ability to respond to small JIT orders as well as bulk quantities.

- *Frequency of delivery.* Ability to deliver JIT requirements according to the frequency desired by the customer.

- *Order completeness.* Percentage of demand filled complete at time of order.

- *Order status information.* Capability to inform customers at any time of their order status.

Ideally, each supply channel should strive to establish performance standards across a wide range of service values. These standards should not only span the entire chain of customers, but they should also span the complete order cycle beginning with *pretransaction* activities, such as inventory availability and order response, progressing through *transaction* activities, such as order fill rate and on-time delivery, and concluding with *posttransaction* activities, such as complaints, returns, and service parts availability.

The final step in managing channel service value is structuring a mechanism that will provide for ongoing performance measurement. Just reviewing the expectations of the customer once a year will hardly provide the kind of metrics necessary for effective service management; there must also be flexible measurement tools in place that provide detailed information on an ongoing basis and that change as marketplace expectations change. Once the means to gauge service performance have been formalized, the results can be measured against the standard and corrective action then taken to eliminate the variance.

Developing effective service value performance measurements involves measuring *both* individual company internal as well as channelwide external metrics. In today's business environment, channel service value management is the key to marketplace leadership. Although the end customer occupies the final position in the supply pipeline, at the gateway stands the channel's chain of internal customers. As such, the quality of products and services available to the end customer directly stems from the service quality that is found as it proceeds down through the supply network. If total channel product quality, on-time delivery, replenishment order accuracy, financial transference, and service support contain unacceptable performance gaps, final distribution points cannot help but respond to the final customer by passing on the excess costs, lack of quality, and poor service. On the other hand, programs designed to maintain high service quality will enable each member along the supply pipeline to jump start and sustain continuous improvement initiatives that can then be communicated to each succeeding level. When viewed as a continuous chain of customers, the supply channel should provide effective quality measurement methods that contain a clear statement of product and service expectations, search to remove redundancies in material movement, delivery, and paperwork, and implement ongoing improvements designed to eliminate costs and provide a window into new market opportunities and trends that would be impossible to achieve if each channel member were to act on their own.

Implementing SCQM

The decision of a channel to initiate a program of SCQM process improvement can be directly traced in one manner or another to emerging marketplace requirements for increased customer value. The SCQM initiatives that result may have a variety of different motives and consist of an array of high-impact changes to productive processes ranging from customer service to purchasing. In any case, no matter the improvement path, the goal is to increase the substance and the perception to the customer of increased value. In the past, the customer was treated as an afterthought: The focus was on producing standardized products, passing them through the supply channel, and hoping that they would match customer requirements. In contrast, today's process-focused organizations seek to understand the needs of the customer first, and only then design flexible, high-quality processes that are capable of continuously delivering value-added solutions to the marketplace.

Types of SCQM Projects

Undertaking the implementation of SCQM projects can be the result of several different factors. They might arise as a response to unexpected, yet strong challenges from competitors that have gained a slight marketplace superiority due to new technology or management methodologies. On the other hand, channel improvement projects might began as a result of the efforts of a single channel member who has decided to overhaul internal processes and now demands that the entire channel do the same. Further, channel members may jointly undertake process improvement in a decidedly low-key and evolutionary manner. As individual company managers and operating staffs slowly begin on their own to quietly improve processes, the management of the processes of the entire channel are infused gradually with the new emphasis on continuous quality improvement.

Then again, the impetus to channel process improvement may come as part of a proactive decision. Some channels might focus on using process improvement as a way to gain marketplace advantage before the competition undertakes a similar initiative. Some might perceive it as a way to cut channel costs and improve on productivities that will allow them to reach new markets or undercut rival channels. Others may have to embark on an SCQM program as an answer to customers who require their suppliers to initiate quality and productivity improvement efforts leading to certification. Still others will use it as a means to reverse decidedly unhealthy trends of growing product and service quality defects. Regardless of the motive, the SCQM project will seek to define several key high-impact actions that will provide the channel with the desired result.

Whatever the motive, an SCQM improvement project normally seeks to accomplish one or a mixture of the following objectives: responding to demands for enhanced customer value, increasing the productivity of a particular process by either cutting costs or accelerating the movement of value through the supply channel, or creating market-winning differentiation. There are many directions a process improvement effort targeted at customer value can take. The following are some of the most important: better marketing communications regarding promotions and deals, new value-added services, inventory availability, increased order processing capabilities, shorter delivery cycles, or enhanced order receipt and follow-up. SCQM projects targeted at improving efficiencies and productivities are normally more focused and narrow in scope than those targeted at impacting customer value. Critical improvement actions at this level can consist of the following [24]:

- Eliminating a process step or combining it with another related process
- Speeding up a process by adding additional resources or eliminating internal cycle time
- Performing processes in a parallel mode that previously were carried out sequentially
- Improving the efficiency within a process through the application of TQM techniques
- Improving process consistency and standardization by reducing sources of variability
- Better coordinating and integrating a process as it flows from one channel member to another

Finally, an SCQM improvement project can be undertaken that is designed to give channel products/services a unique or distinctive appearance to the marketplace. This "differentiation" refers to the capacity of channel processes to customize certain product or service attributes to meet specific customer needs.

When deciding on an SCQM improvement project, channel members will usually focus on a single critical area that includes, perhaps, a set of homogeneous secondary areas. It would be very difficult for a supply channel to undertake an improvement project that considers multiple areas at once. Given the collaborative requirements and complexities involved in an SCQM project, most projects will consist of a single improvement target that will bring the most value to the entire channel network. The steps required to implement an SCQM project are considered in the next sub-section.

Implementing SCQM Process Improvement Actions

The successful implementation of SCQM quality and productivity process improvement efforts is a much more difficult task than similar projects carried on only at the company level. Although lacking none of the prerequisites necessary for company-level improvement projects, projects carried on at the channel level contain several unique requirements. To begin with, the supply channel must identify its critical processes. Although there may be regions of similarity, each supply channel network possesses it own unique set of business processes. Channels may have analogous functions, such as order processing, logistics, and procurement, but the actual processes performed can differ greatly, depending on what industry the channel belongs to and the nature of the customer value it is pursuing. Whereas precisely describing each business process is alone a difficult task within a single company, it is even more complex when a given process activity, say product delivery, is traced across several channel members. Being able to converge what could be radically different ways of conducting business into a single comprehensive process requires a new type of conceptual quality management thinking that seeks to look horizontally across the channel rather than simply considering processes occurring within an individual company.

Second, the requirements for effective leadership to conduct a successful SCQM project are exceptionally demanding. Normally, the leadership role will be filled by a chief executive from one of the channel member companies. Usually, this individual is from one of the largest stakeholders in the channel. The role of this individual is to define the vision and goals for SCQM quality and productivity improvement projects and to establish reasonable expectations concerning the magnitude of the changes involved. In addition, the project sponsor will have to work closely with other channel member executives in guiding them in their efforts to redefine the cultures of their organizations and direct them on the path of continuous cooperative channel process improvement. The role of channel quality leader does not end with the completion of a given project but is really an ongoing position with the goal being a never-ending search for channel value. No better example can be given than the tremendous quality efforts being led today by the automotive industry giants driving the philosophy of continuous quality through multiple layers of channel suppliers. Chrysler, for instance, introduced a supplier management program called Score, and acronym for "supplier cost-reduction effort." The goal of the program is to work closely with suppliers to reduce Chrysler's procurement costs by a targeted 5% of a given supplier's annual billings. Since its inception, the program has netted total savings of $2.5 billion for Chrysler. But more importantly, the demand for savings has cascaded down the supply channel, as each supplier asks of their supplier, in turn, to engineer more productive processes [25].

Third, networks of companies and their work forces must all be aware of their pivotal place in continuous channel process improvement. The key to SCQM success is for everyone in the channel, from the raw-materials supplier to the retailer, to recognize that they are all essential parts of a common quality program. Such a program requires each channel member to fully understand each business process and how resulting inputs and outputs can impact the channel's chain of customers. Creating such a new mind-set requires the existence of open channels of communication to publicize success and share credit among network participants. It also requires all employees regardless of their places in the channel to consider themselves as "process-centered engineers" dedicated not only to performing their jobs efficiently but also to searching for continuous process improvements both in their own companies and across the entire supply network.

Fourth, supply channels should be realistic about the SCQM projects they select. An easy way to begin a project is to build on successes that have already occurred on the company or channel level. By staying close to past quality-focused programs that have been successful, channel members can embark on quality campaigns that are within their capabilities and have a good likelihood of success. Another method is to use the "VIP" method [26]. The "V" stands for *visibility* and refers to a particularly critical process problem or one that offers significant potential for quality improvement. The "I" refers to the *impact* the improvement will have on channel quality and productivity. Finally, the "P" denotes the *probability of success* the improvement project will enjoy. The overall goal of the method is to select successful projects that will engender a sense of enthusiasm and momentum among channel participants. It is important that the concept of continuous channel process improvement not become bogged down. Methods to prevent this occurrence are changing the area targeted for improvement, rethinking rewards systems, and searching for quick successes that will build credibility for the SCQM improvement process.

Fifth, once selected, the progress of SCQM improvement projects must be carefully measured. Each channel system will have to identify key performance measures. Some will be based on the requirements emanating from the channel's chain of customers. These measures will signal whether issues such as the ease of order placement, delivery, or other aspects of customer service performance is the central standard. On the other hand, productivity measures, such as cost reduction, asset utilization or other process targets, are critical to charting the collective internal improvement needs of all channel members. Whatever the actual content of the measures identified, they must reflect the performance of the process across the entire channel, not just for a single company, and be communicated to and be used by all channel members. A useful tool in creating transchannel performance measurement is information technology. Today's computer systems can be easily linked together, thereby facilitating data collecting, transfer, and analysis.

The last step in ensuring successful SCQM improvement projects is to make process improvement a channelwide strategic and operational paradigm. Like most good ideas, the value of continuous channel process improvement is in direct proportion to its becoming a permanent part of the management landscape. Localized or one-shot channel improvement successes will accomplish little in changing a channel's interest from being focused around individual company successes to being channel process focused. In earlier chapters, it was stated that the real advantage offered by SCM was the ability of allied companies to utilize the competencies and productivities of each other to build unstoppable sources of marketplace advantage and to create whole new regions of competitive space. Realizing the benefits of SCM requires the conversion of all channel members to continuous SCQM process improvement. At the very heart of SCM stands a collective commitment to constantly assuring that channel processes are being performed to create customer value and optimize on productivities, to continuously search for new opportunities for improvement, and to effectively translate these opportunities into performance realities [27].

Summary and Transition

A decided focus on quality improvement and customer service value creation has become one of the hallmark characteristics of today's "world-class" organization. For over a decade the typical enterprise has recognized that the only path to competitive advantage in today's global marketplace is to be found in a firm commitment to the implementation of quality management techniques. However, although companies have been quick to adopt QM philosophies to improve *internal* functions, it has only been recently that they have turned their attention to extending QM to their *external* supply chains. Several radical changes in the business environment are spearheading this movement toward supply chain quality management (SCQM). These changes, found in the rising level of customer expectations, product design and manufacturing management, government regulation, technology, and value-add opportunities offered by channel partnerships, and exploration of new methods targeted at reducing total channel costs and accelerating the flow of value through the supply network. Managing these challenges effectively and efficiently has made SCM today's hottest business topic.

As a channel management concept, SCQM can be defined as

> the participation of all members of the supply channel network in the continuous and synchronized improvement of all processes, products, services, and work cultures focused on generating sources of productivity and differentiation and the active promotion of market-winning product and service solutions that provide total customer value and satisfaction.

When this definition is viewed in detail, it is evident that SCQM is concerned with several crucial elements of marketplace success. To begin with, it is focused on the continuous creation of customer value by facilitating the active promotion of channel effectiveness in the performance of critical service quality elements, channel efficiency in providing the desired mix of products and services perceived by the customer as providing superior value for the price, and channel differentiation characterized by the uniqueness of channel product and service combinations. Second, SCQM provides today's supply chain with the ability to leverage the TQM toolkit to maintain an atmosphere for the continuous creation of customer service value.

Third, SCQM requires each channel node to follow a formalized process of product and service quality improvement that provides the best value to the customer. Fourth, SCQM requires the development of a formal methodology requiring interchannel management and operational commitment, definition of the service quality requirements for the entire chain of customers, charting of current levels of performance, defining strategies to improve channel quality to exploit new marketplace opportunities, quality program implementation, and the creation of organizational and performance techniques that provide for continuous channel quality improvement. Finally, SCQM process cannot take place without the creation of effective performance measurements that ensure that the combined resources of the entire channel network are being utilized in the most efficient and value-enriching ways possible to provide market-winning, cost-effective customer value.

Supply chain quality management represents the latest stage in the total quality movement. Unlike most of the TQM and reengineering process improvement movements of the recent past, SCQM marks a decisive shift away from product-based and enterprise-based quality to one that focuses on how the entire supply channel can increase service value for the chain of customers it serves. Although the application of SCQM can take many different forms based on channel circumstances, such as marketplace position, strength of the competition, management culture, and other operational characteristics, they all share a basic set of common attributes. To begin with, they are all centered around promoting business process quality through the application of a formal process improvement methodology that fuses TQM and BPR concepts into a single channel management technique focused on increasing service value. Instead of a one-time project, the SCQM improvement methodology provides supply chains with the capability to sustain process improvement as a permanent management objective.

In addition, all SCQM projects use similar analytical models and quality and productivity measurement techniques. At the heart of quality analysis tools can be found quantitative techniques that are used to track statistical means, variances, process distributions, random sampling, and other methods of identifying process error. The goal of these quality measurements is to provide meaningful data relating to the magnitude of customer service value, channel productivity, and channel process effectiveness. Another productivity measurement tool is benchmarking. The goal of this technique is to analyze the level of product and service quality of industry competitors with an eye not just to match but to exceed the best performance of the process anywhere.

Finally, the ability of SCQM to provide for continuous improvement in service quality and the development of new regions of competitive value is directly dependent on the efforts of the collective efforts of the supply channel network's people resources. As yesterday's "command and control" work culture gives way to work force empowerment and team management styles, quality management has migrated from a product- to a customer-focused paradigm led by teams of *process engineers* ceaselessly searching to remove wastes and add value to the processes they work with daily. The typical supply channel quality organization would consist of a quality management executive board and special channel task forces recruited from each network member. The objective of the channel organization would be to link the knowledge and skills of people across the entire supply network to open new vistas of innovative thinking, achieve levels of customer value previously thought impossible, and establish the framework for sustained quality improvement that wins customers and creates a source of unbeatable competitive advantage.

The central focus of SCQM is the management of customer value. "World-class" supply channel are tireless in their examination of every aspect of their operations in search of untapped sources of quality and customer satisfaction. Activating such a vision means that the entire supply channel must be vigilant in measuring customer perceptions of service quality, identify service shortfalls, and respond to service gaps across the length of the pipeline. Once these gaps have been quantified, cross-channel quality teams can begin to redesign supply point interfaces and value structures, refocusing them on continuous incremental improvements in process and information performance. Channel service leaders see SCQM as one of the fundamental building blocks of marketplace dominance, set levels of service that exceed their customers' expectations and surpass the standards set by the competition, have an action-oriented attitude focused on teamwork and commitment to excellence, and are ceaseless in their endeavor to satisfy the customer at all costs.

Undertaking the implementation of SCQM can be the result of several different factors. A powerful new competitor may challenge a supply channel to reexamine the quality and productivity of its processes. On the other hand, a channel network might decide to use SCQM as a way to gain marketplace advantage by reaching new markets or undercutting rival channels. Regardless of the motive, the implementation of SCQM involves a number of unique requirements. To begin with, the supply channel must identify its critical processes. Second, the project requires exceptionally strong leadership to set the quality vision, spearhead the improvement process, and assist channel members to redefine the cultures of their organizations and keep them on the path to continuous cooperative channel quality improvement. Third, the network of companies and their work forces must all be aware of their pivotal place in maintaining continuous channel process improvement. Fourth, supply channel leaders must be realistic about the SCQM project they select to undertake. Fifth, once an SCQM project has begun, the results must be carefully measured to provide benchmarks of success. The last step is to make SCQM a channelwide strategic and operational paradigm.

Notes

1. The reader is referred to such works as Thomas J. Peters and Robert H. Waterman, Jr., *In Search of Excellence.* New York: Harper & Row, 1982; David A. Garvin, *Managing Quality: The Strategic and Competitive Edge.* New York: The Free Press, 1988; A. V. Feigenbaum, *Total Quality Control.* New York: McGraw-Hill, 1961;

Philip B. Crosby, *Quality Is Free: The Art of Making Quality Certain.* New York: Signet Mentor, 1979; W.E. Deming, *Out of the Crisis*, Cambridge, MA: MIT Press, 1986; Joseph M. Juran, *Quality Control Handbook.* New York: McGraw-Hill, 1979; Shingo Shigeo, *Non-Stock Production: The Shingo System for Continuous Improvement.* Cambridge, MA: Productivity Press, 1988.

2. An excellent summary of quality management issues can be found in Richard J. Schonberger, *Building a Chain of Customers.* New York: The Free Press, 1990, p. 66.

3. See the discussion in C. John Langley, Jr. and Mary C. Holcomb, "Total Quality Management in Logistics," in *The Logistics Handbook,* (James F. Robeson and William C. Copacino, eds.) New York: The Free Press, 1994, pp. 183–186.

4. This excellent phrase was coined by Richard J. Schonberger in his book *Building a Chain of Customers.*

5. These quality attributes have been adapted from the research done by David A. Garvin, *Managing Quality: The Strategic and Competitive Edge.* New York: The Free Press, 1988, pp. 49–60, and Schonberger, *Building a Chain of Customers*, pp. 83–85.

6. Helen L. Richardson, "Speed Replaces Inventory," *Transportation & Distribution* 37 (11) (November 1996), pp. 69–76.

7. Michael Hammer, *Beyond Reengineering.* New York: HarperBusiness, 1996, pp. 6–7.

8. This is Hammer's definition, *Beyond Reengineering*, p. 11.

9. See the discussion in Robert Hall, *The Soul of the Enterprise.* New York: HarperBusiness, 1993, pp. 74–77, Schonberger, *World Class Manufacturing,* pp. 91–102, and Martin Christopher, *Logistics and Supply Chain Management.* Burr Ridge, IL: Irwin Professional Publishing, 1994, pp. 47–69.

10. See the discussion in Patrick M. Byrne and William J. Markham, *Improving Quality and Productivity in the Logistics Process.* Oak Brook, IL: Council of Logistics Management, 1991, pp. 163–176, and Douglas M Lambert, "Logistics Costs, Productivity, and Performance Analysis," in *The Logistics Handbook.* New York: The Free Press, 199 pp. 260–302.

11. For a detailed treatment of this subject, see Byrne, et al., pp. 163–168.

12. For general information relating to benchmarking see: Robert Camp, *Benchmarking: The Search for Industry Best Practices That Lead to Superior Performance.* Milwaukee, WI: ASQC Quality Press, 1989; Robert Camp, "Benchmarking," in *The Logistics Handbook,* pp. 303–324, Francis G. Tucker, Seymour M. Zivan, and Robert C. Camp, "How to Measure Yourself Against the Best," *Harvard Business Review* 87 (1) (January–February 1987), 8–10, and Christopher, *Logistics and Supply Chain Management*, pp. 71–94.

13. The changes to workforce organization have been excellently summarized in Schonberger, *World Class Manufacturing*, pp. 177–200.

14. This definition can be found in Gary Hamel and C. K. Prahalad, *Competing for the Future.* Boston, MA: Harvard Business School Press, 1994, p. 219.

15. Bernard LaLonde and Paul H. Zinszer, *Customer Service: Meaning and Measurement.* Chicago: National Council of Physical Distribution Management, 1976, pp.

203–217, Douglas M. Lambert and James R. Stock, *Strategic Logistics Management*, 3rd ed. Homewood, IL: Irwin, 1993, pp. 111–112.

16. Bernard LaLonde, Martha C. Cooper, and Thomas G. Noordewier, *Customer Service: A Management Perspective*. Chicago, IL: Council of Logistics Management, 1988, p. 5.

17. William A. Band, *Creating Value for Customers*. New York: John Wiley & Sons, 1991, p. 80.

18. Richard J. Schonberger, *World Class Manufacturing: The Next Decade*. New York: The Free Press, 1996, pp. 19–48.

19. Valarie A. Zeithaml, A. Parasuraman, and Leonard L. Berry, *Delivering Quality Service*. New York: The Free Press, 1990, pp. 20–23; Valarie A. Zeithaml, A. Parasuraman, and Leonard L. Berry, "A Conceptual Model of Service Quality and Its Implications for Future Research." *Journal of Marketing* (Fall, 1985) 41–50; David F. Ross, *Distribution: Planning and Control*. New York: Chapman & Hall, 1996, pp. 391–393.

20. Sarah A. Bergin, "Recognizing Excellence in Logistics Strategies," *Transportation and Distribution* 37, (10) (October 1996), p. 48.

21. Band, p. 21.

22. The use of the service matrix concept is more fully explained in Karl Albrecht and Lawrence J. Bradford, *The Service Advantage*. Homewood, IL: Dow Jones-Irwin, 1990, pp. 171–185.

23. Zeithaml et al., *Delivering Quality Service*, pp. 9–13.

24. For a fuller treatment of these improvement areas see Patrick M. Byrne and William J. Markham, pp. 177–179.

25. Justin Martin, "Are You as Good as You Think You Are?" *Fortune* 134 (6), 146.

26. This method is detailed in Byrne and Markham, pp. 244–249.

27. See Hammer's judgment on the importance of the process-centered organization in *Beyond Reengineering*, pp. 10–17.

8

The Work Force and Information Technology

The growth of the concept of supply chain management (SCM) is the direct result of several dramatic changes in the way today's business environment is structured and how companies compete for marketplace advantage. Some of these changes are to be found in the methods by which products are developed, manufactured, warehoused, and sold, the way the enterprise is organized and its productivities measured, and the skills required to manage, motivate, and empower the work force. Other changes have come from without. The global marketplace has rendered obsolete the vision (which was surely never a practical strategy) that single companies could seize and maintain market leadership solely by the strength of their own efforts and precipitated the age of the "virtual" organization and supply chain partnership. The explosion in the various forms of information technology has also acted as the catalyst as well as the foundation of today's revolution in the way customers and suppliers engage in the business of buying and selling. Finally, these changes have altered forever almost century-long organizational models by which companies were run, the structure of the relations existing between management and labor, the methods used to plan and measure competitive success, and the place each company occupied in the business ecosystems of which they were a part.

Of all these changes, none has the immediacy and poignancy as do those that have redefined the values of labor and impacted the organization of work. Except for a very privileged few, most people spend their lives working not only to provide for the physical wants and needs of life but also to pursue their own self-realization as fundamentally creative beings. The radical reengineering and restructuring strategies pursued over the past half-decade by all institutions from commercial businesses to health care have, however, rendered problematic once well-defined and immutable paths of work-life behavior and often substituted doubt and bewilderment and, sometimes, anger in their place. The revolutionary nature of these changes are comprehensive: They have transformed the kinds of work people do, the nature of the jobs they hold, the knowledge and skills they must possess, the methods by which their productive value is measured and rewarded, the careers they pursue, and the role their managers and their organizations play in directing their efforts and activating their personal sense of accomplishment and self-worth. Although the rise of new business philosophies, like SCM and business process reengineering (BPR), have indeed accelerated these changes, they have also established new directions and new opportunities for the rebirth of work and organizational values. What the work life of tomorrow will be has yet to take definitive form; what can be certain is that it will provide infinitely greater sources for personal choice and heightened avenues for self-value and accomplishment.

The pace of the changes driving today's business activities and the capabilities and missions of the work forces who perform them are directly driven by the speed and availability of information technology. It has often been said that we have moved beyond the production era and are now living in the Information Age. Such an assertion is borne out by the growing recognition that the fundamental bases of business value and wealth have subtly but definitely been altered. Driven by the enormous advances occurring in computer architecture and software during the past decade, companies have slowly come to realize that the foundation of economic wealth is no longer to be found simply in materials, manufacturing processes, and labor, but rather in access to accurate information, the speed by which it can be communicated, and the depth of knowledge it provides people in their efforts to make effective decisions. In addition, the value of information and communications technology (ICT) is not restricted to simply being used as a tool to accelerate the speed and productivity of business functions. In fact, the real "information revolution" can be found in the ability of ICT systems to be a fundamental enabler providing companies with the opportunity to continually activate new workplace cultures and structures as well as new visions of how they can leverage the competencies of allied channel partners to create radically new ways of pursuing competitive advantage.

Exploring how changes to the work force and information technologies are impacting SCM is the subject of this chapter. The analysis begins with a review of the migration of work force functions and values from a concern with mass-production manufacturing and mass-distribution organizational paradigms to the revolution in the roles, requirements, and functions of the work force to be found in the "Age of Supply Chain Management". Key topics will revolve around discussing the challenge of SCM leadership, the learning organization, and principles of workforce activation. Next, the chapter moves to a wider perspective with an exploration of the new forms of organization necessary to successfully execute the SCM philosophy. Particular attention is paid to the establishment of effective SCM work-force teams and leveraging the "virtual" supply channel. At this point, the chapter shifts to a review of how ICT is reshaping the tactical and strategic functions of the work force and supply channel networks. Based on the nature and types of ICT structures detailed, the chapter proceeds to analyze how individual companies and whole supply channels can leverage ICT tools to achieve breakthroughs in organizational and supply network capacities and competitive positioning. Finally, the chapter concludes with a comprehensive discussion of the organizational framework necessary to ensure the continuous implementation of ICT in today's SCM environment.

Today's Work-Force Challenge

The ability of companies to realize superior levels of performance is ultimately founded on the capabilities, competencies, and creative visions of the people who comprise their organizations. In the past, management attention on the people organization was focused narrowly on formulating the right management principles necessary to ensure optimal work-force productivities through the maintenance of rigid organizational structures and operations systems and procedures. In contrast, the enabling power of ICT tools and the need for close intersupply channel partner functional integration have dramatically shifted the management of human resources away from a concern with the maintenance of rigid "command and control" organizational hierarchies to a radically new perspective that considers lasting competitive advantage to be attainable only by liberating and activating the skills, discipline, self-motivation, creativity, ability to solve problems, and capacity for learning that is to be found among the individual managers and functional professionals located throughout the supply channel network. Ultimately, what a company or a whole supply channel is really selling their customers is not products and services but the enrichment value of the skills and knowledge possessed by the people who work within the organization and outside in its partner companies and suppliers.

The Changing Character of the Organization

As little as a decade ago, most manufacturing and distribution companies were still being managed as they had been for decades. The organization was divided into a series of management and submanagement hierarchies that spanned the entire enterprise. At the top of the command pyramid resided the top management staff, the firm's leaders who were charged with the task of defining enterprise strategy, setting business direction, and providing the objectives to be pursued by departmental managers. On the next level of the pyramid could be found the company's tactical managers. It was this group's responsibility to translate the business strategy into detailed departmental objectives and budgets. Reporting to this group could be found, at the bottom of the organizational hierarchy, the firm's operations managers, who, in turn, directed the functional supervisors and work force in the performance of daily tasks and activities.

The operations culture that provided the mechanics of the whole structure was a reflection of the rigidity of the "command and control" organizational hierarchy. The values of both management and work force emphasized unity of command, division of work into detailed business functional units whose performance could be easily measured, a strong respect for scalar authority, responsibility, unity of direction, and functional and strategic centralization. This organizational model also promoted the concept that "thinking" was the preserve of managers as one moved up the hierarchy, and "doing" was the responsibility of those found at the broad-based bottom of the pyramid. Functionally, organizational processes moved in two rigidly defined directions. Moving up the structure was information from business activities. As information moved up the chain from layer to layer, middle managers summarized and presented it to top management. Executives, in turn, used these data as the basis for enterprise decisions and mandates that made their way back down the channel. As management directives passed serially through each level, they were used to create local objectives which, in turn, were used to guide the direction of the tasks to be carried out by ground-level supervisors and workers. Each management area was responsible for its own performance and productivity; each area fit neatly into some box on the firm's organizational chart.

Despite its acknowledged deficiencies and growing obsolescence, the multitiered hierarchical organization has served as the basic model for the workplace for almost a century. Developed to respond to the needs of an era characterized by the mass production and distribution of standardized goods, the objective of the whole system was to ensure conformity and predictability in management activities, organizational behavior, and productive functions. Yet, despite its strengths, by the 1980s, no matter how firmly entrenched in the minds of executives, their staffs, and the work force, this business organizational model began to unravel under the pressures of radically new marketplace developments. Some stemmed from the growing challenge of foreign competitors who were able to achieve dazzling productive and quality successes through the use of different management techniques that emphasized operations flexibility, the elimination of nonvalue activities, commitment to total quality and customer value, and the utilization of work force management methods designed to empower, energize, and enable people. Other forces impacting the old management structure could be found in the dramatic change of the customer from being a passive buyer to an active collaborator in productive and distributive processes, the ability of the computer to enable once segregated islands of people to network across hierarchical boundaries, and the use of the productive competencies of business partners that resided out in the supply channel.

The organizational response to these challenges that struck at the very core of competitive survival was, at first, disappointing. It is not that the inefficiencies and inflexibilities of the traditional organization were not unknown to executives. It is just that as competition from global companies intensified in the early 1980s, these deficiencies became glaringly apparent. Utilizing the management toolbox of the traditional model, companies, at first, sought to improve performance by initiating improvement task forces mandated with the mission to shrink costs, increase flexibility, decrease cycle times, and improve quality and service. New computer systems were installed; consultants were hired; management and work force were schooled in the latest quality management techniques. Unfortunately, no matter how methodically the new techniques and acquired skills were applied, the performance gap between old style managed companies and "world-class" competitors only widened.

Finally, by the late–1980s, companies began to understand the nature of their performance dilemma. What they had been concentrating on was applying task management methods to improve what were inherently *process* problems. Instead of concentrating on improving a single unit of productivity, such as labor efficiency, what was really needed was a broader initiative that focused on improving the whole process. Once companies recognized the fundamental relationship between competitive advantage and process improvement, it was obvious that the strategy of applying narrowly focused solutions that simply sought to shore up the departmentalisms of the old hierarchical-based organization was doomed to failure.

Substituting radically new organizational models to meet the challenge of global competition in the early 1990s coalesced around the application of two process-centered improvement approaches. The first response grew out of the Just-In-Time (JIT) movement and was characterized by a commitment to continuous incremental improvement in the firm's internal and external business processes. Centered around a variety of quality management (QM) techniques, the object of this approach was to reshape the attitudes and redirect the functional behavior of everyone in the firm away from narrow departmental task management and toward a broader understanding of how the productive processes that spanned the entire company could be continuously improved.

The second response was more radical. Known today as *process reengineering*, this management method was concerned with "the fundamental rethinking and radical redesign of business processes to achieve dramatic improvements in critical contemporary measures of performance" [1]. Instead of trying to improve existing processes, organizational reengineering sought nothing less than the total dismantling of processes and rebuilding them from the ground level up. Whether through incremental improvement or fundamental restructuring, the demise of the traditional hierarchical management model and the rise of new organizational structures built around whole business processes signaled tremendous changes in the roles and values of the work force.

The Changing Role of the Work Force

Quality management and BPR have provided today's work force with new tools to create superlative business processes that produce customer-winning service value and unbeatable competitive advantage. It is not that the work force that labored under the old pyramid organizational structure possessed less skills and knowledge or lacked motivation and drive. Clearly, the rigid cost and labor standards of yesterday's firm shows people worked hard and understood the nature of their roles. The real problem was not one of effort, but rather of the organization of work. One of the underlying principles of hierarchical management models is the assumption that people cannot function effectively when they report simultaneously to multiple managers nor when they are charged with the performance of parallel tasks. In fact, instead of expanding the scope of labor to encompass whole business processes, the traditional organization sought to divide those processes (such as production and inventory control or order fulfillment) into smaller work units under the control of a single manager who then was responsible for supervising, monitoring, and optimizing all work-cell activities. The goal of the organizational approach was for each departmental unit to successfully complete a common group of tasks constituting their portion of the process and then to pass it on to the next sequential work unit who, in turn, completed their portion and passed it further down the work process [2].

The role of management and the work force in such an environment was to ensure that the work that passed through their departments was performed as accurately and as efficiently as possible. The functions of the work force were a reflection of the splintering of business processes. Workers were segregated into distinct departmental units based on common skills and the nature of work activities. Workers were trained and expected to master the intricacies of their jobs. Salaries and other types of recognition were tied to the performance of specific job classifications. On their part, the function of management was to measure, improve, and enforce the work procedures governing the span of activities proscribed for each individual department. Once tasks were performed, it was management's responsibility to smooth the transfer of work to the next department and report on efficiencies, utilizations, and costs expended. The work culture of both managers and work force focused around following departmental procedures and attaining work unit performance objectives. What was happening in other parts of the company, even sister departments on opposite ends of the process flows, was generally unknown; entrepreneurial behavior was discouraged as being disruptive of the equilibrium established by departmental rules and management control.

The operational structure and culture of the hierarchical organization severely militated against managers and the work force transcending the narrow boundaries of their departments to consider the actual flow of product, service, and information processes as they moved through the enterprise, much less than how they flowed between linked outside channel members. The division and subdivision of process management, rigid "command and control" organizational values, separation of thinking and doing, perpetuation of narrow departmental "turfism", a focus on engineering complexity and individual initiative out of the performance of work tasks—all these aspects of traditional enterprise management destined even the best intended improvement project to failure. Once companies finally began to understand that the source of their competitive problems lay not in the improvement of the performance of narrow departmental tasks but in the splintering of fundamental core business processes, the foundations for the dramatic improvement of enterprise and interenterpise quality and functional processes were in place.

The shift from task-oriented to process-oriented organizations has had a dramatic impact on the basic structure and culture of management and labor. A process perspective means that focusing people exclusively on segmented pieces of a process, no matter how effectively performed, will result in relatively little gain. The old management theory had been that if the segregated parts of a process could be performed to meet optimal standards, then it could be reasonably assumed that the performance of the whole process would, by extension, also be optimized. This fundamental assumption was the basis of the division of labor into small task units that had guided industrial organizations since the time of Adam Smith's *Wealth of Nations*. The migration to process-oriented organizations, however, had rendered this fundamental assumption obsolete. The task-focused organization often required people to function and pursue objectives that were at cross-purposes with other departments, promoted internal finger-pointing and misunderstanding, and sought the optimization of the part at the expense of the whole. In contrast, process-centered organizations require the redirection and elevation of the efforts of people toward the achievement of enterprise-level objectives associated with the effective performance of whole business process that not only traverse the enterprise from one end of the business to the other but also serve as the link joining together the supply channel networks of which each firm is so closely integrated.

Principles of Work Force Activation

Today's business literature has awakened to the transformation in the nature of work, and there are many excellent books [3] exploring the shift in the role of the work force as it migrates from the industrial to the postindustrial world, from being industrial workers to process engineers. All of them point to the same conclusion: The old multitiered hierarchical organization is woefully insufficient to handle today's business requirements for global competition, ever-increasing flexibility of processes and people resources, and creation and continuous development of forms of "virtual organizations" composed of alliances of supply chain partners. The implications for the work force can only be considered as staggering. Instead of laboring in narrow, simple, task-oriented jobs, people now find that their work lives are multidimensional, spanning departments, oftentimes connecting them with counterparts in other companies out in the supply channel. As old artificial departmental demarcations lose their meaning, people increasingly find themselves coalescing into temporary process teams based on work competencies. Instead of being governed by a "command and control" management style where performance is measured by activity, they are now empowered to make decisions previously considered the preserve of management and they are measured by results. In place of training to achieve optimal task performance, the emphasis has switched to educating people to increase their insight and understanding of the converging enterprise processes of which they are a part. Finally, managers, formerly the guardians of authority and knowledge, have had to change from being taskmasters and scorekeepers to coaches and leaders.

In addition to changes in the nature of work, the *value* of the work force has also been dramatically altered. In the age of mass production, the value of labor, whether line or managerial, was to be found in the volume of productivity performed as defined by departmental standards. Today, the value of labor is to be found in the level of the skills, knowledge, expertise, and information that the people *within* the firm can utilize to provide enrichment value to customers and supply channel partners. In the old hierarchical organization, an enterprise's competitive assets were defined as plant, equipment, and machinery, with people and inventory considered as variable assets that could be manipulated to optimize the firm's competitive assets. In contrast, today's best organizations, such as Federal Express, Texas Instruments, Hewlett-Packard, US Robotics, and others, have come to recognize that it is the skills and competencies of their people that is their secret to marketplace success.

At this point, many readers may object to the glaring discrepancies between such a statement as "people are a company's most valuable asset" and the reality of mass layoffs, maneuvering of highly skilled (and highly paid) people out of their jobs and the hiring of lesser qualified (and cheaper) workers in their place, and continual erosion of jobs to outsourcing experienced by the work force everyday. One response is to point a finger at companies (and especially the ignorance, incompetency, and arrogance of their management staffs) that mistake BPR and QM with a narrow-minded and eventually self-destructive preoccupation with incrementalist cost cutting and subservience to Wall Street. Managers that focus solely on short-term gains that leave the company denuded of its best talent and capabilities for innovation have lost sight of the very purpose for its existence. These companies are doomed: Without the ability to evolve and to fight for their place in the business ecosystems they inhabit, they will be quickly devoured by more resilient competitors or even by once-friendly channel partners hungry to strengthen their place in the struggle for marketplace survival.

Another response is to take a hard look at what workers should be doing and thinking about when it come to their jobs. Although old-fashioned hierarchical organizations did stifle the growth of the individual, they also provided workers with the prospect of lifetime jobs. It is still amazing to talk to retired people today and hear how they spent 30–40 years or more with GM or General Electric. Those days are gone and will never return. If the nature of work and the value of the labor of people have changed, today's work force must be prepared to take the bad with the good. In reality, people do not have "jobs" (jobs belong to companies who have the power to change or eliminate jobs at their discretion); they have *careers* [4]. A career consists of the skills, aptitudes, and values a individual possesses, based on personal learning, development, and mastery of a range of productive capabilities that provide for self-realization and personal accomplishment. Whether people choose to actualize their careers in one company or several companies is immaterial—their careers and how they are managed are now squarely the responsibility of the individual, existing apart from any particular company. This realization is at once exhilarating as well as frightening. Today's work force enjoys virtually unlimited opportunity to exercise their personal sense of creativity and entrepreneurialism and to sell their knowledge and capability at their own discretion to the highest bidder at any time. However, the new-found freedom is not without its cost: Unlimited opportunity comes at the expense of security. In the end, the responsibility for the nurturing of successful and rewarding careers no longer rests within the organization but in the capabilities and personal sense of worth and value that individuals contribute to the companies by whom they are employed.

Because the responsibility for career success belongs to the individual, what are the key attributes characteristic of process-centered professionals? One of the best ways to illustrate the shift in work force values from a task-oriented to a process-oriented organization is to contrast the following critical elements:

- *Skills.* In the task-oriented organization, the work force is expected to perform a set of well-defined tasks, each of which can be optimized. The goal of both management and staff is to construct standardized tasks and provide for their continuous monitoring to ensure that all productive activities are indeed optimized. As a result, each job category has been rigidly defined and workers are discouraged from deviating from work standards developed by the company's corps of production and quality engineers. The authority to respond to the possibility of process variation has been removed from line control.

    In contrast, in the process-oriented organization, both managers and line staff see their functions as part of entire business processes that are inherently complex and changing. As such, effective decision making regarding daily productive activities has been pushed down to the lowest possible level. In place of a unyielding adherence to optimizing task standardization, the work force is constantly on the search for ways to continuously improve work methods and flows. Because of their close involvement with process performance, the work force is closely involved in the definition of plant layout, the organization of individual and team functions and roles, requirements for training and education, and criteria for rewards and recognition.

- *Knowledge and Learning.* In the task-oriented organization, the work force is considered a variable cost whose primary role is to perform functional tasks. In fact, one of the prime goals of management is to continually search for ways to reduce the level of skill required to perform activities and to structure jobs so that relatively anyone could step in and run the task. Managers closely oversee that each task is correctly performed. The work force, in turn, is trained only in the standard method: Experimentation with job tasks is discouraged as potentially disruptive of the departmental process.

  Recognizing that an intelligent and self-motivating work force is at the center of enterprise success, the process-oriented organization continually searches for ways to enhance the skills of its people in order to constantly pursue opportunities for improvement. In contrast to "deskilling" their workers, the goal of the organization is to create a dynamic atmosphere where learning, experimentation, and problem solving occur. Rather than considering change as disruptive of jobs bounded by standardized procedures, the process-centered organization encourages people to take appropriate risks and think about how to do things differently. Process-centered organizations continuously seek to promote learning and thinking environments that develop vital core competencies that can be applied as both internal and supply chain resources providing new vistas for innovation and sustained success. Finally, people migrate away from considering themselves as workers performing the tasks of a department—engineering, sales, accounting, and so forth—to professionals possessed of *skills* that can be utilized across traditional business functions.

- *Communications.* In hierarchical management structures, communications is normally confined to departmental information regarding tasks, priorities, and performance reporting. Workers are isolated in their own departments and rarely pass or receive information from workers across the enterprise. Job classifications and rigid process standards provide workers with detailed operations procedures covering every possible activity and eventuality. Information passing through layers of planners and supervisors is critical in ensuring the execution and control over the many complex systems used to conduct operations. Departmental supervisors are measured on how closely workers have adhered to specific procedures and whether they can meet the daily task requirements of plans transmitted from upper management.

  In the modern process-oriented enterprise, neither the skills of the work force nor the productive processes of the firm are rigidly etched in stone; they are "virtual" resources, available to meet company or supply channel resource needs at any time. Because all processes are so closely integrated, connectivity of information both inside and outside the enterprise is essential if companies and channel systems are to be able to continually reconfigure channel resources, competencies, ideas, and people to realize radically new marketplace opportunities. Through the power of information technologies, people have the capability to communicate directly with other people without having to filter through hierarchical organizational structures. This peer-to-peer networking capability enables people anywhere in the supply channel to access the knowledge of others and facilitates the creation of cross-functional and cross-enterprise teams who can converge their competencies and resources to work on marketplace opportunities. Finally, the availability of networked information removes the superior-subordinate relationship of pyramid management styles, pushes decision making down the organization, and accentuates the value-added capabilities and excitement of shared visions, insight, and unique talents of people linked across space and time.

- *Management Control.* The underlying management basis of hierarchical organizations is the concept of "command and control." The creation of strategic and operations plans are the preserve of management. It is the purpose of all levels of corporate management to carry out these strategies, thereby attaining the organization's goals. The criteria for judging these management actions are the efficiency and effectiveness by which the work force achieves these overall goals. These forms of direct control include such performance techniques as variance analysis, direct supervision, and detailed procedures. Task control is transaction oriented. Rules to be followed in accomplishing tasks are prescribed as part of the management control process; task control ensures that the rules governing performance are followed. Although task control activities are considered "scientific" (based on optimization studies), managers, as they know more than workers about what should be done and how to do it, must constantly supervise actual activities. A clear understanding of management's responsibility, authority, and prerogatives is the pillar on which organizational productivities are based.

  Process-oriented organizations, on the other hand, perceive the value of the work force as consisting in their ability to continuously create and invent new products and processes. Pyramid organizations seek to separate people into small portions of much larger processes and rigidly control every activity. Work force visions and the natural spirit of innovation are considered of little importance and actually threaten established procedures. In contrast, the work force in a process-oriented environment is expected to exercise new avenues for creative thinking and entrepreneurialism that anticipates how a product or a service may be constantly improved to meet future requirements. Because process-centered workers can see the functioning of the entire process, even when it extends outside company boundaries, they can apply their insight and knowledge to relentlessly search for whole new regions of product and service quality and uniqueness that also acts as a catalyst to inform and inspire the visions of other channel members. The activation of process-oriented cross-functional, cross-enterprise teams is made possible by the communication of experiences, training, and understanding through computerized networking tools, such as e-mail, the Internet, and computer-to-computer linkages. In such integrated environments, rigid control has been replaced by continual learning and problem solving at all levels.

Work Force Competencies and SCM

It has become apparent in the mid-1990s that industry leaders can no longer hope to maintain their positions of marketplace strength by simply concentrating on strategic agendas focused around quality, time to market, and customer responsiveness. These values provided companies with competitive advantage in the 1980s. Today's complex global business environment, on the other hand, requires companies not only to do the basic things well but also explore new ways of proactively shaping the marketplace by continuously regenerating themselves, their core strategies, and their capabilities and productive resources. Uncovering tomorrow's opportunities requires more than a single vision; it requires the collective perspective of many groups spanning both the enterprise and related channel network systems. Companies that separate themselves functionally and geographically are unlikely to gather the intellectual energy and innovative resources necessary to understand and influence the forces shaping tomorrow's marketplace.

Transcending the limitations of traditional organizational models requires that the work force be capable of accumulating talents, skills, and competencies necessary to realize emerging opportunities. Traditional strategies emphasized the necessity of matching existing resources with targeted business plans. In contrast, keeping competitive in a global business environment marked by the blinding speed of today's marketplace and the short cycle times of product shelf life, service factors, and operating and planning information means that existing company resources will be, sometimes substantially, out of alignment with strategic aspirations. Much can be done *inside* the enterprise to shrink the gap by ensuring that employees are defocused away from vague policy statements and short-term operational goals and are provided with a sense of strategic direction (as opposed to "control") and competitive meaning that fosters initiative and creativity. By demonstrating the vital role of each member of the work force in attaining clear competitive challenges, such as creating new sources of customer satisfaction or dramatically better products and services, individual employees can develop personalized goals that are supportive of their own objectives as well as those of the enterprise. In place of a narrow concern with numerical scorekeeping, work force consciousness shifts to capability building, to utilizing advantage-building management tools, such as total quality management (TQM), reengineering, systems modeling, and teamwork methods, in order to understand and continuously improve the critical processes of the enterprise.

Still, regardless of the strength of *internal* resources, companies have increasingly come to realize that it is only by utilizing the collective work-force competencies to be found in the supply chain management systems to which they belong can they hope to sustain the level of marketplace innovation necessary for competitive leadership. This premise is based on several key points. To begin with, as companies are increasingly conceived of as collections of productive resources focused around key market-satisfying processes rather than as repositories of wealth, the unique capabilities and knowledge of their work forces becomes a more important source of supply chain value than the products they sell. In fact, it can be said that the particular suite of products and services offered by a company is simply the manifestations of bundles of past skills and innovative thinking possessed by its productive competencies. In reality, the value of these competencies is to be found not in past accomplishments, but rather in their future capability to create new and exciting marketplace offerings that can enrich not only their customers but also their supply channel partners. Think of the competencies offered by Motorola (communications), Microsoft (information software), and 3M (adhesives, substrates, and other materials) that are continuously utilized by their channel partners to achieve even broader ranges of product and service opportunity.

Second, exportable work-force competencies enable groups of companies to converge specialized resources to exploit new opportunities or solve critical problems that would be impossible if attempted from the standpoint of a single company. In fact, if it can be said that a company can only progress when it utilizes its learning experiences as a springboard to more effective action, then the ability to exploit the cumulative experiences of whole channel systems in search of critical ideas can generate dramatically new methods of improvement and innovation. Through alliances, joint ventures, licensing, and franchise, companies can gain access to new competencies and resources, and then by internalizing them, they can create new regions of workforce skills and capabilities. In reverse, some firms seek to export their innovative competencies to fund tomorrow's competitive breakthroughs. For example, companies such as Samsung, Cannon, and Sharp sell OEM components and finished goods to Hewlett-Packard, Eastman Kodak, Philips, and others as a way of financing their leading-edge research in imaging, video technology, and flat screens [5]. Whether it is in sharing development risks, utilizing resources (for example, when software companies contract relatively low-cost programmers in India and China), or networking designers and marketers via communications technologies, the goal of the process is to gain access to new resources, thereby gaining control of next-generation innovation without having to invest in core competency building.

Third, the mere possession of or capability for converging resources from several sources does not, by itself, guarantee success. Leveraging channel competencies requires their skillful blending in ways that compound the value of each resource. Integrating several channel competencies to solve a particular problem unleashes opportunities to create unique fusions of often quite different philosophies and techniques. Besides blending competencies, companies must also be able to balance the competencies they possess or acquire through contract with partners. Regardless of the composite of the mixture, the synthesis must possess proportional and complementary capabilities that must be continually rebalanced over time to ensure equilibrium, the even distribution of risks and rewards, and the maximization of time to payback. The effective application of these elements is particularly important to companies engaging in SCM. The ability to continuously blend and balance internal resources with the spectrum of shifting competencies to be found out in the supply channel places a significant burden on strategic foresight teams. Besides searching for the right match of complementary competencies that will generate high-value processes, companies must also be careful to shield their own work forces from being raided by competitors or even erstwhile partners seeking to build internal capabilities [6].

Supply chain management is a management philosophy that can assist companies acquire critical work force skills, talents, and business vision by enabling naturally linked supply network partners to integrate complementary competencies focused around common strategic objectives. The work-force component of SCM provides a clear solution to the imbalance between available resources and strategic market aspirations today's most aggressive market leaders often experience. Finally, SCM provides managers with new opportunities to creatively leverage the wide range of resources that can be gathered from the typical supply channel. Chapter 4 detailed how effective competitive strategies provided the overall direction for the firm and its supply network partners; this chapter has focused on how managers can effectively coalesce channel resources to achieve order-of-magnitude breakthroughs in products and services that provide superlative value to the customer and, by extension, to their supply chain partners, in the search for overwhelming marketplace advantage.

New Enterprise Organizational Forms

The dramatic changes occurring in the roles, capabilities, and goals of today's work force could not have occurred without complimentary changes in the way companies are organized. In fact, as a general principle, it can be said that changes occurring to the work values and operating activities of a company's people resources will have a reciprocal impact on the forms of organization to which they belong, and vice versa. Today's relationship between labor and organizational structures and values is evidence of this principle. The development of global markets, the creation of "virtual" products and services, the networking capability provided by modern information and communications technologies, and the utilization of work force competencies found outside the boundaries of the enterprise have rendered obsolete the operational norms and organizational models that have governed the way companies have been run for almost a century. As the work force struggles to understand and create for themselves new labor values and behavioral expectations arising from the reformulation of the meaning of work, so too have today's companies struggled to redefine the fundamental bases of the way they are organized around productive processes, job requirements, the role of management, performance measurements, rewards, and their ultimate importance of organizational forms to the success of the enterprise.

The Basis of Organizational Structure

In the past, the organizational structure of a company was considered as a collection of *responsibility centers.* A responsibility center was defined as a work or task unit headed by a manager who was responsible for its activities. Responsibility centers within a company were arranged in a hierarchy. At the lowest level in the organization could be found responsibility centers that performed the daily operational activities of the firm, such as the weld shop, assembly, packing, inventory control, and order processing. The next higher level in the organization consisted of departments or division staff and management people charged with enterprise planning and budgetary responsibilities. Finally, from the perspective of top management, the whole corporation was considered a single responsibility center. The assumptions governing the operating model of this organizational structure were simple. To begin with, productive processes were best managed by dividing and subdividing them into small work/task units. Centered around the development of rigid standards necessary to achieve the effective utilization of mass-production functions, the productivity of these work units was best measured and controlled by narrowly defined job descriptions and task activities. Decisions were made by management at the top of the organizational pyramid, and then carried out by departments and work units which were expected to meet the operating guidelines assigned to the function. The role of the individual worker was to optimize and conform to the work unit performance standards developed for each possible task.

Although each responsibility center had its own particular objectives, presumably these objectives were in alignment with and supportive of the strategies of the whole organization. Resolving potential conflict of goals among individuals, responsibility centers, and the corporation was the central purpose of management. It was their responsibility to encourage and guide *goal congruence*, to ensure that the many objectives of the participants composing the organization were consistent with the strategies of the organization as a whole. Management endeavors directed at achieving cooperation and ameliorating conflict were assisted by the maintenance of specific rules and procedures. These "rules," defined as codified sets of business practices, guidelines, job descriptions, customs, operating procedures, manuals, and codes of ethics, formalized the management control environment and specified the work culture and values of all employees [7].

During the past decade, the basic principles underlying the traditional organization began to unravel. The content of these dramatic changes, detailed above in the discussion of today's work force, posed fundamentally new challenges as well to organizational structure. These challenges can be detailed as follows:

- In the traditional organization, the work force and their managers performed the tasks assigned to their departments and then passed the work to the next serial node in the process. In contrast, today's marketplace requires organizations to be based not around tasks, but whole processes. In a process-centered organization, there are no departments as such and no work unit managers whose job it is to ensure adherence to standardized tasks and procedures. The goal of the organization is to secure and assign the proper resources, regardless of where they originate, to form focused teams charged with both running and continuously improving the firm's productive processes. Because the flow of work is no longer compartmentalized into distinct, self-contained work units, the traditional organization loses its meaning and functional importance and gives way to "virtual" organizations that are created to accomplish whole work processes and then are dispersed when the work objectives are completed.

- Unlike the fairly static business environment of the age of mass production, today's dynamic and constantly changing market landscape requires organizations that are opportunity based. Such organizations must be considered as pragmatic, flexible, constantly metamorphosing in order to be responsive to the shifting needs of the customer. As product life cycles decline and development and distribution costs accelerate, organizations will be responsible for entire process flows and be required to amass a wide range of competencies and resources, sometimes from outside company boundaries.

- The requirements of a process-focused organization have demanded radically new organizational roles for the work force and management. Because process organizations wax and wane around specific job processes, the work force and management must possess the skills and be capable of moving from one process to another. This requires heightened requirements for employee development and knowledge that extends beyond the narrow unit task training of departmentalized organizations. Managers must migrate from taskmasters and scorekeepers to process owners and coaches responsible for allocating resources to productive processes as well as mentoring employees to ensure the long-term availability of knowledgeable and innovative people. The goal is to develop "world-class" *centers of excellence* containing superlatively trained and flexible people from which successful process teams are to be assembled.

- Process-centered organizations rely more on the knowledge and skills of their people than on the functions they perform. Therefore, enabling people and organizations to remain focused in a fluid work environment where teams are configured and reconfigured to meet specific challenges becomes of the utmost importance. Meeting this need requires the existence of information networking and knowledge integration. Responding to today's complex matrix of market, governmental, competitive, and supply channel needs requires companies to move far beyond considering their employees purely as task performers and more as repositories of specialized skills and knowledge. Being able to bring together these productive factors through information networking facilitates the task of skill mastering, reinterpreting new workplace patterns, and continuously improving whole business processes in search of tomorrow's competitive success.

- The challenges posed by the global marketplace have increasingly forced today's process-centered organization to utilize the knowledge, competencies, and resources to be found in their supply chain networks. Forming "virtual organizations" composed of targeted people both from within the enterprise and from channel partners provides companies with unique opportunities to maximize on resources and achieve order-of-magnitude synergies of significant productive and innovative power. The task of engineering and directing these SCM forms of organization stands at the forefront of today's most critical management challenges. Leveraging the enormous reservoirs of skills and knowledge to be found in the supply chain permits companies to be adaptable and opportunity-focused so as to provide unassailable sources of customer satisfaction and value.

The difference between yesterday's task-focused department and today's process-focused teams is dramatic. In the past, companies managed change by unfreezing the organization, developing new standards and procedures, implementing these changes, and then once again freezing the organization. Today, the speed of change and the tremendous risks associated with the loss of competitive advantage have required companies to create fluid organizations capable of continuous change and of using knowledge teams composed of people from any place in the supply chain network continuum. A summary of the changes to the different elements required to capitalize on supply chain organizations is as follows:

- *Organizational Agenda*: Progression from bureaucracy, stifled initiative, bounded growth, and lack of customer focus to internal and supply channel interlinkages fostering the use of core competencies that enable collective solutions to realize process breakthroughs and customer-centered opportunities.

- *The Work Force*: From task-center to process-center jobs; from performance centered around meeting rigid operations standards to continuous improvement in productive processes; from jobs focused around narrow skill sets to jobs requiring continuous learning, innovation, and flexibility.
- *Organizations*: Transition from hierarchical defined departmental hand-offs of tasks to "virtual teams" of people required to perform whole processes and determine their own measurements.
- *Work Teams*: From narrow work unit teams to cross-functional, even cross-enterprise teams.
- *Managers*: End of authoritarian-based management styles and rise of managers as coaches and facilitators. Everyone is vested with the responsibility for the traditional roles once held by the management class.
- *Leadership*: End of "command and control" methods of leadership and emphasis on teamsmanship and facilitating the growth of team and individual sources of innovation and entrepreneurialism.

Defining Today's Process-Focused Organization

The basis of today's process-focused organization can be found in the unfolding of two concepts. The first is the assumption that people (both the work force and management) have the capacity for multitasking. This means that the human organization is capable of performing multiple processes in parallel by being able to easily switch from participating in one process to another. This is possible because the increment of labor value is not task performance but rather the unrestricted application of highly skilled, knowledgeable people who can use new forms of information, work efficiently with others, and possess the critical problem-solving skills necessary to accomplish the objectives of multiple teams functioning concomitantly. The second assumption is that today's most competitive form of organization is the "virtual organization" composed of multiple teams than span enterprise functions and that have the capability to also include the human resources of channel partners existing outside the boundaries of the company. The primary objective of these multiprocess-focused, often geographically dispersed, teams is to transform the multitude of customer-centered themes arising out of the marketplace into practical, measurable product, process, and service strategies. By integrating capabilities from virtually any node in the supply channel network, their goal is to swiftly and completely turn customer-winning opportunities into market-satisfying solutions.

Although the process-focused organization can take many forms (partnership, joint venture, strategic alliance, supplier-subcontractor, cooperative agreement, outsourcing contract, and integrated channel process team) based on the nature of the competitive strategy pursued, they all contain several similar characteristics. To begin with, all process-focused organizations are driven by comprehensive strategic plans. As detailed in Chapter 4, these plans reveal the strategic goals and missions of the organization that provide not only clear direction but also critical corporate values and cultural norms. In addition, these plans help define the organization's competitive visions, the operational targets to be pursued, and the critical linkages binding together the functions both internal and external to the firm. Clear business strategies make visible the requirements of productive processes and enable the creation of targeted cross-functional, cross-channel virtual teams necessary to accomplish process objectives. Process-focused teams also have the responsibility to constantly challenge and enrich the enterprise's strategic vision.

The second critical element of the processed-focused organization is the existence of *centers of excellence*. The capability to assemble virtual process teams is dependent on the availability of large numbers of highly skilled, reliable, and knowledgeable workers who have the ability to rapidly adapt to change, possess the ability to perform parallel tasks, understand and can use the newest information technologies, and can work efficiently with others. Although formally the organization is still broken down into the traditional departments of marketing, sales, engineering, manufacturing, distribution, finance, and human resources, they no longer act as independent functional silos but rather are to be conceived of as pools of resources to be drawn from in the creation of actual process teams. Centers of excellence do not perform tasks—that is the role of the actual process team. Department mangers will still hire, train and educate, and develop individuals within their particular career discipline; however, their roles will change from taskmasters and bosses to career mentors and coaches charged with the continuous development of the company's repository of work force skills and competencies. The ability to easily tap into the resources of supply chain partners to create truly interchannel process teams further accentuates the value not just of enterprise-level, but also of channel-level *centers of excellence* [8].

The third critical element is the capability of individual enterprises and whole channel systems to activate and network targeted individual people resources to create teams focused around a specific process. Today's continuously advancing world of information and communications technology (ICT) provides the capacity to link both internal and external resources and removes dependence of physical and temporal proximity of team members. Teams can function in "real time" through the use of such tools as electronic mail, fax, computer notebook files, distributed databases, technical data interchange, and a host of other networking systems designed to enrich communications and data transfer. The effective use of the Internet is perhaps the most fertile area for team networking. The use of the "web" not only enables process teams to define new ways of developing interrelationships but provides access literally to the knowledge and skills of the entire world to form almost limitless collections of informal teams clustered around the networking infrastructure. ICT tools enhance and make practical the use of centers of excellence and multiple process teams by liberating the knowledge and problem-solving talents to be found vertically within the enterprise and horizontally among supply channel partners.

Visualizing the process-focused SCM team network reveals a complex and open-ended transchannel organization. Figure 8.1 is an attempt at portraying how such a framework would look. The first observation that must be made is how radically different the operating paradigm is from traditional corporate structures. In fact, it is not an organizational chart at all, but rather a model meant to

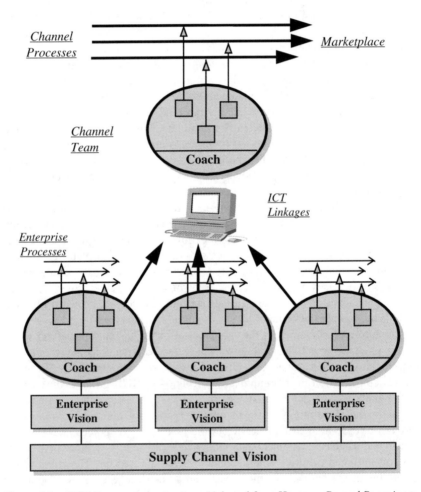

Figure 8.1. SCM Process team structure. (Adapted from Hammer, *Beyond Reengineering,* p. 126 [3.])

demonstrate the components of a multienterprise process-focused team network. It is noteworthy that the traditional trappings of the pyramid organization— closed departmental silos, fixed hierarchies, and narrow definitions of authority— are completely absent from the model. Instead, the model illustrates how individual enterprise teams converge to activate and orchestrate value from the effective performance of both enterprise-level and channel-level work processes.

At the base of the model are the individual enterprise and combined supply channel business strategies. The goal of these strategies is twofold: first, to realize new opportunities for competitive advantage resulting from innovative thinking and the convergence of core competencies; second, to execute operational objectives centered around products, value-added services, and customer value. Moving up the structure can be found the *centers of excellence* that exist within each company. Within the ovals can be found each enterprise's management (coaches) and functional departments, staffed by skilled professionals. Enterprise teams are formed by deploying people from a center of excellence who are then assigned, assisted by ICT tools, to specific work processes. Each process can be said to have an "owner" who is responsible for defining the objectives, making decisions, and setting daily task priorities. At the very top of the model can be found the *channel-level centers of excellence* composed of selected individuals or whole enterprise teams that have been networked by ICT to perform the work of channel processes presided over by a channel-level process owner.

The structuring of SCM process teams is only one of the ways traditional hierarchies and process team models can be contrasted. Some of the other differences can be found in the way the work force is utilized and how process tasks are actually performed. Instead of being assigned to a narrowly defined department, workers in a process-centered organization are moved frequently in an attempt to match process requirements and individual skills and aptitudes. A purchaser will remain attached to the purchasing department where career development and evaluation will occur. However, individual purchasers will find that they are often assigned to a multitude of possible process activities that span other departments, such as a major new engineering project or a channelwide supplier quality management program. In addition, although process owners can be said to direct process activities, they are, in fact, not "bosses" at all. Process owners and coaches offer guidance and direction; it is up to the individual professional to actually achieve process results and to unearth new areas for improvement. Thus, SCM process teams can be thought of as changing, constantly mutating collections of coaches, process owners, and skilled professionals that coalesce to perform the work of specific processes, and then are dispersed after specific objectives are achieved. The overall direction of enterprise and supply channel process objectives is the responsibility of top management, who determines cumulative long-term vision and assists in resolving goal conflicts based on individual enterprise and channel business strategies.

Developing SCM Process Teams—Product Flow Team Example

The development of process teams that span the entire supply channel have become increasingly important to competitive survival. The ability to link the marketing, product, and information process flows that traverse the channel from producer, through wholesaler, and ending with the retailer will not only provide effective mechanisms for process performance and review but will also enable focused teams to search for innovative ways to constantly improve the channel's operating activities. For example, as Martin explains [9], the creation of *product flow teams* (PFTs), charged with the responsibility for ensuring the application of the most cost-effective and timely methods for the movement of product through the supply pipeline system, enable channel partners to network and integrate their inventory planning, warehousing, and delivery functions. Such a product flow team should consist of professionals who perform similar functional roles in their respective companies. The competencies and skills of the people from each segment of the channel system should be complimentary and supporting. As is illustrated in Figure 8.2, the professionals working in sales at the manufacturing site, at the wholesaler, and at the retailer should all belong to the same SCM process team. The same can be said for similar teams involved in transportation, purchasing, and other functional areas. In fact, all of these subteams could arguably be combined into one unified "channel customer fulfillment" team.

As discussed above, the creation of such SCM process teams by no means requires constituent channel members to embark upon a program of radical organization reengineering. People can still remain within enterprise-level organizational structures. All that is required is for member companies to recognize the importance of channel process teams and deploy their employees whenever possible to take advantage of the best opportunities for creating total channel process value. What will be required is the decision of each channel member's management staff to collectively commit to cross-channel team building, borderless sharing of power between channel constituents, empowerment of team members, and freedom of teams to develop and implement new avenues for process improvement without the typical "noise" of management politics and the worth-

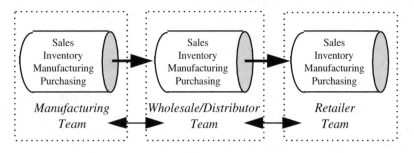

Figure 8.2. SCM Product Team.

less exercise of management authoritarianism. Although channel process teams are usually spearheaded by a dominant channel player who is searching for ways to leverage the supply channel capabilities, the leadership of peripheral channel members in forming their own channel process teams can provide novel approaches to new sources of competitive advantage that will benefit the entire channel.

Creating SCM process-focused teams requires the collective commitment of all business partners that are touched by the channel process. There are six steps that supply channel process owners can follow in their efforts to create effective SCM teams:

1. *Develop Objectives.* No SCM process team effort can hope to be effective without clear definition of the processes, objectives, and benefits to be gained. In this step, multicompany process owners will need to distinctly define the nature of the channel process and concisely agree upon the objectives to be attained and the benefits to be expected.

2. *Confirm Commitment.* As is the case with internal process teams, unless all members who touch the process are fully committed to providing critical resources, participating in joint process activities, and continuously searching for improvements that will provide for the collective improvement of the whole process, the SCM process teams that are formed will fall short of their goals. For example, if quick delivery has been identified as critical to channel marketplace leadership, then all channel constituents who have the responsibility for moving product down the supply pipeline will need to form a joint team targeted at continuously finding ways to improve delivery processes.

3. *Develop SCM Team.* In this step, all channel process members will have to identify and assign the appropriate people resources to the SCM process team. These professionals should possess the technical and team-building skills necessary to ensure effective cross-channel process management.

4. *Develop Network Links.* The effectiveness of a specific SCM process team can be measured by the strength of the tools available for interchannel communication. ICT tools, such as direct computer-to-computer linkup, EDI, fax, computer notes, and others, will assist in drawing geographically dispersed SCM team members into a closely networked "virtual" organization capable of decisive decision making as well as creating the infrastructure necessary for the daily processing of cross-company transactions.

5. *Team Management Skills.* Unlike enterprise-level teams, where there is a clear process owner who sets overall team objectives and is responsible for daily activities, SCM teams will often have multiple leaders, each

with different agendas. This fact will require SCM teams to develop new ideas about how to construct solutions that will meet the needs of each channel member while seeking optimal total channel solutions.

6. *Performance Measurement.* No SCM process team can hope to succeed without establishing effective performance measurements. Although gathering and assessing channel data is a complex affair requiring the active participation of all the companies composing the process team, concise metrics will enable process team leaders to gauge the effectiveness of current objectives and expose alternative directions for the development of new team strategies.

Benefits of SCM Process-Focused Teams

For the past half-decade, manufacturers and distributors have been undergoing a fundamental transformation. The vertically integrated business structures of the past have been disappearing as companies concentrate more on their own core competencies, draw in the expertise of channel alliances to attain critical skills and aptitudes, and outsource peripheral products and service functions to supply channel partners. Companies, like heavy equipment manufacturer Case Corporation, have disbanded many internal logistics functions and created an alliance of third-party service providers to fill in the gap. Case's channel partner, Fritz Companies, Inc., acts as lead systems integrator for the alliance, handling standard air and ocean shipments and providing services such as customs brokerage, duty-drawback, cargo insurance, warehousing, packing, crating, pick-pack, documentation, and information management. GATX Logistics, provides warehousing services, operates a central materials receiving center and several parts depots, and performs all related value-added services including sequencing, kitting, returnable container management, dunnage removal and disposal, and plant support transportation. Finally, Schneider Logistics Inc. manages the shipment of all inbound material to Case's manufacturing plants from nearly 2000 suppliers worldwide, handles all outbound shipments in North America, and manages inbound shipments of parts to Case's nine parts depots and its North American dealer network [10].

Regardless of whether the SCM initiative is marketing, logistics, or product design based, today's leading companies have all come to recognize the fundamental principle of SCM: *Supply chain cooperation can enhance competitiveness.* Consider these examples of cooperative organizations: Sematech and the Microelectronics and Computer Consortium (MCC); the clothing industry consortium {TC$_2$}; USCAR, pooling the resources of the Big Three automakers; CAM-I, Computer-Aided Manufacturing, International; and, the New England Suppliers Institute (NESI). At the same time, there has been an explosion in cooperative collaboration among companies considered as traditional rivals. Witness Apple's partnering with Sony to bring the Macintosh PowerBook to market and with

Sharp to manufacture the Newton. IBM, Motorola, and Apple jointly developed the PowerPC chip and operating system. Similarly, IBM and Toshiba partnered to develop the flat-panel display in their PC products; IBM, Toshiba, and Siemens jointly funded a 256-megabyte DRAM chip production facility, a $1.5 to $2 billion cost that none of the partners could afford individually [11].

Such examples of cooperative partnership underscore the simple principle that forming collaborative intercompany process teams, regardless of how they are formed or what their objectives are, can provide unique sources of marketplace advantage. It also provides for a redefinition of what is meant by a "supply chain." Most texts and articles conceive of a supply chain as the merging of logistics functions designed to smooth the transfer of product, the performance of channel transactions, the elimination of wasteful pipeline costs, redundancies, and cycle times, and the acceleration of channel information necessary for effective decision making. In contrast, a supply chain should really be defined in its broadest perspective as the active converging of the competencies and skills of the people resources of companies, sometimes even deadly rivals, to realize order-of-magnitude breakthroughs in some aspect of product/service development, manufacture, or deployment. SCM, therefore, must be understood as the process of founding, empowering, and continuously reinventing channel process-focused teams in the search for innovative ways to realize radically new forms of competitive advantage.

The utilization of SCM process-focused teams can be said to provide companies with the following advantages [12]:

1. *Sharing Infrastructure Competencies.* The single greatest advantage of SCM process teams is that they permit allied companies to draw on the intellectual and physical resources of channel partners to undertake projects and provide levels of process value unattainable if they were to act solely on their own. Each individual company team brings to the channel unique capabilities and skills that will enrich and facilitate realization of exceptional opportunities for marketplace value.

2. *Leveraging Channel Infrastructures.* By utilizing SCM process teams, companies have the ability to tap into unique infrastructure competencies, critical human and technology resources, and share the risk of new product/service, manufacture, or distribution systems development, the costs of which could not be borne by a single company acting in isolation. One SCM team, for example, could be constructed to facilitate the design of a new product; another team could be created to continuously search for cost reduction in the transportation of products as they traverse the supply channel.

3. *Reducing Process Cycle Times.* One of the fundamental attributes of SCM process teams is their ability to perform multiple functions in parallel. The goal is to converge the talents of people, physical resources,

and ICT systems found in the supply chain to reduce the time it takes for a process to be completed. In a business environment characterized by the continuous shrinking of product and service life cycles, the ability of SCM teams to focus a wide spectrum of resources on a particular project is critical to exploiting the high-profit window of opportunity before competitors catch up. Reducing cycle times for product design, customer order fulfillment, transport of products, and other actions will result in decreased costs, increased service, faster product design time to market, and increased quality.

4. *Increasing Process Capabilities.* Today's business environment is filled with opportunities for companies both large and small to create highly productive partnerships that can provide distinct sources of competitive advantage. These partnerships enable companies to acquire additional ranges of capability that are outside the physical resources of existing plant and scope of operations. A highly visible example is the explosion in the use of third-party logistics services. In the Case Corporation study discussed above, success is predicated on blending the core competencies of Case's own staff with those of their third-party logistics partners.

5. *Expanding Market Reach.* The fundamental concept of a supply channel is the ability of a firm to link with other firms to create a continuous pipeline for the flow of goods and services. Through the development of SCM process teams, individual companies can leverage the intellectual and physical resources of channel partners to better serve existing customers and to gain access to new markets.

6. *Developing Unique Solutions.* The convergence of knowledge and physical resources that occurs when an effective SCM process-focused team is created enables whole channel systems to develop combinations of products and services that provide unique customer solutions. The SCM process team concept provides supply channel networks with the ability to assemble and constantly restructure a targeted critical mass of skill-based capabilities and to create radically new products and services that consistently outperform the competition and offer customers unique value-based solutions.

SCM and Today's Information and Communication Technologies

The origins and continued development of the SCM concept is directly dependent on the capabilities of today's information and communication technologies (ICT). In fact, realization of almost all of the management and operational aspects of SCM discussed in this book—the networking of geographically dispersed process teams, the integration of channel strategies and operations, communications technology providing connectivity between companies, planning systems that facilitate inventory management integration across the supply channel pipeline, and others—would be impossible without effective ICT systems. SCM provides such a critical management and operational approach for competitive advantage because it is inherently intertwined with the networking power to be found in today's computerized information and communications systems. As capabilities of ICT tools expand, there can be little doubt that the integrative and infomating capabilities of SCM to provide fresh competitive perspectives will likewise expand.

Basis of Information Processing

As a principle, it can be stated that the operational and strategic capabilities of today's SCM-focused organization must be in balance with the functional dimensions of the information technologies available. Simply, the organization's ability to create, collect, assimilate, access, and transfer information must be in alignment with the velocity of the activities necessary to effectively execute the management of supplier, customer service, manufacturing, logistics, and financial functions. Changes in the speed of the velocity can come from two directions: Advancements in the capabilities of information technologies or changes in the internal and external business environment, or sometimes both, can render the previous equilibrium untenable and stimulate the rate of growth in either or both dimensions.

Historically, the capability of the enterprise not only to control physical events but also to leverage data to achieve operational optimization and exploit the internal and external linkages between activities has been determined by the information processing technologies in existence at the time. In the past, the velocity of information in the business environment was inhibited by limitations in information processing. Data could be collected, assimilated, and passed on to other business functions only as fast as human efforts, assisted by crude forms of automation, could process it. These limitations stood as the basis for the hierarchical organizations discussed earlier. Levels of management sought to gain meaning from the flood of information affecting the organization by dividing it into much smaller functional units from which homogeneous streams of data could be extracted and summarized. The data could then be passed up the organizational structure to specialized corporate staff planners, who, in turn, passed strategic and operational directives down through the departmental hierarchy. The time that was expended in manipulating information flows, planning, action, recognition of variance, replanning, and corrective action was covered by increased pipeline inventories, long lead times, and long delivery schedules.

With the advent of the computer, capable of handling information and communications data in volumes and at speeds previously thought unimaginable, the cumbersome information processing constraints of the past were lifted, revealing new horizons for the application of information and obsoleting many of the older management methods and organizational processes structured around previous information systems. Finely tuned corporate bureaucracies with narrowly defined policies, procedures, and job descriptions increasingly fell out of alignment with the new realities and opportunities of a marketplace activated by technology. Instead of a rigid structure dominated by "command and control" management styles, the ability of companies to integrate internal business functions and network with supply channel partners inaugurated a new era of individual employee empowerment and the activation of intercompany process teams capable of performing critical activities and making fundamental decisions without the interference of management.

Integration and Networking [13]

At the core of the concept of SCM stands two critical dimensions of information management. The first is the existence of a technical infrastructure that links computer systems and people. The word commonly used for this process is *integration*. One of the problems with this dimension is understanding exactly what it means. As Savage has pointed out [14], there has been a great deal of confusion concerning the definition of the word "integration." It is often erroneously used synonymously with *connectivity* and *interfacing*. Connectivity means connecting processes together, such as when a telephone system connects customers and order processing functions. Interfacing means bringing information from one system and presenting it for input to another, such as occurs in an EDI transaction. Although both terms provide for the assembly and transmission of information, neither connectivity nor interfacing change the way the organization works. If the organization is really nothing more than a collection of independent functions, neither connectivity nor interfacing will have much impact on converging or unifying the strategic and operational activities within the enterprise or outside in the supply channel system.

In contrast, integration calls for the elimination of the ideological, strategic, and performance barriers characteristic of the old hierarchical organization. Integration means to come in touch, or to be in touch with itself. Organizationally, integration means leveraging information tools that bring operational functions together, both on the enterprise and the channel level, by facilitating ever-closer coordination in the execution of joint business processes. Integration focuses on activating the creative thinking of highly skilled professionals within and between enterprises. Integration attempts to bring into alignment the challenges and opportunities offered by information technologies and the cultures and capabilities of tightly linked SCM process-centered organizations.

The second dimension at the core of the SCM philosophy is *networking*. In the past, manual and early computer system architecture permitted only hierarchical communication. As each information node completed its tasks, the output was then available for the next information node, which, in turn, passed its output to the next downstream node. With the advent of today's minicomputers and client/ server architectures, the process of communicating information has shifted from processing hierarchies to connecting different computers and their information databases together in a network. The growing availability of open-system softwares and arrangements of local-area and wide-area networks is targeted at solving the problem of the dissimilarity of hardware operating systems. The advantage of networked systems is that people can now communicate information directly to other people in the network. This peer-to-peer networking enables companies to leverage the capabilities, skills, and experience of people by integrating and directing their talents around focused tasks. What is more, the establishment of process-focused teams can occur not only inside the enterprise but also can be extended to people performing similar functions throughout the supply channel network.

Integration is the process of linking business functions together; networking, on the other hand, is the activation of those links by enabling and empowering people to cut across functional and company barriers and interweave common and specialized knowledge to solve a wide range of competitive problems. Integration and networking are complimentary activities that can be combined and defined as the *integrative process.* According to Savage, this process

> puts us in touch with the whole, with one another, with customers, and with suppliers in ever-changing patterns of relationships. It also puts us in touch with our own wills, emotions and knowledge. The integrative process is a process of human networking: networking our visions and knowledge so we can take decisive action in concert with other efforts. [15]

This integrative process is the driving force in the acquisition of computerized technology and is, in turn, governing the development of its topology.

Technology Architectures

The ability of today's supply channel organizations to integrate and network information is achieved through three major technology architectures. One of the key breakthroughs in modern information processing is the use of shared databases. The development of shared databases solves one of the fundamental problems that had long plagued the effective transfer of information in the enterprise: How to establish a single source of data that could be easily accessed by a community of users to facilitate meaningful decision-making and operations execution. In hierarchical organizations, each department defined its own repository of information resulting from decision and execution processes and then, after summarizing the results, passed it on to the next sequential node in the hierarchy. As could be expected, critical information that needed to be communicated throughout the enterprise was often incomplete, misunderstood, and lacked timeliness. The existence of a common database, on the other hand, enabled organizations to escape from the serial hand-off of information characteristic of the era of batch processing. The power of today's ICT systems to provide real-time, interactive, and integrative access to information enables process-centered professionals both within and outside the enterprise to significantly compresses time out of productive functions and speed the creation of marketplace value.

The power of today's database architectures has been enhanced by the second major ICT enabler—the ability to network databases. In the past, companies were restricted to using information software that worked on a single hardware platform. Trouble arose when databases that existed on multiple computer systems needed to be integrated. The development of computer-to-computer networking has solved this problem. This breakthrough in intraenterprise and interenterprise ICT database integration has taken several forms. To begin with, companies can network mainframe and minicomputers in remote branch sites or with business partners found in the supply channel. This solution, where all the layers of computing (screen input or query, the application area for database edit and update, the database, reporting, and the data warehouse) is on a single-tier, is termed a *fat client*. Another possible configuration is to utilize powerful (client) desktop personal computers (PCs) with a small server computer to keep the desktops talking to each other. Still another uses a multitiered architecture with powerful clients and database servers capable of networking the external channel. The future holds the promise of inexpensive network computers where users can timeshare to run their software or use the Internet as a form of electronic commerce permitting anyone with access to the web to gain entry to and perform transactions such as purchasing and funds transfer [16].

Database networking makes key data such as customer information, inventory status, shipment delivery, and transportation status transparent to users at any point in the supply channel. Today, the ongoing development of *open-systems computing* is providing companies with the opportunity to run software independent of hardware operating systems, thereby shrinking user's dependence on having to pass data to centralized processors and enabling the exploration of a whole new dimension of PC software applications. By linking channel structures and partners across business functions, geographical regions, and marketplace segments, the availability of shared information provides everyone in the supply channel with an important new source of competitive advantage [17].

The third key ICT tool enabling channelwide integration and networking is electronic data interchange (EDI). Although database architectures now provide the channel with the ability to store, access, and enter common data, EDI technologies constitute one of the most widely used forms of physical linkage. EDI provides for the computer-to-computer exchange of business transactions, such as customer orders, invoices, and shipping notices. As is the case with the implementation of any information technology tool, the success of an EDI system is only as good as the implementation effort and the ability of the channel information nodes to adapt to new organizational structures and values provided by the real-time sharing of data throughout the channel network. The various benefits of EDI are as follows:

- *Increased Communications and Networking.* By enabling channel partners to transmit and receive up-to-date information regarding customers and processes, the entire supply channel can leverage the integrative nature of today's ICT technology for competitive advantage.

- *Streamlining Business Transactions.* EDI can significantly shrink cycle times in a wide spectrum of transaction processing activities.

- *Increased Accuracy.* Because business information is transferred directly from computer to computer, the errors that normally occur as documents are transferred from function to function and business to business are eliminated.

- *Reduction in Channel Information Processing.* EDI provides for the removal of redundant paperwork flows that simply add time and cost.

- *Increased Response.* EDI enables channel members to shrink processing times for customer and supplier orders and to provide for timely information that can be used to update planning schedules throughout the channel.

- *Increased Competitive Advantage.* EDI enables the entire supply channel to shrink pipeline inventories, reduce capital expenditure, and improve on return on investment and to actualize continuous improvements in customer service and productivity.

Identifying ICT System Solutions

The opportunities to be found in the SCM concept, coupled with the explosion of computerized tools in the 1990s, have provided whole supply chains with an array of organizational and information technology choices targeted at increasing total channel productivity and serviceability while reducing total costs. The ability to properly align information technology tools and targeted breakthroughs in organizational capacities requires companies to match the objectives of the enterprise with the capabilities of the technology available. In weighing the decision to acquire new technology, managers must find answers to such questions as the following:

- What is the nature of the information requirements *internal* and *external* to the enterprise necessitating a change in the manner by which information is currently processed?

- What level of channel networking is required?

- Of the many business areas, where should ICT be applied first, and which areas will provide the largest competitive boost for the expense?

- What impact will the implementation of new information technology have on the customer service, logistics, and financial functions of the current organization, and of the supply channel as a whole?

- What will the implementation of new technology cost in terms of resources and operational trauma to the existing organization? How will channel allies react to requirements for increased information networking?

- How is new technology to be integrated with existing technology?

- What new resources will be required to operate the new technology?

- What new opportunities for competitive advantage will be available to the enterprise and to the supply channel in general?

In responding to these and other questions, each company in the channel network must first thoroughly understand the scope and nature of possible technical solutions available. As was detailed in Chapter 2, ICT systems for SCM can be divided into four general areas. In the first area can be found a wide range of *business system solutions* ranging from integrated corporate-level enterprise requirements planning (ERP) to workstation specific products, such as freight rating/routing, warehouse management, and shop-floor data-collection software. The objective of these packages are to provide accurate and "real-time" information to enable intraenterprise and interenterprise networking across functional boundaries. The wide range of *communications technologies* constitutes the second area of ICT systems for SCM. In this area can be found technologies as commonly used as the telephone and the fax, and as complex as EDI, satellite, and digitally transmitted data. In the third area can be found a diverse mixture of *material-handling equipment*. In this category can be found technologies such as automated vehicles, bar code readers, and radio frequency (RF). The application of these tools permit whole supply channels to automate processes in an endeavor to increase output rates, increase information accuracy, reduce material handling, and accelerate information transfer. The final area of SCM ICT systems is the expanding use of the *Internet*. Although the Internet is used today as primarily a way of providing information concerning products and services, the prospect for true "electronic commerce," where firms will actually buy and sell by "surfing the supply chain" is looming in the immediate future [18].

The utilization of ICT systems is focused around leveraging the tremendous opportunities available when the entire supply channel system is closely networked together. In the today's most forward thinking companies, SCM process teams are already designing and operating global channelwide information networks focused around data collection, communications, and data management to support the design and operation of their channelwide physical networks of vendors, carriers, distribution centers, plants, and customer facilities. For example, take Polycon Industries's use of a combination of ICT systems to satisfy Ford's requirements for the delivery of sets of sequenced car bumpers 4 hours after they are ordered. The company is a tier-one supplier that engineers and builds complete bumper assemblies including fascia, support rails, fog lamps, and so forth. The order management process begins when a car is put on the line at Ford's Oakville, MI plant. Via a continuous modem connection, Polycon receives broadcast orders into a production system running on a PC normally 4 hours prior to the time the bumper is to be fitted to the automobile. The PC application then parses the order to retrieve style, trim, color, and the sequence number of the vehicle on the line. The bumpers must be delivered in a precise sequence to match the vehicles on the line. Each of these order "packets" is then input into the Polycron's client-server production system. From here, each order is transmitted on a radio frequency (RF) broadcast system to hand-held receivers with display terminals used by warehouse operators. Upon receipt, the warehouse operator picks the proper bumper and loads it on one of three feeding conveyors. Bumpers are then sent to a central conveyor where they are finally sequenced, verified via a computerized shipping workstation, bar code labels generated and applied to the product, and Ford notified of the impending delivery. Finally, the inventory is reduced and the data input into the plant's MRP II system. The system also monitors the production system, calculating "real-time" yield, utilization, downtime, and other measurements necessarily to spot capacity trouble spots. Polycon's ICT solution permits the company to achieve internal process and planning optimization and provides the ability to interface with external suppler and customer applications like Ford's broadcast system [19].

The range of information integration provided by ICT systems, such as that used by companies like Polycon Industries, can only be expected to grow as the year 2000 approaches. The major requirements of 21st-century SCM process management will require robust ICT systems that will provide solutions to the following challenges:

- Ability to provide tailored, value-added services that will focus on specific, predetermined service targets established to maximize life cycle profitability.

- Creation of long-term partnerships enabling suppliers to operate in individualized markets and fulfill requirements for low-quantity, fast delivery while maintaining economies characteristic of mass markets.

- Continuing evolution of ICT systems. Among these tools will be enhancements to *data collection* through invisible and miniature bar codes, holograms, and voice recognition, *data storage* through data compression and use of powerful analytical tools for data retrieval, *data communications* through fiber optics, digitized transmission, and EDI via Internet and Intranet, and, finally, *data processing* through the use of multiple processors banded together to attack complex computing problems, embedded expert systems, and object-oriented, integrated software applications [20].

- Increased transaction capability permitting financial recognition and funds transfer at the time of purchase.

- Greater use of either third-party service providers or industry wholesale distributors who will assume increasing responsibility for traditional logistics functions.

- Expansion of ICT tools for product and service scheduling, production planning, and delivery to match contracted commitments.

- Heightened focus on reverse logistics functions.

Selecting ICT Systems for SCM

The implementation of ICT systems can assist companies realize a number of critical SCM operational and strategic opportunities. Several key enablers come to mind. To begin with, effective ICT systems can lower cost and improve productivity on any level of the value chain. Historically, technology has always been applied to reduce the cost of activities subject to repetitive processing. The recent explosion in ICT, however, has also empowered the enterprise not only to search for cost and productivity drivers through activity automation but also to expand the information content of the activities themselves. In this sense, ICT can not only help individual channel members to function more effectively, it can also exploit advantages in competitive scope. Second, ICT systems can enhance strategic differentiation. ICT has the power to leverage cost, logistics differentiation, and strategies focused around a channel network's combined product/service mix. By shortening the order processing and delivery cycle and providing customers with the ability to customize product requirements while reducing lead times and increasing the speed of information transfer through EDI, companies can improve product positioning. ICT can also enhance the performance of SCM value-added services, such as inside delivery, installation, kitting, and crating, that increase marketplace differentiation. Finally, ICT can broaden competitive scope. The effective implementation of targeted ICT systems can provide the entire supply channel with the opportunity to broaden the competitive scope of its operations and, by extension, alter the reach of its competitive advantage. Besides increasing the ability to coordinate activities on a global scope, ICT can assist individual companies to penetrate new markets and offer value-added services beyond former capabilities.

Although ICT can provide entire channel systems with the ability to achieve order-of-magnitude breakthroughs in productivity and competitiveness, companies must be careful to match ICT solutions with overall strategic business scope. Perhaps the first action to be undertaken is identifying clearly the range of the business problems to be solved. This step will significantly narrow the array of possible solutions. An effective requirements definition will greatly assist companies in avoiding critical errors, such as buying technology that is overkill, does not address the critical issues, or narrows future technology options because of hardware or software compatibility limitations. The following principles will greatly assist supply network members in leveraging ICT systems for competitive advantage.

1. *Assess Information Intensity.* Before an ICT solution can be effectively chosen, companies must evaluate the existing and potential information intensity of its products and operations processes. The objective is to determine the *breadth* of the information required both to run the supply channel network and to manage product and service processes. The former includes the number of customers and suppliers in the channel, size of marketing and selling information, number of product variations and size of product variety, and length of cycle times. The latter includes product related issues such as manufacturing complexity, requirements for buyer knowledge and training, and the ability of the product to service alternative uses. ICT solutions should enable the enterprise to leverage product and process information content to achieve marketplace leadership.

2. *Performance.* There are several critical performance issues that companies must resolve. Will the proposed ICT solution increase ICT availability and access to key data? Does the proposed ICT system's relative costs and benefits meet or exceed those of other possible solutions? How quickly can internal people and outside SCM process teams learn the software?

3. *Industry Impact.* Companies must closely examine the impact of the introduction of new ICT on the marketplace's competitive forces. ICT can dramatically alter a firm's bargaining power with suppliers and buyers, its ability to offer new and substitute products, and its ability to fight off new as well as old competitors. Effective ICT strategies can enable a company to seize marketplace leadership and force competitors to follow.

4. *Search for Ways ICT Can Increase Competitive Advantage.* By targeting activities that represent a large proportion of supply channel costs, are critical to marketplace differentiation, or compose critical links internally and externally within the supply channel, ICT can assist companies to identify new avenues for sustainable competitive advantage and explore opportunities to increase competitive scope. Some key questions here are: Does the ICT system permit entry to new market segments or invade the preserve of strong rival channels? Will ICT enable the entire channel to compete globally or to exploit interrelationships with other industries?

5. *Analyze System Scalability.* One of the most critical factors in selecting an ICT solution is confirming the ease by which an ICT system can adapt to change. Among the questions to be answered are: Can the system be operated, updated, and maintained with ease and efficiency? Can the system be integrated with new or existing technologies? Does the system possess the ability to easily change the number and location of users who are using it with minimal cost and operational disruption? To be competitive, today's SCM computing environment requires coexistence among all kinds of different architectures, platforms and hardware. ICT coexistence seeks to unify the computing environment found anywhere in the supply channel from mainframe to client/server and beyond in the search for unique sources of channel competitive advantage.

6. *Develop a Long-Term Plan That Seeks to Continuously Leverage New ICT.* The entire supply channel system must be diligent in instituting a formal methodology for the ongoing review of new ICT applications, the strategic alignment of business opportunities and new ICT products, the investments necessary to implement new hardware and software, and the impact ICT will have on supply channel linkages. Successfully enhancing competitive positioning can no longer occur simply by entrusting the exploration of new information technologies to MIS functions. Tomorrow's successful companies will require the participation of all functional levels, both within the organization and outside in the partner channel, if ICT tools are to be effectively utilized.

In the final analysis, the success of any ICT solution is measured on how closely the resources and capabilities of those enterprises comprising the supply channel network can be merged and focused on customer satisfaction. Leading-edge companies recognize that effective ICT systems provide them with enormous potential to fuse the diversity of the supply channel into a unified supply network system focused on creating unbeatable sources of marketplace advantage through the timely and accurate transmission of information.

Summary and Transition

The ability of companies to realize superior levels of performance is ultimately founded on the capabilities and creative visions of the people who comprise the organization and the robustness of the information and communications technologies used to run both internal and external supply channel processes. In today's business environment, the growing importance of core competencies and work force empowerment and the virtual explosion in information technologies has created a revolution in organizational structures and values and activated radically new ways of leveraging productive processes in the pursuit of global competitive advantage.

In the past, management attention on the people organization was focused narrowly on formulating the right management principles necessary to ensure optimal work-force productivities through the maintenance of rigid organizational structures and operations systems and procedures. The organization of the typical company was divided into a series of management and submanagement hierarchies that spanned the entire enterprise. At the top of the command pyramid resided the top management staff, the firm's leaders who were charged with the task of defining enterprise strategy, setting business direction, and providing the objectives to be pursued by departmental managers. On the next level of the pyramid could be found the company's tactical managers. It was this group's responsibility to translate the business strategy into detailed departmental objectives and budgets. Reporting to this group could be found, at the bottom of the organizational hierarchy, the firm's operations managers, who, in turn, directed the functional supervisors and work force in the performance of daily tasks and activities.

Despite its historical strengths, by the late 1980s this organizational model had become increasingly untenable. Efforts aimed at substituting radically new organizational models to meet the challenge of global competition coalesced around the application of two process-centered improvement approaches. The first, converging around a group of quality management (QM) techniques, focused on reshaping the attitudes and functional behavior of everyone in the firm away from narrow departmental task management and toward a broader understanding of how the productive processes that spanned the both the company and the supply channel could be continuously improved. The second and more radical response, business process reengineering (BPR), sought nothing less than the total dismantling of existing processes and rebuilding them up from the ground level. Whether through incremental improvement or fundamental restructuring, the demise of the traditional organizational model and the rise of organizational structures built around whole business processes signaled tremendous changes in the roles and values of the work force.

The implications for the work force of the end of the traditional hierarchical organization have been staggering. Instead of laboring in narrow, simple, task-oriented jobs, people now find that their work lives are multidimensional, spanning departments, oftentimes connecting them with counterparts in other companies out in the supply channel. As old artificial departmental demarcations lose their meaning, people increasingly find themselves coalescing into temporary process teams based on work competencies. Instead of being governed by a "command and control" management style where performance is measured by activity, they now find themselves empowered to make decisions previously considered the preserve of management and are measured by results. In place of training to achieve optimal task performance, the emphasis has switched to educating people to increase their insight and understanding of the converging enterprise processes of which they are a part. Finally, managers, formerly the guardians of authority and knowledge, have had to change from being taskmasters and scorekeepers to coaches and leaders. In addition to changes in the nature of work, the *value* of the work force has also been dramatically altered. In the age of mass production, the value of labor, whether line or managerial, was to be found in the volume of productivity performed as defined by departmental standards. Today, the value of labor is to be found in the level of the skills, knowledge, expertise, and information that the people *within* the firm can utilize to provide enrichment value to customers and supply channel partners.

In addition to leveraging internal people resources, companies have increasingly come to realize that it is only by utilizing the collective work-force competencies to be found in the supply chain management systems to which they belong can they hope to sustain the level of marketplace innovation necessary for competitive leadership. This premise is based on several key points. To begin with, as companies are increasingly conceived of as collections of productive resources focused around key market-satisfying processes rather than as repositories of wealth, the unique capabilities and knowledge of their work forces becomes a more important source of supply chain value than the products that they sell. Second, exportable work force competencies enable groups of companies to converge specialized resources to exploit new opportunities or solve critical problems that would be impossible if attempted from the standpoint of a single company. Finally, the mere possession of or capability for converging resources from several sources does not, by itself, guarantee success. Leveraging channel competencies requires their skillful blending in ways that compound the value of each resource. The work-force component of SCM provides a clear solution to the imbalance between available resources and strategic market aspirations today's most aggressive market leaders often experience.

A summary of the changes to the different elements required to capitalize on supply chain organizations is as follows:

- *Organizational Agenda:* Progression from bureaucracy, stifled initiative, bounded growth, and lack of customer focus to internal and supply channel interlinkages fostering the use of core competencies that enable collective solutions to realize process-breakthroughs and customer-centered opportunities.

- *The Work Force:* From task-centered to process-centered jobs; from performance focused around meeting rigidly defined operations standards to continuous improvement in productive processes; from jobs composed of skill sets to jobs requiring continuous learning, innovation, and flexibility.

- *Organizations:* Transition from hierarchical-defined departmental hand-offs of tasks to "virtual teams" of people required to perform whole processes and determine their own measurements.

- *Work Teams:* From narrow work unit teams to cross-functional, even cross-enterprise teams.

- *Managers:* End of authoritarian-based management styles and rise of managers as coaches and facilitators. Everyone is vested with the responsibility for the traditional roles once held by the management class.

- *Leadership:* End of "command and control" methods of leadership and emphasis on teamsmanship and facilitating the growth of team and individual sources of innovation and entrepreneurialism.

The origins and continued development of today's view of role of the work force, the structure of the organization, and the utilization of management philosophies such as BPR and SCM are directly dependent on the capabilities to be found in the information and communication technologies (ICT) available to companies. In fact, such organizational breakthroughs as the networking of geographically dispersed process teams, the integration of channel strategies and operations, communications technology providing connectivity between companies, planning systems that facilitate inventory management integration across the supply channel pipeline, and others would be impossible without effective ICT systems. SCM is such a critical management and operational approach for competitive advantage because it is inherently intertwined with the networking power to be found in today's computerized information and communications systems. As capabilities of ICT tools expand, there can be little doubt that the integrative and infomating capabilities of SCM to provide fresh competitive perspectives will likewise expand.

As a principle, it can be stated that the operational and strategic needs of the organization must be in balance with the dimensions of the information necessary for effective action and decision making. In the past, the velocity of the business environment was inhibited by limitations in information processing. With the advent of the computer, capable of handling information in volumes and speeds previously thought unimaginable, the heavy information processing constraints of the past were lifted, revealing new horizons of information and obsoleting many of the older methods and organizational processes and structures. Today, the explosion in information technologies has provided companies with the ability to integrate once-isolated business functions into unified supply channel organizations. This ability to network information by activating the links to empower people to cut across functional barriers and interweave common and specialized knowledge to solve enterprise problems is at the heart of the SCM concept.

When exploring the application of ICT solutions, implementers must ask several critical questions focused around the proper alignment of information technology tools and perceived increases in enterprise productivity and service-ability. Perhaps the most important decision is to identify clearly the scope of the business problems to be solved. This step will significantly narrow the range of possible ICT solutions and ensure that the effort is focused around core business issues. Equally as important is charting the effect an ICT implementation will have on the organization and its capabilities. In fact, the more encompassing the implementation, the more levels of learning and adjustment are required to utilize it. This alignment of the organization with the proposed ICT system impacts the enterprise in three ways. To begin with, the integrative process requires managers to restructure the culture and capabilities of their organizations around values promoting continuous process improvement and teamwork. Second, integrative ICT systems enable the organization not only to rethink traditional enterprise information flows but also to leverage new information tools such as graphics, workstation technology, and network-to-network computer integration. Finally, the effective application of new information technologies requires a redefinition of the goals and skills of the enterprise's people resources.

Notes

1. This definition is from Michael Hammer and James Champy, *Reeingineering the Corporation.* New York: HarperBusiness, 1993, p. 32.

2. For an expanded analysis of these points see Charles M. Savage, *Fifth Generation Management: Integrating Enterprises Through Human Networking.* Burlington, MA: Digital Equipment Corporation, 1990, pp. 65–146, and Michael Hammer, *Beyond Reengineering.* New York: HarperBusiness, 1996, pp. 1–17.

3. Currently some of the best books are Hammer and Champy, *Reengineering the Corporation*; Hammer, *Beyond Reengineering*; Steven L. Goldman, Roger N. Nagel, and Kenneth Preiss, *Agile Competitors and Virtual Organizations.* New York: Van Nostrand Reinhold, 1995; and also by the same authors *Cooperate to Compete.* New York: Van Nostrand Reinhold, 1996; James P. Womack, *Lean Thinking.* New York: Simon & Schuster, 1996; James Champy, *Reengineering Management.* New York: HarperBusiness, 1995; William H. Davidow and Michael S. Malone, *The Virtual Corporation.* New York: HarperCollins, 1992; and Peter M. Senge, *The Fifth Discipline.* New York: Doubleday, 1990.

4. This crucial point is elaborated upon by Hammer, *Beyond Reengineering*, p. 50.

5. Gary Hamel and C.K. Prahalad, *Competing for the Future.* Boston, MA: Harvard Business School Press, 1994, p. 182.

6. See the excellent discussion on build enterprise competencies in *Ibid.,* pp. 163–193.

7. For an excellent summary of the purpose and functionings of the tradition organization see Robert N. Anthony, *The Management Control Function.* Boston, MA: The Harvard Business School Press, 1988.

8. Critical discussion of the *center of excellence* concept can be found in Savage, *Fifth Generation Management*, pp. 214–215, and Hammer, *Beyond Reengineering*, pp. 116–126.

9. Andre Martin, *Infopartnering*. Essex Junction, VT: omeno, 1994, pp. 136–138.

10. Leslie Hansen Harps, "Case Corp Constructs Logistics Model of the Future," *Inbound Logistics* 16 (10) (October 1996), 25–32.

11. These business alliances are identified in Goldman et al., pp. 209–210.

12. See the discussion in *Ibid.*, pp. 210–220.

13. This section was drawn from David F. Ross, *Distribution: Planning and Control.* New York: Chapman & Hall, 1996, pp. 714–715.

14. Charles M. Savage, *Fifth Generation Management*, p. 70.

15. *Ibid.*, p. 71.

16. Richard Brown, "Configurable Network Computing," *APICS: The Performance Advantage* 6 (12) (December 1996), 36–39.

17. Christopher Gopal and Harold Cypress, *Integrated Distribution Management*. Homewood, IL: Business One Irwin, 1993, p. 173.

18. See the informative article by Laurie Joan Aron, "Surfing the Chain," *Inbound Logistics* 16 (9) (September 1996), 40–44.

19. Mike Ngo and Paul Szucs, "Four Hours," *APICS: The Performance Advantage* 6 (1) (January 1996), 30–32.

20. These trends can be found in Richard L. Dawe, "Tackle 21st Century Technology Today," *Transportation and Distribution* 37 (10) (October 1996), 112–120.

9

Implementing Supply Chain Management

It has been the central theme of this book that supply chain management (SCM) represents nothing less than a radically new strategic management philosophy enabling today's enterprise to realize the significant opportunities for competitive advantage to be found in the global marketplace of the late 1990s. Similar to other management concepts such as Just-In-Time (JIT) and Quality Management (QM), SCM can be described from several perspectives as an implementable technique, a management process, and a business philosophy. Beginning as an aspect of integrated logistics management centered around linking the common logistics functions to be found among supply and customer channel partners in search of throughput and cost advantages, SCM has evolved from a purely operational tactic to a universal strategic philosophy that seeks to converge the productive and innovative capabilities of enterprises linked together in a supply chain into a single, unified competitive force. The fundamental value of SCM is *cooperation*, and it is manifested in the willingness of allied chains of companies to link their strategic objectives and fundamental operational processes to create unique, borderless, market-satisfying resources that are invisible to the customer yet capable of quickly massing critical competencies and physical processes to form uncopyable sources of competitive advantage. In the past, companies relied on the development of fixed channels of supply where standardized mass-produced products would be distributed based on the least-cost principle. Today, market leadership belongs to those supply channels that can activate concurrent business processes and core competencies among their members, merge infrastructure, share risk and costs, jointly leverage design and productive processes, and anticipate tomorrow's opportunities for radically new products and competitive space.

This concluding chapter seeks to discuss the management and organizational elements necessary to effectively implement SCM. Although it is true that any actual SCM implementation will be unique due to particular business circumstances, productive and distributive processes, sets of core competencies, and strategic visions of the supply channel networks involved, there are, nevertheless, a number of formative principles that can be found in any application of the SCM concept. The chapter begins by revisiting these basic principles of SCM. Next, the discussion shifts to a detailed review of the requirements necessary for a successful SCM implementation. The chapter then concludes with a series of questions managers can use when determining how SCM can help their businesses achieve competitive success and what direction their implementations should take.

SCM Revisited

In Chapter 1, an attempt was made to construct a definition of SCM that could be used as a benchmark for the discussion to follow. Although it was stated that SCM, as is the case with any management philosophy, was still in the process of evolving, the concept could be broken down into three closely integrated dynamics. The first perceives SCM narrowly as an operations management technique that enables companies to move beyond merely integrating logistics activities to a position where all enterprise functions—marketing, manufacturing, finance, and logistics—can be closely integrated to form the foundation of a unified business system. SCM at this level enables the firm to connect and synchronize the day-to-day performance of operational activities, such as inbound logistics, all types of processing functions, outbound logistics, and marketing, planning, and control support activities, that are necessary to ensure the continuous alignment of departmental tactical objectives, the optimization of all operations functions, and the continuous creation of customer service value from the enterprise perspective.

In the second dynamic, SCM is described as a channel management method that seeks to extend the concept of integrated logistics to the performance of complimentary logistics activities by suppliers at the input end and by customers at the output end of the supply pipeline. The objective of SCM at this level is to closely interface, if not merge altogether, the logistics functions of an organization with the identical functions performed by logistics counterparts found in outside supply channel partners. Examples could be as simple as providing inventory planners with the ability to look directly at their suppliers' inventories via computer-to-computer linkup, or as complex as the partnerships some companies form with third-party logistics service providers who perform a matrix of logistics activities, such as warehousing, rate collection and billing, transportation, and others, working in close collaboration with their customers' logistics operations staffs. This dynamic supports the concept that in today's business environment, no company exists independently nor possesses by itself all of the competencies and knowledge necessary to maintain market leadership. Instead, by networking logistics functions, supply channel partners, in effect, acknowledge that they are inextricably bound with networks of other channel partners in the pursuit of common and mutually supportive competitive objectives.

The third, and final, dynamic seeks to define SCM in an entirely new light as perhaps today's most powerful strategy for leveraging the enormous capabilities and capacity for innovation to be found when the individual companies comprising a supply chain system are fused into a single competitive entity. Such an understanding of the role of SCM is described as constituting a distinct break with contemporary definitions. Basically, current treatments define SCM primarily as an *operations management activity* focused on accelerating the flow of inventory and information through the supply channel network, optimizing and aligning internal operations functions with those of outside channel partners, and providing the mechanism to facilitate continuous channelwide cost-reduction efforts and increased productivity. However, although these concepts are clearly a part of the operations dynamics of SCM, they are by themselves insufficient when viewing SCM from a strategic perspective. The operational elements of SCM provide today's enterprise with the ability to survive in the struggle for marketplace advantage: SCM, however, also provides a company with a distinct strategic perspective enabling the construction of a shared competitive vision with its supply channel partners, the formation of coevolutionary and mutually beneficial alliances, the opportunity to exploit the productivity of channel processes and technologies, and the ability to leverage core competencies and critical resources beyond the capacities of any single organization to lead market direction, spawn new associated businesses, and explore radically new opportunities.

Based on these three dynamics, SCM can be defined as follows:

> SCM is a continuously evolving management philosophy that seeks to unify the collective productive competencies and resources of the business functions found both within the enterprise and outside in the firm's allied business partners located along intersecting supply channels into a highly competitive, customer-enriching supply system focused on developing innovative solutions and synchronizing the flow of marketplace products, services, and information to create unique, individualized sources of customer value.

Throughout this book, these different facets of SCM have been discussed in detail. SCM has been described in one way or another as all of the following:

- *A Method.* The central focus of SCM is the management of customer value. Value to the customer is measured by how closely perceptions of product and service quality are matched by actual performance. In today's business environment, satisfying the end customer can only take place when the entire supply channel, from materials supplier to retailer, are linked closely together in the pursuit of innovative ways to actualize service value. Accomplishing this objective requires companies to rigorously explore the implementation of channel-level quality management and business process reengineering techniques in an endeavor to search for methods of continuously improving the productivity and cost-effectiveness of internal and supply channel work processes. Regardless of the industry or marketplace orientation, at the very heart of SCM stands a collective commitment on the part of all channel network members to constantly validate that channel processes are being performed that optimize functional productivities and create superlative customer value.

- *A Concept.* Whereas SCM represents a radically new approach to enterprise management, the body of knowledge surrounding it is not purely the product of some management theory. The concept of SCM, in fact, is connected with and in many ways is the product of the significant changes that have occurred in modern logistics management. Over the past 30 years, logistics has progressed from a purely operational function to become a fundamental strategic component of today's leading manufacturing and distribution companies. As logistics has evolved through time, the basic features of SCM can be recognized, first in their embryonic state as an extension of integrated logistics management, and then latter as a full-fledged business philosophy encompassing and directing the productive efforts of whole supply chain systems. This "intellectual" heritage validates SCM and provides it with recognizable concepts and operating norms practiced widely today.

- *A Philosophy.* Although it is concerned with the pursuit of high-quality operations processes that continually add value to the customer, the real power of SCM is to be found when it is perceived as a management philosophy. Today, SCM has evolved from being purely a set of logistics performance tools to an enterprisewide, even channelwide, operating philosophy. SCM is a boundary-spanning, channel-unifying, dynamic, and growth-oriented philosophy of enterprise management. SCM requires companies to search for competitive advantage by looking beyond the frontiers of their own organizations. As a unifying force, SCM enables whole supply channel systems to act as a single competitive entity. SCM is also a dynamic, open-ended management philosophy focused on the continuous process of shaping and reshaping intracompany and intercompany performance, information technology tools, products and services, and organizational and personal excellence to exploit the ever-changing contexts of marketplace opportunities. Finally, SCM is, above all, a business philosophy that enables individual companies, as well as allied channel network members, to achieve high levels of productivity, profit, and growth.

- *A System.* SCM provides all types of businesses with a comprehensive system to respond effectively to the enormous changes occurring in today's business climate brought about by marketplace globalization, the enabling power of technology, the development of channel alliances and "virtual" organizations, continued organizational reengineering, and work force empowerment. SCM presents companies with the operational and strategic mechanisms to meet customers' requirements for tailored combinations of products, services, and information that will provide unique value and a solution to their needs. It enables producers to construct affordable agile design, manufacturing, and delivery processes providing superior customer-winning quality. It provides for the convergence of information and communications technologies (ICT) and empowerment-based management techniques stimulating productivity and opening new vistas for competitive innovation. It also enables firms to leverage the opportunities to be found in the opening of the global marketplace and the growth of strategic channel alliances. Finally, SCM fosters the creation of superlative logistics functions that support customer service objectives, reduce cycle times, stimulate productivity, and reduce total supply channel costs.

- *A Process.* Preeminently, SCM can be described as a business philosophy centered around the effective management of supply channel processes. The term "supply channel processes" refers not to a concept but rather to the actual physical business functions, institutions, and operations strategies that characterize the way a particular channel system moves goods and services to market through the supply pipeline. The objective of SCM is to solve the following critical channel process problems: effectively managing and transporting channel inventories, facilitating marketing, product, and information exchange, increasing supply channel efficiency, facilitating financial and distribution processes by providing capital and credit and postponement functions, reducing channel complexity by standardizing channel activities, providing key alliances, assembling new competencies, exploring new opportunities to be found in global supply channels, and enabling the growth of channel specialization whereby performance of critical elements of supply channel management, such as exchange, transportation, or warehousing, are invested in channel partners specializing in that function.

- *A Strategy.* SCM represents such a revolutionary and powerful competitive management paradigm that it should be considered as a fundamentally new source of competitive advantage. Effective SCM strategies provide today's enterprise with radically new opportunities to create marketplace advantage by leveraging supply chain partnerships, information and communications technologies, and the knowledge and innovative capabilities of the entire channel work force. SCM strategies require firms to identify themselves and the basis of their competitive advantage less by the products and services they offer to their customers and more by the processes that they use to create marketplace value. Real growth and expansion today, the kind that consistently preempts the competition and increases marketshare, occurs when companies move beyond a pre-occupation with internal capabilities and focus around strategies that continuously search to converge their own strengths with the productive resources and innovative knowledge of their supply channel partners.

- *A State of Mind.* As a major business philosophy, the effective application of the principles of SCM requires broad acceptance by the staffs of all supply channel members. In fact, leveraging the capabilities and creative vision of the people who comprise the supply channel is, without a doubt, the essential key to realizing superior performance through SCM. Achieving "buy in," however, requires companies to tackle the difficult challenges to the traditional concepts of management and work occurring as a result of enterprise reengineering, the growth of "virtual" organizations, application of quality management techniques, and team-based organizational styles. Activation of organizations ready to implement SCM requires environments that foster the generation of ideas which will continually improve quality and productivity, management policies and practices that support individual initiative and entrepreneurialism, and new workplace values enabling the convergence of specialized resources to be found among channel members to exploit new opportunities or solve critical problems that would be impossible if attempted from the standpoint of a single company.

Supply chain management has become today's most important management concept for competitive advantage because it provides whole supply channel systems with the capability to move beyond cost incrementalism and process reengineering and toward a new competitive vision based on supply channel cooperation, the sharing of critical core competencies, and a commitment to coevolutionary business processes. The strategic and operational dynamics of SCM provide allied companies with both long-range market foresight as well as the day-to-day operational principles necessary to leverage the competitive possibilities of a global marketplace bursting with coalitions, partnerships, and consortia of all kinds. Leveraging this new environment for competitive advantage means that companies will have to wisely seek out channel partners, devote critical resources when merging channel partner competencies, identify centers of potential innovation, and engineer productive, creative relationships that utilize the capabilities of the very best companies to structure superlative channel system product and delivery processes of unsurpassed marketplace value.

Implementing SCM

As is the case when implementing any major business philosophy, tackling SCM presents a formidable undertaking. A flood of questions comes to mind: Where does one start? What kinds of resources will be needed? How much will it cost? How much time will it take? What will an SCM implementation do to the company's organizational structure, to basic cultural values, to traditional operations processes, to the way the work force performs their jobs? Unfortunately, there is no one best answer. SCM is based on the assumption that a company's basic business strategies, processes, organizations, and people are evolutionary in nature, rather than stagnant functional entities. SCM also assumes that the marketplace is inherently dynamic, that companies work best when they merge their productive resources with the competencies of their supply partners, and that it is continuous innovation and foresight that drives competitive advantage today. Choosing an SCM implementation path, therefore, requires a company *and* its supply chain partners to conscientiously and decisively explore a multitude of alternatives and make choices depending on the type of business, the manufacturing and distribution processes used, and the level of sophistication of the channel environment.

Implementation Guidelines

There is no "proven path" to SCM implementation. There are so many facets to SCM, both operational and strategic, that any given implementation can take an infinite variety of different forms, progress through radically different stages, and result is several different outcomes. Broadly speaking, however, the contents of an implementation project should focus around the three dynamics of the SCM concept described in the previous section. The first SCM dynamic (*total enterprise functional integration*) is the easiest to implement. This approach seeks the full integration of all enterprise functions, ranging from finance to logistics, as well as related business units composing the internal supply channel, into a single business system. Achieving this objective requires not only the close synchronization of all daily operational and planning processes, but also the removal of departmental biases and the establishment of strategic congruence and consensus. At this level, an SCM implementation could utilize some common management techniques, such as Just-In-Time (JIT), Total Quality Management (TQM), and Business Process Reengineering (BPR), to break down remnants of "silo" management styles and refocus the entire enterprise on pursuing continuous process improvement objectives, organizational and work force transvaluation, and convergence of departmental strategies centered around common customer-focused goals.

The second area of an SCM implementation (*channel operational integration*) seeks to expand on the integrative foundations of the first dynamic. This approach utilizes the same management techniques and pursues similar operational goals as the first dynamic. The difference between the two can be found in the fact that the second dynamic requires the implementation of the principles of SCM operations integration with business partners existing beyond the borders of the organization. This dynamic represents more than an expansion in scope: Channel operations integration requires companies to develop clearly defined and communicable objectives, advanced negotiating skills, knowledge of critical business processes, and the ability to work and make decisions that span the operations of multiple channel partners. In addition, implementation of this dynamic requires companies to create and empower effective intrachannel and interchannel process teams. These process teams assume fundamental agreement of operational objectives both within individual organizations and between channel members. It is the strength of these teams and the circumference and depth of their functional congruence that defines the level of SCM integration.

Implementing the third dynamic of SCM (*channel strategy development*) requires a significantly larger effort than what is required of the operational dynamics. Whereas the first two dynamics focus on the operations side of SCM (order management, warehousing, and transportation), implementation of the third dynamic requires companies to continuously search for opportunities to create *strategic* initiatives with business partners that enable the coevolution of radically new methods of providing the chain of customers with value, merging complimentary channel capabilities, the joint development of whole new business processes and technologies, the structuring of new forms of vertical integration and economies of scale, and the leveraging of core competencies found within or among associated enterprises. An SCM strategy, such as merging the knowledge, physical and technical resources, and innovative skills of a consortia of allied companies to constantly create market-wining products, requires a high level of cooperation among management and work force, concise definition as to objectives and process performance, a strong commitment on the part of all companies involved, and superlative information networking capabilities. Implementing the operational dynamics of SCM also requires whole supply channels to focus on achieving short-term marketplace advantage through the application of cost and cycle time reduction techniques. Strategic SCM, on the other hand, enables implementers to search for long-term solutions constructed around market foresight and the ability to coalesce the unique core competencies to be found among supply channel partners in the pursuit of radically new marketplace advantages.

How the three dynamics of SCM are to be implemented is unimportant: The key is understanding that the true value of SCM can only be attained when all three dynamics are implemented simultaneously. If only the *operational* dynamics are implemented, then SCM is really nothing greater than integrated enterprise resource planning or integrated logistics. On the other hand, an implementation of the strategic dynamic of SCM that lacks the presence of strong intrafunctional and interchannel operational integration also cannot succeed. One approach is to begin at the first dynamic (intraenterprise functional integration) and, with this as a base, proceed to expand the operational elements of SCM into the supplier and customer channel network. With these processes in place, implementing effective shared supply channel strategies can then occur. Other methods might favor a pilot approach where certain operational processes might be implemented within the organization and with key channel partners before expanding to the entire supply channel, or a parallel approach where all three SCM dynamics are implemented at the same time.

Ten-Step Game Plan for SCM Implementation Success

The decision to implement SCM is normally undertaken in response to a variety of possible marketplace or competitive conditions. Some are purely operational in nature, such as declining customer satisfaction, massive quality gaps, changes in productive techniques, availability of information technology, arrival of new channel partners, and company functional reengineering. Others are the result of the rise of more critical strategic issues, such as the entry of new competitors into the marketplace, changes in governmental regulation, opening of new markets (e.g., passage of NAFTA and the opening of China), and massive earnings loss due to poor performance. In any case, once a company has decided that focused changes are necessary to ensure corporate survival, they must be careful not to fall into the reengineering/incrementalist trap in the mistaken assumption they are implementing SCM. Management initiatives centered around such techniques as benchmarking the best, creating focused task teams to study the situation, streamlining the organization, and people and cost-reduction efforts are all part of the old paradigm. Such efforts are reactive—they are based on responding to the lead already held by the competition and will result in little net gain.

Before companies can begin, therefore, the task of effectively applying the SCM concept and creating a targeted implementation, it is imperative that their management staffs understand that the real goal of SCM is to fundamentally increase the value-added capabilities possessed by the firm's core competencies, supported by its channel partners, to create and sustain continuous customer satisfaction. This is a competitive-winning philosophy that seeks to not just to catch but to beat the competition. Achieving such a strategic perspective requires companies and the supply chain partners they deal with to ask fundamental questions regarding the basic operating and strategic objectives of their businesses. Are the products and services that are offered capable of providing customers with unique market-winning solutions? Has the company nurtured the critical core competencies necessary to achieve continuous innovation and quality? What channel partners have been assembled to provide the level of human and physical resources the firm needs to pursue its strategies? Does everyone in the company understand that their primary objective is to continuously reduce the gap that exists between the firm and the supply chain network and the customer?

Once companies and their supply chain systems have formulated such critical strategic requirements, they can begin the process of meaningfully implementing SCM. The steps in this process (Fig. 9.1) can be described as follows:

1. *Create an Effective SCM Education Program.* The changes expected to occur both on the individual company as well as on the channel

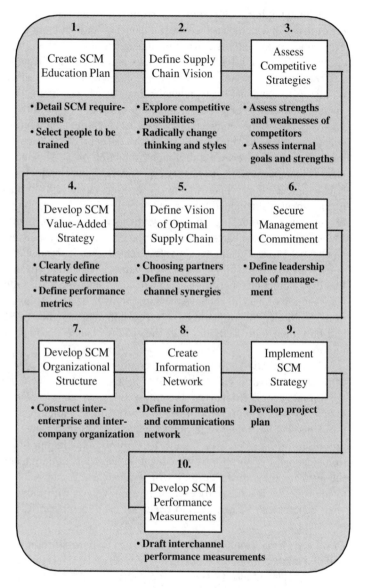

Figure 9.1. SCM implementation process.

level as a result of an SCM implementation will not provide sources of competitive advantage without the education of the channel's managers and work force. In fact, a comprehensive SCM implementation is so extensive and requires organizational structures so different from traditional practices that it will impact everyone in the channel. As a result, maintaining momentum will require a significant educational effort that will encompass the whole channel. Although literature and, even more importantly, trained consultants and educators are hard to come by, there are a few excellent books and articles available [3].

There are three broad groups of people that will need to be educated on the operational and strategic elements of SCM. The first is the top management staffs of each participating channel organization. The goal is to ensure they understand SCM goals and operating objectives and to get them to "buy in" to the SCM concept. Next, each SCM process team must receive a thorough grounding on SCM principles, team management skills, and the use of networking technologies. The goal is to teach these channel teams how to merge concepts of quality management, value-added management, and process innovation practiced within the boundaries of their own firms with the identical efforts found among their channel allies. Ultimately, it is the responsibility of channel process teams to activate targeted channel strategies and convert them into ongoing sources of channel competitive advantage. Finally, all other members of the channel work force who act as process team support or who do not have the opportunity to be directly involved in the SCM effort also should have a thorough understanding of SCM and how their daily activities will impact the channel's competitive mission.

Whereas effective training in the operational and strategic elements of SCM is critical, it is imperative that it be focused and include the appropriate people from across the entire supply chain. Poorly trained people will often confuse priorities and divert their efforts to solving problems that have very little impact while more critical value-added processes are excluded from review. Effective training requires people to understand the purpose of the education and training they receive, the need for the knowledge they will acquire, how it can be applied to improve the business processes they are involved with today, and what benefits can be expected to occur for each member of the supply chain.

2. *Develop a Defined Supply Chain Vision.* The creation and communication of a market-winning competitive vision shared not just by individual companies but also by whole supply chains is absolutely essential before any SCM project can begin. The process of "visioning" or "foresight" enables companies to look into the future and visualize what they would like to be. Visioning provides companies with specific goals and strategies on how they plan to identify and realize the opportunities they expect to find in the marketplace. It is important to note that visioning is not synonymous with dreaming or wishful thinking. In contrast, the kind of visioning that creates new markets, pre-

empts the products and service strategies of the competition, and continuously searches for new forms of customer value is grounded on what is possible, practical, and realistic.

Companies that are content with traditional marketplace approaches lack the ability to effectively vision. Because they fail to utilize customer-wining business philosophies, such as SCM, the basis of their strategic plans, goals, and priorities are turned inward, are often confused and lost in silo management styles, and depend on incrementalism for process improvement. Achieving success through effective visioning requires companies to completely overhaul their traditional ways of thinking and the competitive values they currently possess. One of the most important changes is broadening the *opportunity horizon* [1] of the firm. This is accomplished by perceiving the company as a collection of process and people core competencies that have the potential to infinitely mutate into sources of distinct competitive advantage. Another change is fostering the ability to see the future not just through the window of a single company but through the prism of supply chain interfunctional and interprocess strategy debate. SCM strategic visioning occurs when the marketing, product design, manufacturing, and distribution functions of each channel member are closely integrated internally within each firm and outside with the analogous functions to be found along the supply chain network.

3. *Assess Competitive Strategies.* The successful implementation of SCM requires companies to be constantly in the process of assessing the strengths and weaknesses of their strategies. Strategic assessment enables whole supply channel systems to formulate answers to three critical questions [2]:

- Where are we? Answering this question requires individual companies to map their value-added processes beginning with internal marketing and sales, product/process design, inventory planning and manufacturing, and distribution and postsales service, and then move to an exploration of the processes that are performed in conjunction with supply chain partners. Accomplishing this task can be a daunting experience due to existing departmental barriers and traditional performance systems, not to mention the difficulty in gaining the cooperation of channel partners. By making each intracompany and intercompany process visible, opportunities to increase critical competencies and new areas of competitive value become apparent.

- Where do we want to be? The strengths and weaknesses identified in the firm's and the channel's value-added processes must be matched to marketplace needs and compared to the capabilities possessed by competitors. Understanding customer requirements means that the whole supply chain must not only be able to meet the needs of the existing marketplace but also possess the foresight to continually unearth radically new dimensions of products and services through the design and closer integration of channelwide value-added processes. In addition, an effective process assessment should permit companies to position themselves relative to competitor channels. By using a simple matrix, companies can plot current performance dimensions against those of the competition.

- How do we get there? Once current value-added processes and customer and competitor dimensions have been identified, companies must then assess the capability of existing resources to achieve competitive targets. This objective is accomplished by selecting one or a matrix of value-added strategies. Supply channels could choose to compete as a low-cost or high-quality producer, on the basis of delivery performance, flexibility, and responsiveness, or by offering innovative products and services. Achieving even one of these value-added strategies normally requires today's enterprise to integrate the competencies of supply chain partners.

Continual strategic assessment is needed because neither the marketplace nor the competition is static. SCM requires entire supply networks to rigorously investigate channel processes to ensure that they are providing superlative customer service, effectively utilizing channel productive competencies, and providing a source of unbeatable competitive advantage.

4. *Develop an SCM Value-Added Strategy.* Once a thorough competitive assessment has been completed, supply chain members will need to engineer a comprehensive value-added strategy that will guide the selection and improvement of functional processes that, in turn, will provide competitive advantage to both individual firms and the supply chain as a whole. The creation of a comprehensive SCM value-added strategy has two purposes. First, it provides for the establishment of clear, strategic direction for the supply channel network and illustrates the awareness and dedication of supply channel members to the channel strategy. Second, a detailed channel strategy will define the performance metrics and feedback structure against which the success of the strategy can be measured and consistency established, thus providing the performance review mechanisms characteristics of coherent SCM value-added strategies.

   As Figure 9.2 illustrates, such a strategy rests on the identification of critical intrafunctional and interenterprise value-added processes. It

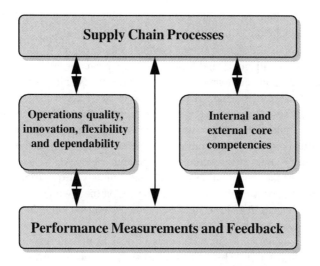

Figure 9.2. SCM value-added strategies.

is the strength and potential for innovation of these product design, manufacturing, and delivery processes that forms the core of SCM competitive operations and strategic objectives. Next, the second piece of the SCM value-added strategy requires focusing attention on developing and improving those critical value-added operational functions that link the supply network. Critical objectives to be realized are increasing the intensity of intercompany operations integration, quality, capacity for innovation, cost reduction, and expansion of flexibility and dependability.

The third area of an effective SCM value-added strategy is concerned with deepening the scope of internal and channel competencies and physical and technological resources. This area also includes the creation of alliances with only the marketplace's best companies in an effort to attain missing or incomplete capabilities. It is only by developing and leveraging the human and technological resources to be found across the width of the supply chain that companies will find tomorrow's competitive edge. Finally, the last area of an effective SCM value-added strategy is the creation of the appropriate performance measurements and feedback mechanisms that provide meaningful data on the strategic impact of channel processes. The goal is the creation and buy-in by the collective channel network of initiatives aimed at achieving the continuous improvement, not only of operational processes, but also of strategic processes arising from the knowledge and skill base of the people organization and the ability of the entire supply chain to activate customer-winning innovation and competitive advantage.

5. *Develop a Vision of What the Optimal Supply Chain Would Look Like.* Once overall goals and strategies have been formulated, individual supply chain companies will then have to establish the criteria to be possessed by channel partners necessary for the realization SCM objectives. The process begins by drawing on the results of the strategic assessment. The goal is to focus attention on gaining the support of partners who can lend critical competencies to assist in the improvement of targeted operational and strategic objectives. This may mean looking for channel assistance to make already strong processes even stronger; or it might mean searching for partners who possess "world-class" capabilities in an area of extreme weakness targeted for outsourcing.

    Choosing the right partners is critical to the success of the SCM implementation. The first place to start is with the array of existing channel partners. The existence of complementary competencies to support internal weaknesses may already be present in the current channel, but have just not been thoroughly exploited. If the right partners cannot be found among current channel members, then companies existing in other business systems, even traditional competitors, might be brought in to the channel alliance. In any case, the channel system that is created must be able to generate the synergies required to move the entire channel to the next competitive level. If the wrong partner is selected, an incredible amount of energy and resources can be expended in a very short time with no real payback. In addition, the business partners that are really needed may no longer be available, having signed on with the competition.

6. *Secure Management Commitment.* Without the full support and commitment of the top management teams of each supply chain member any SCM implementation will be fragmented and achieve only minimal success. The real role of management is to provide effective *leadership* that galvanizes the SCM effort and gains the attention of and inspires everyone associated with the implementation. Management support necessary for the implementation of SCM is complicated by the fact that it requires the full communication and participation of management in actively promoting SCM both within the boundaries of the organization and outside in the supply channel. Internally, management must be prepared to lead TQM, process reengineering, and core-competency-building efforts targeted at continuously improving all aspects of the firm's operational performance. Externally, mangers must be prepared to work closely with their supply channel allies. This means that they must develop consensus with their channel counterparts to facilitate

intercompany operational linkages and tackle the more difficult strategic issues, such as joint product design, database networking, and the establishment and guidance of SCM process teams.

A clear directive to proceed should be issued by all channel top management teams concurrently to ensure that the scope and objectives of the SCM initiative is conveyed to everyone throughout the supply chain network. Arranging for visits to the facilities of channel partners, celebrating events such as computer network linkups, or even moving staff to work directly in the plant of a channel ally will promote recognition of the importance of the SCM linkage and facilitate operational performance and continued innovation. Ultimately, it is the responsibility of the top management SCM team to continually reinforce the importance of the implementation and to maintain its visibility and priorities.

7. *Develop SCM Organizational Structure.* One of the keys of a successful SCM implementation is the development of an effective interchannel organization. Regardless of the depth of comprehensiveness and degree of innovation to be found in the channel's strategies, no SCM initiative can hope to succeed without the existence of strong interchannel organizations that enable the merging and positioning of the critical competencies and resources necessary to perform supply channel processes. Effective SCM organizations require several elements. To begin with, the internal structures of companies must be characterized by organizations that are flat, provide for people empowerment, and are cross-disciplinary and cross-departmental. SCM initiatives simply will not be successful if the organization has layers of bureaucracy and functional silos. Such an environment will only provide internal and SCM teams with conflicting objectives resulting in suboptimal results. In contrast, effectively organized companies strive to maintain open communication lines, deepen internal and channelwide process understanding, and continuously eliminate crippling departmental barriers to cross-functional and cross-channel activities.

In addition to the creation of superlative *internal* organizations, SCM relies on the configuration and empowerment of cross-channel process teams targeted at achieving channel strategic objectives and continuously creating innovative sources of customer value. The interchannel process team is the most effective organizational form for managing SCM activities. SCM process teams ensure that each channel member has representation and can bring the special competencies and skills of their organizations to the channel effort. SCM process teams also assist companies to maintain channel strategic priorities and ensure that channel resources are both adequate and focused on the right strategies. Finally, SCM process teams seek to network performance across the breadth of the entire channel through the merging of information and communications technologies.

8. *Create the Information Network.* The combination of information and communications tools that can be used to network a supply chain together depends on the scope of the SCM strategies and the requirements for interchannel integration. The system support tools to be implemented should provide supply chain members with the following capabilities:

    - Enable the design of information infrastructures that will provide for the close integration of channel systems and support effective process solutions for all operational activities, including product design, purchasing, manufacturing, logistics, customer order management, customer service, and finance

    - Provide a mechanism for enabling and monitoring the application of quality management and reengineering principles to channel processes in the pursuit of continuous improvement

    - Foster a learning environment where the skill levels, competencies, and potential for empowerment of people on each level of the supply chain can be increased

    - Develop intracompany and intersupply channel organizations focused on the design of networked process teams whose mission is to continuously search for ways of increasing customer value

    - Serve as a interchannel database system from which channel members can retrieve accurate and timely information for operations and strategic planning, as well as for the effective performance of daily activities.

When implementing information and communications technologies (ICT), companies must be careful to ensure that the various architectures, platforms, and hardware that are to be found both within the enterprise and outside in the supply chain can be effectively linked together. The capability of channelwide ICT systems to "coexist" enables companies to design a unified computing environment that supports everything from mainframe to client/server and beyond. Although some companies may select different architectures to support specific applications, coexistence enables supply chains to run the same applications across several environments. Such a computing environment provides for the integration of existing systems with those of localized systems such as EDI and freight rating/scheduling applications. In addition, coexistence permits companies to migrate legacy applications to new technology architectures without losing tried and true business functionality or having to reinvest in a new system implementation [4].

9. *Implement the SCM Strategy.* The actual implementation of SCM operational and strategic objectives is perhaps the greatest single challenge a firm and its channel partners can undertake. An SCM implementation project requires a great deal of time, effort, and often expense on the part of the chain of companies involved. From the first, however, it is important to note that an SCM implementation is not to be confused with finite projects, such as implementing a new enterprise information system or designing a new product family. Although an SCM implementation begins at a certain moment in time, it should serve as the foundation for the continuous pursuit of opportunities for the realization of new value-add interchannel processes and strategic innovations that will enable the entire network alliance to achieve ongoing marketplace leadership. In this sense, an SCM implementation must be considered as the gateway through which supply chains progress as they search for radically new ways to capture emerging markets and invent whole new regions of competitive space.

In order to guarantee continuous reinvention of the supply chain's competitive dynamics, the SCM implementation process should consist of the following elements (Figure 9.3):

• SCM Value-Added Strategy. As discussed earlier, the SCM value-added strategy is the driving force that not only provides a vision of the competitive goals of the channel network but also establishes clear performance targets to guide the implementation effort. The

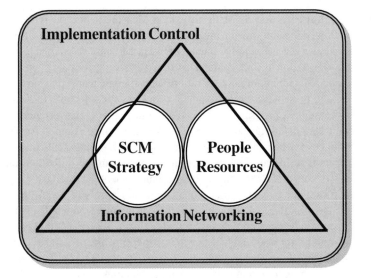

Figure 9.3. Elements of SCM implementation.

convergence of the SCM strategies originating from different regions in the channel permeates and provides direction for each dimension of the implementation effort. Finally, it can be said that SCM strategies are constantly evolving as whole supply chains react to changing marketplace and technology conditions and search for fresh ways to preempt the competition.

- People Resources. The ability to harness the knowledge, skills, and creative power of the people resources to be found along the supply chain and to focus them on the performance and generation of value-creating processes is at the core of continuous SCM implementation success.

- Information Networking. Also to be found at the core of the continuous implementation of SCM are the information and communications tools that are used to network the channel system's people resources. Information networking enables interchannel process teams to integrate ideas and activities across geographical space and time.

- Implementation Control. Surrounding all SCM implementations efforts is the need for interchannel management and control. Effective SCM implementation control ensures that channel resources are focused on the best competitive opportunities and that the most advantageous value-creating strategies are constantly being pursued at all levels in the supply chain.

10. Develop Effective SCM Performance Measurements. As is the case with any meaningful business activity, the ongoing success of SCM strategies for any supply channel network can only be gauged through the development and application of effective performance measurements. The intercompany operations and strategic teams that have been created to manage supply channel processes must be able to develop measurement programs and benchmarking metrics that provide relevant data detailing the actual level of performance attained for SCM initiatives whether operational or strategic. Drafting such measurements is a difficult task because they must span the individual businesses constituting the entire supply chain. One of the foremost dangers is formulating measurements that focus too narrowly on the performance of individual companies to the detriment of the whole channel. Fashioning channel processes around performance metrics that provide advantage for one channel partner at the expense of another clearly would result in the eventual failure of the suite of SCM strategic initiatives being pursued by the supply chain as a whole.

Just as individual organizations need them, supply chain networks also require performance measurements that are calculable, relevant, and promote quality process improvement. Such measurements must include both process and results measurements. Process measurements will provide *internal* metrics relating to performance issues such as cycle time reduction, cost reduction, increases in learning curves, and levels of work force competence. Results-oriented measurements will focus on *channel-level* metrics associated with increased sales, increased profitability, increased customer satisfaction, reduced product design time to market, capability to penetrate existing and new markets, and instances of new product, service, and process innovation.

SCM Implementation Pitfalls

Similar to any business undertaking, some SCM implementation projects are doomed to failure. One of the most difficult aspects of implementing strategic SCM is the fact that not some but all channel partners must be part of the effort and they must all pursue SCM objectives with the same degree of vigor. The problem with all process improvement projects, and SCM is nothing more than a process improvement project on a very large scale, is that all elements constituting the process must be improved simultaneously and in synchronization. Simply improving only a few parts of the process will result in little overall improvement. Some of the common reasons why SCM implementations fails can be listed as follows:

- Lack of Vision. The SCM project is conceived too narrowly either as to scope of channel partner participation or as to the processes to be improved. Just as bad is the other extreme, where the SCM vision is so broad that it is little more than dreaming or wishful thinking.

- Priorities. Often companies can undertake projects that are so comprehensive and contain so many initiatives that priorities become jumbled and meaningless. When this occurs, process teams become confused as to which project to devote their attention to with the result that often projects of lesser importance are considered first, and more critical processes improvements are postponed. In addition, creating too many priorities often results in critical resources being spread too thin, rendering it virtually impossible to achieve with any degree of success original SCM initiatives.

- Lack of Customer Focus. As can happen with any project, SCM process teams can lose sight of the real objective—to increase customer value. A lack of understanding of the customer-focused nature of the SCM initiative can result in implementers improving channel processes that add little or no value to the customer. Customer satisfaction is a moving target and it changes constantly. A successful SCM implementation occurs when process teams are working on objectives that are in alignment with customer expectations and have generated processes that enable the entire supply chain to continuously meet and exceed them.

- Choosing the Wrong Channel Partners. Pivotal to the ongoing success of SCM is the requirement that companies choose only the best supply chain partners. Selecting the wrong partner can divert the focus of the SCM effort and actually discourage other members. The best supply chain partners are those who possess comparable operational and strategic values and who are prepared to commit to converging people and physical resources in the search for new forms of competitive advantage.

- Lack of Management Commitment. Many SCM implementations fail, not because of the organizational and technical requirements but because of a distinct lack of management commitment. Requirements for consistent and directed leadership on the part of management is even more crucial when it is considered that all the management groups participating in a SCM initiative must all be fully committed. If even one management group in the coalition provides only half-hearted support, other management teams will soon slowly withdraw their involvement until the project either dies or achieves only a fraction of its competitive-enhancing possibilities.

- Lack of Measurements. Being able to measure the benefits of an SCM implementation is absolutely critical to its ongoing success. Utilizing erroneous performance metrics, metrics that are weighted to the advantage of one channel partner over the others, metrics that promote attainment of short-term over long-term objectives, or simply no valid measurements at all are all contributors to SCM implementation failure. The best performance measurements are those that demonstrate continuous operating and customer-focused process improvement and which provide calculable results illustrating the enrichment value of the SCM competitive alliance.

Critical Questions for Management

Implementation of SCM concepts can provide companies gathered together in supply chains with the capability of designing highly competitive, cross-channel operational process teams driven by strategies that continually seek to create new sources of customer value. At this point, now that the power of SCM has been described in detail, it is possible for readers to look at their own companies and supply channel relationships and gauge how closely they are leveraging SCM for competitive advantage. What effect could the implementation of SCM have on your company? What are the areas that you could identify that would benefit by SCM integration or reengineering?

Listed below are two sets of questions that readers can use to benchmark their SCM performance improvement potential. The first set of questions relates to strategic SCM requirements; the second set relates to operational SCM requirements. A simple YES or NO will be sufficient to answer each question. Be thoughtful and accurate in recording your response.

Strategic Questions

1. When viewing the industry, does your company continually search for more innovative ways of making a profit and satisfying customers than does the competition?

   YES   NO

2. Is your company normally in the forefront in utilizing new technologies for competitive advantage?

   YES   NO

3. Does your company look toward process, product, and service innovation to maintain marketplace advantage?

   YES   NO

4. Have the design and marketing formulas used to determine product and service priorities been consistently changed over time to meet new competitive requirements?

   YES   NO

5. Does management consider improvement projects as consisting mainly of efforts to continually reengineer strategic processes rather than core processes?

   YES   NO

6. Is the culture of your company centered around continuously creating new opportunities and new businesses?

   YES   NO

7. Does your company management perceive SCM as purely an extension of logistics management with relatively little impact on strategic business or marketing planning?

   YES   NO

8. Does top management consistently seek to form close, mutually beneficial relationships with suppliers and customers?

   YES   NO

9. Has your company often willingly shared power and compromised with supply chain partners to achieve mutually beneficial results?

   YES   NO

10. Does your company consciously seek to exploit the processes, technologies, and core competencies that can be found in your supply chain partners?

    YES   NO

Operations Questions

1. Do you spend most of your creative time as an strategic architect designing the future rather than reengineering the work tasks you are involved with on a daily basis?

   YES   NO

2. Is the human resources focus of your company's management centered around improving the core competencies of the people organization versus staff reduction?

   YES   NO

3. Does your company actively promote the creation of process flow teams both within the enterprise and between your organization and those of supply chain partners?

   YES   NO

4. Has your company reduced cycle times and pipeline inventories by at least 50% over the past 3 years?

   YES   NO

5. Can your company claim at least 98% customer order on-time delivery and reliability today?

   YES   NO

6. Has your company's production and delivery costs declined by 20% or more over the past 3 years?

   YES   NO

7. Have the number of active supplier's been reduced by at least 50% and the delivery lead times of those remaining vendors shrunk also by at least 50% during the past 3 years?

   YES   NO

8. Has your company's cost of quality declined by 50% over the past 3 years?

   YES   NO

9. Does your company search for new ICT tools that facilitate productivity and more closely integrate internal business functions?

   YES   NO

10. Has your company actively sought to promote interfunctional and intercompany process teams?

    YES   NO

Analysis

It would be difficult today to find a company that could legitimately score a YES on all of the above questions. The management view that sees the whole supply chain as a boundless source of competitive advantage is in many aspects just too new for many companies to have expended considerable thought and developed the necessary action plans. Also, it would be quite normal for today's best companies to score higher on the operations elements of SCM than on the strategic elements. Reengineering internal and external processes have long been recognized as sources of cost savings and competitive advantage. In any case, your answers to the 20 questions detailed above can help you position your company and make visible a program to implement the necessary elements of operational and strategic SCM. Tally your "YES" answers and then review below the recommendations associated with your current business position.

18 to 20 YES answers: Your company is one of a small group of marketplace leaders that is committed to competitive success through the application of SCM operations and strategic concepts. Maintaining leadership for these companies means constantly searching for new pathways to expand the SCM concept.

15 to 17 YES answers: Your company is one of a group of high performers who have consciously sought to apply SCM to gain a position of leadership over most of your competitors. The majority of your NO answers most likely appear in the strategic elements area with only a few NOs to be found in the operations area. The NO answers indicate areas of significant risk that must be tackled in order to maintain competitive leadership. You may be currently comfortable in a market niche, but that confidence is transitory and can change dramatically as the competition or the marketplace changes.

14 to 16 YES answers: Your company has significant competitive weaknesses. Most of the questions relating to SCM strategy have been answered NO, as has the interchannel aspects of the SCM operations questions. Your company is most likely just barely able to maintain its competitive positioning and you are constantly having to expend considerable money and effort to catch up to marketplace leaders.

13 or less YES answers: Your company is currently in serious strategic and operational trouble. Your company's perception of itself is narrowly focused around creating internal strengths through cost incrementalism and reengineering. Such companies view suppliers and customers adversarially and constantly seek to squeeze concessions from their channel partners. Companies in this class must rapidly refocus their efforts to exploring and prioritizing those SCM strategic and operational principles that will assist them in regaining market leadership.

Summary

Tomorrow's marketplace leaders must be uniquely positioned and possess razor-sharp organizations and business processes that continuously create unbeatable sources of customer advantage. As the pace of the marketplace quickens, the scope of competition widens, and the demand for unique product and service solutions and ever-shorter delivery times accelerates, companies have no choice but to seek out and develop ever-closer relationships with their supply chain partners. In today's environment, no company can achieve a position of marketplace leadership without intensifying their commitment to and dependence on their channel network allies. No company today can possibly gather the depth of expertise and level of knowledge to preempt the competition than can be gained when the core competencies and physical resources of several supply chain partners are converged to solve specific competitive problems or exploit key marketplace opportunities. No company can afford to ignore the tremendous enabling power of information and communication technologies that have the power to network critical competencies across global space and time. Above all, no company can rest on the old paradigms of hierarchical management and organizational culture, but must consciously seek to tap into the tremendous repository of innovation and entrepreneurialism to be found in their work forces.

Today's customer simply does not care that a product is the result of a joint engineering effort among several companies, that the components of a product come from several suppliers, that the product was finally configured somewhere in the supply pipeline and passed through several distribution points. What the customer cares about is that the product possesses absolute quality, is accompanied by "world-class" services, represents the very latest in product design, is available at a competitive price, and provides a targeted solution that assists in opening new vistas for competitive advantage and the creation of radically new market-places where the competition has not yet dared to tread.

Notes

1. This crucial term is explored in more depth in Gary Hamel and C. K. Prahalad, *Competing For the Future.* Boston, MA: Harvard Business School Press, 1994, pp. 79–116.

2. See the discussion on strategic assessment found in Stanley E. Fawcett, "Using Strategic Assessment to Increase the Value-Added Capabilities of Manufacturing and Logistics." *Production and Inventory Management Journal* 36 (2) (Second Quarter 1995), 33–37.

3. Among the foremost can be found William C. Copacino, *Supply Chain Management: The Basics and Beyond.* Boca Raton, FL: St. Lucie Press, 1997; Christopher Gopal and Harold Cypress, *Integrated Distribution Management.* Homewood, IL: Business One Irwin, 1993; David F. Ross, "Meeting the Challenge of Supply Chain Management." *APICS: The Performance Advantage* 6 (9) (September 1996), 38–42; and Lisa Harrington, "How to Join the Supply Chain Revolution." *Inbound Logistics* 15 (11) (November 1995), 20–24.

4. See the comments in Richard Brown, "Configurable Network Computing." *APICS: The Performance Advantage* 6 (12) (December 1996), 36–39.

Index